Contents

Preface, vii

UNIT I--WHOLE NUMBERS, 1

    Chapter 1   Addition, Multiplication, Powers, and Square Roots, 3

        Pretest, 4
        1-1   Reading Whole Numbers, 5
        1-2   Rounding Off Numbers, 9
        1-3   Addition Facts, 13
        1-4   Adding Whole Numbers, 15
        1-5   Multiplication Facts, 19
        1-6   Multiplying by One Digit Numbers, 21
        1-7   Multiplying by Numbers Having More Than One Digit, 25
        1-8   Handling Zeros in Multiplying Whole Numbers, 29
        1-9   Powers, Roots, Comparisons, 33
        1-10 Applications Involving Whole Numbers, 37
        Summary, 41
        Practice Test A, 43
        Practice Test B, 44
        Supplementary Exercises, 45

    Chapter 2   Subtraction, Division, and the Order of Operations, 49

        Pretest, 50
        2-1   Subtraction Facts, 51
        2-2   Subtracting Whole Numbers, 53
        2-3   Division Facts, 57
        2-4   Remainders, Dividing by Zero, 61
        2-5   Dividing by One Digit Numbers, 65
        2-6   Dividing by Numbers Having More Than One Digit, 69
        2-7   Zeros in the Quotient, 75
        2-8   The Order of Operations, 79
        2-9   Primes, Divisibility, and Factoring into Primes, 83
        2-10 More Applications Involving Whole Numbers, 87
        Summary, 91
        Practice Test A, 93
        Practice Test B, 94
        Supplementary Exercises, 95

Unit I Exam, 98

UNIT II--FRACTIONS AND DECIMALS, 99

Chapter 3 Fractions, 101

Pretest, 102
- 3-1   The Meaning of Fractions, 103
- 3-2   Equivalent Fractions, 107
- 3-3   Mixed Numbers, 113
- 3-4   Multiplying Fractions, 117
- 3-5   Dividing Fractions, 121
- 3-6   Multiplying and Dividing Mixed Numbers, 125
- 3-7   Fractional Parts of Numbers, 129
- 3-8   Adding and Subtracting Like Fractions, 133
- 3-9   Adding and Subtracting Unlike Fractions, 137
- 3-10 Finding the Least Common Denominator, 141
- 3-11 Adding and Subtracting Mixed Numbers, 145
- 3-12 Complex Fractions, 149
- 3-13 Comparing Fractions, 153
- 3-14 Applications involving Fractions, 157

Summary, 161
Practice Test A, 163
Practice Test B, 164
Supplementary Exercises, 165

Chapter 4 Decimals, 169

Pretest, 170
- 4-1   Reading and Writing Decimals, 171
- 4-2   Rounding Off Decimals, 175
- 4-3   Adding Decimals, 179
- 4-4   Subtracting Decimals, 183
- 4-5   Multiplying Decimals, 187
- 4-6   Dividing Decimals, 191
- 4-7   Dividing Decimals--Rounding Off Answers, 195
- 4-8   Multiplying and Dividing by Numbers That End in Zeros, 199
- 4-9   Changing Between Fractions and Decimals, 203
- 4-10 Comparing Decimals, 207
- 4-11 Operating With Both Fractions and Decimals, 211
- 4-12 Applications Involving Decimals, 215
- 4-13 Applications Involving Perimeters and Areas, 219

Summary, 223
Practice Test A, 225
Practice Test B, 226
Supplementary Exercises, 227

Unit II Exam, 230

UNIT III--RATIO, PERCENT, MEASUREMENT, AND GRAPHS, 231

Chapter 5   Ratio, Proportion, and Percent, 233

    Pretest, 234
    5-1   Ratios, 235
    5-2   Proportions, 239
    5-3   More on Solving Proportions, 243
    5-4   Applications Involving Proportions, 247
    5-5   Percents, 251
    5-6   Converting Decimals and Fractions to Percents, 255
    5-7   Percent of a Number, 259
    5-8   Percentage Problems, 263
    5-9   Applications Involving Percents, 267
    5-10   Applications Involving Interest, 271
    Summary, 275
    Practice Test A, 277
    Practice Test B, 278
    Supplementary Exercises, 279

Chapter 6   Measurement and Graphs, 283

    Pretest, 284
    6-1   Reading Measuring Devices, 285
    6-2   The U.S. Customary System of Measurement, 289
    6-3   The Metric System, 293
    6-4   Conversions Within the Metric System, 297
    6-5   Conversions Between the Metric and U.S. Systems, 303
    6-6   U.S. and Metric Temperatures, 309
    6-7   Statistical Graphs, 313
    Summary, 319
    Practice Test A, 321
    Practice Test B, 322
    Supplementary Exercises, 323

Unit III Exam, 326

UNIT IV—INTRODUCTIONS TO ALGEBRA AND GEOMETRY, 327

Chapter 7 Introduction to Algebra, 329

Pretest, 330
7-1   From Numbers to Letters, 331
7-2   Signed Numbers, 335
7-3   Adding Signed Numbers, 339
7-4   Subtracting Signed Numbers, 343
7-5   Multiplying and Dividing Signed Numbers, 347
7-6   The Order of Operations—Signed Numbers, 351
7-7   What Are Equations?, 355
7-8   Solving Equations Using the Addition Property, 359
7-9   Solving Equations Using Multiplication/Division Properties, 363
7-10 More on Solving Equations, 367
7-11 Word Problems, 371
Summary, 377
Practice Test A, 379
Practice Test B, 380
Supplementary Exercises, 381

Chapter 8 Introduction to Geometry, 385

Pretest, 386
8-1   Basic Geometric Objects, 387
8-2   Angles, 391
8-3   Triangles, 395
8-4   Other Common Geometric Shapes, 399
8-5   Perimeters and Areas, 403
Summary, 407
Practice Test A, 409
Practice Test B, 410
Supplementary Exercises, 411

Unit IV Exam, 413

Final Exam, 415

Answers, 417

Index, 464

Preface

<u>To the Student</u>

The intent of this book is to help you develop your math skills. It has been written so that you can learn by actually working with mathematics. The book not only shows you how to solve problems, it also attempts to explain the "why" of the solution. The following features of the book will give you many avenues for learning mathematics.

1. **Objectives:** A statement of the skills you should acquire is given at the beginning of each unit and chapter.

2. **Pretests:** At the beginning of each chapter there is a test that you can take and correct. The purpose of the pretest is to help you determine which sections of a chapter you need to study in detail and which ones you can review quickly.

3. **Explanations and Examples:** The book has readable explanations and detailed examples to show you the "how" and "why" of solving the problems of arithmetic.

4. **Problems:** In addition to explanations and examples, each section has problems for you to solve. By checking your answers against the given detailed solutions, you can determine your understanding of the concepts in each section.

5. **Exercises:** Each section of the book has two sets of exercises to give you practice on the skills covered in each section.

6. **Summaries:** At the end of each chapter you will find a summary of the important concepts and terms used in each chapter.

7. **Practice Tests:** At the end of each chapter and unit you will find practice tests to help evaluate your progress, and to check your mastery of the entire book, there is a final exam on pages 415-416.

8. **Answers and Solutions:** In the Answer Section, pages 417-463, are all the answers to pretests, exercises, chapter tests, unit exams, and the final exam. You will also find detailed solutions to many of the odd numbered exercises.

It is my hope that this book will help you develop your mathematics skills. Remember that math is not a spectator sport. You must participate in the solution of its problems in order to learn it. Even if you have experienced difficulty with mathematics in the past, I am confident that you can master the skills covered in this book if you keep a positive attitude and work at it earnestly. I want to encourage you to develop your math skills and give you best wishes for success.

Ronald Staszkow

To the Instructor

This book has been written for those students who need to acquire a better understanding of and facility with arithmetic, and who desire an introduction to algebra and geometry. The book has enough explanations, examples, solved problems, exercises, and tests to meet the needs of both individualized (lab based) and lecture formats.

In this 3rd edition the word problems have been integrated into each chapter, rather than being contained in a separate chapter. Sections on primes and factoring, reading measuring devices, and the order of operations with signed numbers have been added. The factoring method is used to find the least common denominator, rather than the lattice method of previous editions.

The 3rd edition also offers the following:

1. a non-threatening format that encourages the student to work in the textbook itself, but leaves the textual material intact even if exercise sets are removed.

2. explanations of the "how" and "why" of arithmetic that are concise, readable, and contain many examples and solved problems.

3. over 2500 problems contained in Set A and Set B Exercises at the end of each section. All answers for those problems and selected solutions are contained in the text's Answer Section.

4. over 1000 supplementary exercises for which there are no answers in the text. These are ideal for quizzes, homework, or class work.

5. pretests, summaries, chapter tests, unit exams, and a final exam that give an excellent means for students to review and test their understanding of the material.

6. brain busters in each unit that will challenge the student.

7. a **Resource Book** that contains:
   a. ten forms of each chapter test, unit exam, and final exam.
   b. a sample study guide for using the text in self-paced, individualized programs.
   c. a computer generated testing system for Apple II computers that generates endless forms of chapter, unit, and final exams.
   d. answers to all supplementary exercises.
   e. a diagnostic exam to determine proficiency in basic arithmetic.

Acknowledgements

I would like to thank Virginia Tebelskis for her continuing help, and Joseph Ehardt and the Ohlone Art Department's Laser Writer for the graphics work.

Ronald Staszkow

Unit I

Whole Numbers

Numbers are all around us. Just take a minute to look back over your day. Try to remember how many times you worked with numbers. Did you do any shopping? Did you pay any sales tax? Did you write a check or balance your checking account? Did you measure ingredients for a recipe or lumber for a bookshelf?

Whether you are at home, in a grocery store, driving along a freeway, using a credit card, or reading the sports page, mathematics is there. In our everyday lives and in many occupations, it is important to understand and be able to work with numbers.

In this unit you will learn to operate with the whole numbers (0, 1, 2, 3, 4, 5, 6, 7, 8, 9, 10, 11, 12, 13, and so on). When you have finished this unit, you should be able to do the following:

1. perform addition, subtraction, multiplication, and division of whole numbers.
2. read and round off whole numbers.
3. use the "equals" sign, "greater than" sign, and the "less than" sign.
4. compute simple powers and square roots.
5. use the proper order of computation when different operations are combined in a problem.
6. factor whole numbers into primes.
7. solve word problems that involve operations with whole numbers.

Unit I

Brain Buster

Your boss offers to pay you according to this scheme:
  You get $1 for the first day,
          $2 for the second day,
          $4 for the third day,
          $8 for the fourth day, etc.
          (your daily wage doubles each day).

If you worked 20 days that month,
  a. how much would you make on the 20th day?
  b. what would your total wages be for the month?

(Hint: You would be a millionaire!)

Chapter 1

Whole Numbers: Addition, Multiplication, Powers, and Square Roots

After finishing the first chapter you should be able to do the following:

1. read and round off whole numbers.
2. add whole numbers.
3. multiply whole numbers.
4. calculate powers and simple square roots of whole numbers.
5. use the =, >, and < signs.
6. solve word problems using addition and multiplication.

On the next page you will find a pretest for this chapter. The purpose of the pretest is to help you determine which sections in this chapter you need to study in detail and which sections you can review quickly. By taking and correcting the pretest according to the instructions on the next page you can better plan your pace through this chapter.

Chapter 1 Pretest

Take and correct this test using the answer section at the end of the book. Those problems that give you difficulty indicate which sections need extra attention. Section numbers are in parentheses before each problem.

ANSWERS

(1-1)   1. Write using numbers and commas: six hundred two million, sixty-seven thousand, five.

1. _____

(1-1)   2. Write the following number in words: 42,107

2. _____

(1-2)   Round off 542,950 to the nearest:

3. _____

             3. hundred   4. ten thousand

4. _____

(1-4)   5. 41 + 1278 + 6

5. _____

(1-4)   6. 5007 + 37 + 4716 + 5 + 87,998

6. _____

(1-6)   7. 356 x 7

7. _____

(1-7)   8. 6359 x 486

8. _____

(1-8)   9. 248 x 95,000,000

9. _____

(1-8)   10. 806 x 4,000,009

10. _____

(1-9)   11. 2^3 (1-9)   12. $\sqrt{49}$

11. _____

12. _____

(1-9)   Replace the ? with =, >, or <:

13. _____

             13. 0 x 9 ? 9 x 1   14. 4^2 ? $\sqrt{64}$

14. _____

(1-10)   15. How much would a family of 2 adults and 4 children pay to attend a baseball game, if adults pay $7 and children pay $4?

15. _____

1-1 Reading Whole Numbers

All numbers can be formed using the **digits** 0, 1, 2, 3, 4, 5, 6, 7, 8, and 9. This can be done since a digit's position in a number gives it a specific meaning called its **place value**.

For example:

In 576
  the 5 means 5 hundreds (500)
  the 7 means 7 tens (70)
  the 6 means 6 ones (6).

Thus 576 is read "five hundred seventy-six."

In 4,208
  the 4 means 4 thousands (4000)
  the 2 means 2 hundreds (200)
  the 0 means 0 tens
  the 8 means 8 ones (8).

Thus 4,208 is read "four thousand, two hundred, eight."

BEFORE HE PULLED HER TOOTH, THE DENTIST <u>NUMBER</u> WITH NOVOCAINE.

The following is a chart for the place values in our number system:

Value of Each Group		
billions	millions	thousands

6	7	8	,	3	1	0	,	4	5	6	,	1	2	7
hundred billions	ten billions	billions		hundred millions	ten millions	millions		hundred thousands	ten thousands	thousands		hundreds	tens	ones

In the United States large numbers are separated into groups of three digits and read according to this place value chart.

For example: In the number 17832536 you would start from the right hand side of the number and mark off every three digits with a comma as follows:

17,832,536

Then you read each group of digits that was marked off, followed by the value for that group as noted on the place value chart.

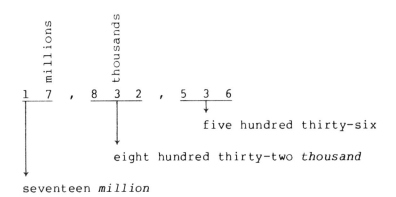

Thus you read it "seventeen million, eight hundred thirty-two thousand, five hundred thirty-six."

Remember, when placing commas to separate the digits, always start from the right hand side of the number and work your way to the left, marking off every three digits.

Note: The word "and" is not used when writing the word equivalent for whole numbers. It is used to indicate the placement of a decimal point, as we shall see later. For example, 5008 is read "five thousand, eight." It is *not* read "five thousand and eight."

Note: Compound numbers from 21 to 99 are written with a hyphen, such as twenty-one or ninety-nine.

PROBLEM 1: Write 10,487 in words.   Answer: ten thousand, four hundred eighty-seven

PROBLEM 2: Write 2,003,500 in words.   Answer: two million, three thousand, five hundred

PROBLEM 3: In the number 325,467
            what does the:

                                             Answers:
            5 represent?   5 thousands (5000)
            4 represent?   4 hundreds (400)
            2 represent?   2 ten thousands (20,000)

PROBLEM 4: Write using numbers: Answers:
            a. four hundred twenty-two million, ten   a. 422,000,010

            b. thirty-five billion, twenty-seven   b. 35,027,002,006
               million, two thousand, six

6

EXERCISE 1-1   SET A

NAME _____ DATE _____

WRITE IN WORDS: ANSWERS

1. 852   1. _____

2. 4,256   2. _____

3. 17,109   3. _____

4. 3,057,010   4. _____

5. 14,100,700   5. _____

6. 946,003   6. _____

7. 1,357,926,183   7. _____

WRITE USING NUMBERS AND COMMAS:

8. seven hundred forty-five   8. _____

9. fifty thousand, sixty-eight   9. _____

10. one hundred five thousand, six   10. _____

11. forty million, thirty-six   11. _____

12. five million, seven thousand, two hundred
    thirty-eight   12. _____

13. twelve billion, fifteen million   13. _____

14. eighty-nine billion, eighty-nine   14. _____

15. three hundred thirteen million, seven
    hundred ten thousand   15. _____

16. seven hundred twelve million, four hundred
    twenty-two   16. _____

IN THE NUMBER 5,876,492:

17. What does the 5 represent?   17. _____

18. What does the 9 represent?   18. _____

19. What does the 8 represent?   19. _____

EXERCISE 1-1   SET B

NAME _____   DATE _____

WRITE IN WORDS: ANSWERS

1. 925   1. _____

2. 6,172   2. _____

3. 13,402   3. _____

4. 6,026,050   4. _____

5. 17,200,400   5. _____

6. 196,002   6. _____

7. 4,307,916,452   7. _____

WRITE USING NUMBERS AND COMMAS:

8. nine hundred sixteen   8. _____

9. thirty thousand, forty-three   9. _____

10. two hundred seven thousand, four   10. _____

11. fifty million, seventy-two thousand, forty   11. _____

12. four million, six thousand, three hundred
    eighty-five   12. _____

13. eighteen billion, twelve million   13. _____

14. seventy-nine billion, thirty-five   14. _____

15. one hundred forty-four million, seven
    hundred thirty thousand   15. _____

16. nine hundred thirteen million, five hundred
    fifty-three   16. _____

IN THE NUMBER 1,827,645:

17. What does the 2 represent?   17. _____

18. What does the 8 represent?   18. _____

19. What does the 1 represent?   19. _____

1-2 Rounding Off Numbers

If the number of people living in your city were 178,952 the road sign would probably read as in the illustration on the left.

The sign gives the approximate population of the city. It is just an estimate of the number of people living there. We say that the population has been rounded off. When you round off a number, it is close to the original number. Rounded numbers are easier to visualize and simpler to work with than the original numbers. Rounded numbers are also used to estimate answers in complicated math problems.

Rounding Off to the Nearest Ten

If you were to count by tens you would say 10, 20, 30, 40, 50, 60, 70, etc. These are called the multiples of ten.

To round off a given number to the nearest ten, you must find the multiple of ten that the number is closest to.

For example: To round off 37 to the nearest ten, the answer would be 40, since 37 is between 30 and 40 but is closer to 40. We say 37 rounds off to 40.

On the other hand, 32 rounds off to 30, since it is closer to 30. We say 32 rounds off to 30.

Now 35 is halfway between 30 and 40. In this case you round off to the higher multiple of ten. Thus 35 rounds off to 40.

PROBLEM 1: Round off 86 to the nearest ten

Answer: 90
Since 86 is between 80 and 90 but is closer to 90.

PROBLEM 2: Round off 723 to the nearest ten.

Answer: 720
Since 723 is between 720 and 730 but is closer to 720.

PROBLEM 3: Round off 2695 to the nearest ten.

Answer: 2700
Since 2695 is halfway between 2690 and 2700, round off to the higher multiple, 2700.

Rounding Off to the Nearest Hundred

The multiples of a hundred are 100, 200, 300, 400, 500, 600, and so on.

To round off a given number to the nearest hundred, you must find the multiple of a hundred that the number is closest to.

> For example: 768 rounds off to 800 since 768 is between 700 and 800 but is closer to the higher multiple of a hundred, 800.
>
> 725 rounds off to 700 since it is closer to the lower multiple of a hundred, 700.

Rounding Off to Other Places

You round off to other places in our number system using the same ideas as we used in rounding off to the nearest ten or hundred. You round off to the higher multiple if a given number is more than half the way between two multiples. You round off to the lower multiple if it isn't. An easy way to decide this is as follows:

> Look at the digit to the right of the place you are rounding to. If that digit is 5 or more, round off to the higher multiple. Otherwise, round off to the lower multiple.

Example 1: Round off 28,752 to the nearest thousand.

$$28{,}752 \doteq 29{,}000 \quad (\doteq \text{ means "is approximately equal to."})$$

thousands place ↑ — digit to the right ↳ is more than 5

Thus you round off to the higher multiple of a thousand by changing the 8 to a 9, getting 29 thousands (29,000).

Example 2: Round off 413,207,593 to the nearest million.

$$413{,}207{,}593 \doteq 413{,}000{,}000$$

millions place ↑ — digit to the right ↳ is less than 5

Thus you round off to the lower multiple of a million by leaving the 3 unchanged, getting 413 millions (413,000,000). Notice: All the digits to the left of the millions place are unchanged while those to the right are replaced with zeros.

PROBLEM 4: Round off 865,000 to the nearest ten thousand.

Answer: 870,000
Since the digit to the right of the ten thousands place is a 5, round off to the higher multiple by adding one to the ten thousands place.

EXERCISE 1-2   SET A

NAME _____   DATE _____

ROUND OFF TO THE NEAREST TEN: ANSWERS

1. 53   1. _____
2. 76   2. _____
3. 25   3. _____
4. 506   4. _____
5. 1473   5. _____
6. 5895   6. _____
7. 6997   7. _____

ROUND OFF TO THE NEAREST HUNDRED:

8. 549   8. _____
9. 872   9. _____
10. 650   10. _____
11. 14,736   11. _____
12. 27,864   12. _____
13. 179,950   13. _____
14. 1,538,276   14. _____

ROUND OFF 215,749,538 TO THE NEAREST:

15. ten   15. _____
16. hundred   16. _____
17. thousand   17. _____
18. ten thousand   18. _____
19. hundred thousand   19. _____
20. million   20. _____
21. ten million   21. _____

EXERCISE 1-2   SET B

NAME _____   DATE _____

ROUND OFF TO THE NEAREST TEN: ANSWERS

1. 72   1. _____

2. 48   2. _____

3. 65   3. _____

4. 307   4. _____

5. 2582   5. _____

6. 4795   6. _____

7. 3996   7. _____

ROUND OFF TO THE NEAREST HUNDRED:

8. 748   8. _____

9. 562   9. _____

10. 850   10. _____

11. 25,604   11. _____

12. 37,589   12. _____

13. 79,950   13. _____

14. 4,356,495   14. _____

ROUND OFF 927,563,742 TO THE NEAREST:

15. ten   15. _____

16. hundred   16. _____

17. thousand   17. _____

18. ten thousand   18. _____

19. hundred thousand   19. _____

20. million   20. _____

21. ten million   21. _____

1-3 Addition Facts

EVEN CAFETERIA WORKERS SPEND TIME ADDITION OUT FOOD.

If you drove 5 miles to the store and from there you drove 3 miles to school, how far did you drive? To find the total miles driven you add the two distances and get the answer, 8 miles. We would say:

$$5 \text{ plus } 3 \text{ equals } 8$$

$$\text{or } 5 + 3 = 8.$$

The names given to each part of an addition problem such as that are:

```
addend   addend    sum
   ↓        ↓       ↓
   5    +   3   =   8
```

This first operation with whole numbers is the basis for the other operations. Addition must be learned well before progress can be made in developing your math skills.

The chart below gives the basic addition facts that you should know.

+	0	1	2	3	4	5	6	7	8	9
0	0	1	2	3	4	5	6	7	8	9
1	1	2	3	4	5	6	7	8	9	10
2	2	3	4	5	6	7	8	9	10	11
3	3	4	5	6	7	8	9	10	11	12
4	4	5	6	7	8	9	10	11	12	13
5	5	6	7	8	9	10	11	12	13	14
6	6	7	8	9	10	11	12	13	[14]	15
7	7	8	9	10	11	12	13	14	15	16
8	8	9	10	11	12	13	14	15	16	17
9	9	10	11	12	13	14	15	16	17	18

To find the answer to any addition problem, such as 6+8, find the 6 on the left hand side and go along that row until you reach the 8 at the top of the chart. That number (14) will be the answer to 6+8.

PROBLEM 1:   Find 7+9 using the chart.   Answer: 16

PROBLEM 2:   Find 4+7 using the chart.   Answer: 11

Since you can not carry that addition chart around with you at all times, you must learn the basic addition facts by working with them. On the next page you will find rows of problems along with their answers. These rows of problems can help you become better at addition. You might also find the use of "flash cards" helpful in mastering the basic addition facts.

Self Drill in Addition

Directions

1. Cover the answers for a row.

2. Quickly figure out and write down your answer to each problem.

3. Check your work.

4. Repeat this for each row until you can do each one quickly and correctly.

```
  1    4    2    5    5    1    3    8    5    6
 +0   +2   +1   +2   +0   +3   +3   +1   +1   +3
 ──   ──   ──   ──   ──   ──   ──   ──   ──   ──
  1    6    3    7    5    4    6    9    6    9


  4    2    4    6    7    7    9    3    1    9
 +0   +2   +3   +2   +0   +3   +0   +1   +1   +3
 ──   ──   ──   ──   ──   ──   ──   ──   ──   ──
  4    4    7    8    7   10    9    4    2   12


  0    7    2    1    9    8    3    0    7    6
 +0   +1   +0   +2   +1   +2   +0   +3   +2   +0
 ──   ──   ──   ──   ──   ──   ──   ──   ──   ──
  0    8    2    3   10   10    3    3    9    6


  8    9    6    2    0    8    5    3    4    0
 +3   +2   +1   +3   +2   +0   +3   +2   +1   +1
 ──   ──   ──   ──   ──   ──   ──   ──   ──   ──
 11   11    7    5    2    8    8    5    5    1


  1    3    4    5    3    6    2    6    9    4
 +4   +5   +6   +5   +6   +4   +6   +6   +4   +4
 ──   ──   ──   ──   ──   ──   ──   ──   ──   ──
  5    8   10   10    9   10    8   12   13    8


  1    5    7    8    2    8    1    5    6    7
 +6   +4   +5   +4   +5   +6   +5   +6   +5   +4
 ──   ──   ──   ──   ──   ──   ──   ──   ──   ──
  7    9   12   12    7   14    6   11   11   11


  0    4    9    0    2    9    0    3    8    7
 +5   +5   +6   +6   +4   +5   +4   +4   +5   +6
 ──   ──   ──   ──   ──   ──   ──   ──   ──   ──
  5    9   15    6    6   14    4    7   13   13


  1    9    4    3    4    6    8    1    2    2
 +7   +8   +7   +9   +9   +7   +7   +8   +9   +7
 ──   ──   ──   ──   ──   ──   ──   ──   ──   ──
  8   17   11   12   13   13   15    9   11    9


  5    3    6    7    5    7    3    5    4    9
 +8   +7   +9   +8   +9   +9   +8   +7   +8   +9
 ──   ──   ──   ──   ──   ──   ──   ──   ──   ──
 13   10   15   15   14   16   11   12   12   18


  9    0    8    6    0    7    8    1    2    0
 +7   +8   +8   +8   +9   +7   +9   +9   +8   +7
 ──   ──   ──   ──   ──   ──   ──   ──   ──   ──
 16    8   16   14    9   14   17   10   10    7
```

1-4 Adding Whole Numbers

Now that you have learned the basic addition facts, you can proceed to the addition of larger whole numbers. Let me explain how that is done.

> To add whole numbers you add the digits in the ones place, tens place, hundreds place, etc. of each whole number.

For example: To add 74 and 23 you place the numbers above each other, lining up respective columns.

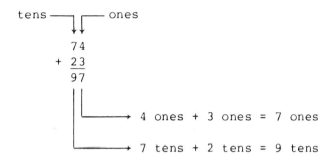

Example 1: 544 + 32 + 3 = ?

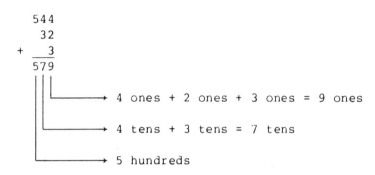

To check your work, add the numbers in a different order.

PROBLEM 1:   1573 + 2312 = ?   Answer: 3885   Check:

                                                    1573   2312
                                                 + 2312   + 1573
                                                    3885   3885

PROBLEM 2:   641 + 324 + 13 = ?   Answer: 978   Check:

                                                     641   13
                                                     324   641
                                                  +   13   + 324
                                                     978   978

                                47
If you try to add + 25 in the same manner as the above problems, you will discover that when you add the digits in the ones column, you get 12. Since you can not have a two digit number in the ones place, what do you do?

Example 2:   47 + 25 = ?

```
  1 ←———— number carried
  47
+ 25
  ──
  72
```

→ 7 ones + 5 ones = 12 ones (1 ten and 2 ones). Put just the 2 in the ones place and carry the ten to the tens column.

→ 1 ten + 4 tens + 2 tens = 7 tens

Example 3:   287 + 158 + 39 = ?

```
  12 ←———— numbers carried
  287
  158
+  39
  ───
  484
```

→ 7 ones + 8 ones + 9 ones = 24 ones (2 tens and 4 ones). Put the 4 in the ones place and carry the 2 to the tens column.

→ 2 tens + 8 tens + 5 tens + 3 tens = 18 tens (1 hundred and 8 tens). Put the 8 in the tens place and carry the 1 to the hundreds column.

→ 1 hundred + 2 hundreds + 1 hundred = 4 hundreds

PROBLEM 3:   3576 + 687 = ?   Answer: 4263

```
  111
  3576
+  687
  ────
  4263
```

PROBLEM 4:   1872 + 54 + 137 + 34,567 = ?   Answer: 36,630

```
    122
   1872
     54
    137
+ 34567
  ─────
  36630
```

PROBLEM 5:   276 + 89 + 2345 + 9 + 500 = ?   Answer: 3219

```
   122
   276
    89
  2345
     9
+  500
  ────
  3219
```

16

EXERCISE 1-4   SET A

NAME _____   DATE _____

ADD THE FOLLOWING: ANSWERS

1.   45   2.   76   1. _____
   + 23   + 12
                                                2. _____

3.   356   4.   729   3. _____
   + 24   + 46
                                                4. _____

5. 5072   6. 8906   5. _____
   + 946   + 398
                                                6. _____

7.   426   8.   716   7. _____
    5382   2053
   + 90   + 84   8. _____

9.   5   10.   7   9. _____
      76   83
     130   5200   10. _____
   + 256   + 194

11. 76 + 128   11. _____

12. 405 + 21 + 7   12. _____

13. 1400 + 7 + 322 + 17   13. _____

14. 576,276 + 8,006 + 475   14. _____

15. 52 + 9000 + 876 + 43 + 4   15. _____

16. 5,000,300 + 852,176 + 3,820   16. _____

17. 18 + 196 + 45 + 2463 + 757 + 95   17. _____

EXERCISE 1-4   SET B

NAME _____   DATE _____

ADD THE FOLLOWING: ANSWERS

1.   62   2.   34   1. _____
   + 15   + 25
                                                             2. _____

3.   756   4.   243   3. _____
   + 17   + 49
                                                             4. _____

5. 4706   6. 9821   5. _____
   + 385   + 359
                                                             6. _____

7.   25   8.   987   7. _____
    4186   1063
   + 385   + 45   8. _____

9.   4   10.   7   9. _____
     706   56
    3819   820   10. _____
   + 26   + 754

11. 84 + 257   11. _____

12. 809 + 28 + 9   12. _____

13. 4500 + 296 + 7 + 63   13. _____

14. 321,506 + 875 + 5,321   14. _____

15. 94 + 7000 + 326 + 43 + 6   15. _____

16. 6,000,200 + 763,918 + 3,457   16. _____

17. 25 + 283 + 76 + 4076 + 656 + 87   17. _____

1-5 Multiplication Facts

If you knocked down 9 pins in six consecutive frames of bowling, how many total pins did you knock down? One way to find the answer is to add 9 for each of the six frames. That is:

$$9+9+9+9+9+9 = 54$$

There is another way to think of that problem. What you have is six nines. Instead of adding nines together, we say six times nine is 54 (6x9 = 54). In mathematics, repeated addition is called multiplication. For example:

$$7 \times 8 \text{ means } 8+8+8+8+8+8+8 = 56$$

$$3 \times 12 \text{ means } 12+12+12 = 36$$

Each part of such a multiplication problem has a name.

SOME WONDER IF MATH IS <u>FACTOR</u> FICTION.

```
factor     factor         product
  ↓          ↓               ↓
  6    x    9       =       54
```

If you were asked to find 637 x 15,218 by repeated addition, you would have quite a job on your hands. That problem could be done much quicker, if you knew the basic multiplication facts and how to apply them.

Your next objective in developing your mathematics skills is to master the basic multiplication facts. They are displayed on the chart below. Since every chapter from here on relies heavily on having them memorized, it is extremely important that you take time to learn them well. The self drill in multiplication on the next page will help make that a little easier. You might also find the use of "flash cards" helpful in mastering the basic multiplication facts.

To find the answer to any multiplication problem, such as 6x8, find the 6 on the left hand side and go along that row until you reach the 8 at the top of the chart. That number (48) will be the answer to 6x8.

x	0	1	2	3	4	5	6	7	8	9
0	0	0	0	0	0	0	0	0	0	0
1	0	1	2	3	4	5	6	7	8	9
2	0	2	4	6	8	10	12	14	16	18
3	0	3	6	9	12	15	18	21	24	27
4	0	4	8	12	16	20	24	28	32	36
5	0	5	10	15	20	25	30	35	40	45
6	0	6	12	18	24	30	36	42	[48]	54
7	0	7	14	21	28	35	42	49	56	63
8	0	8	16	24	32	40	48	56	64	72
9	0	9	18	27	36	45	54	63	72	81

Self Drill in Multiplication

1	4	2	5	5	1	3	8	5	6
x0	x2	x1	x2	x0	x3	x3	x1	x1	x3
0	8	2	10	0	3	9	8	5	18

8	9	6	2	0	8	5	3	4	0
x3	x2	x1	x3	x2	x0	x3	x2	x1	x1
24	18	6	6	0	0	15	6	4	0

Directions:

1. Cover the answers for a row.
2. Quickly figure out and write down your answer to each problem.
3. Check your work.
4. Repeat this for each row until you can do each one quickly and correctly.

1	3	4	5	3	6	2	6	9	4
x4	x5	x6	x5	x6	x4	x6	x6	x4	x4
4	15	24	25	18	24	12	36	36	16

1	5	7	8	2	8	1	5	6	7
x6	x4	x5	x4	x5	x6	x5	x6	x5	x4
6	20	35	32	10	48	5	30	30	28

0	4	9	0	2	9	0	3	8	7
x5	x5	x6	x6	x4	x5	x4	x4	x5	x6
0	20	54	0	8	45	0	12	40	42

1	9	4	3	4	6	8	1	2	2
x7	x8	x7	x9	x9	x7	x7	x8	x9	x7
7	72	28	27	36	42	56	8	18	14

5	3	6	7	5	7	3	5	4	9
x8	x7	x9	x8	x9	x9	x8	x7	x8	x9
40	21	54	56	45	63	24	35	32	81

9	0	8	6	0	7	8	1	2	0
x7	x8	x8	x8	x9	x7	x9	x9	x8	x7
63	0	64	48	0	49	72	9	16	0

6	7	9	8	7	6	9	7	8	8
x7	x8	x8	x6	x9	x9	x7	x8	x9	x8
42	56	72	48	63	54	63	56	72	64

6	6	8	9	7	9	7	8	6	9
x6	x8	x7	x9	x8	x7	x6	x8	x9	x8
36	48	56	81	56	63	42	64	54	72

1-6 Multiplying by One Digit Numbers

Now that you have learned the basic multiplication facts, you can proceed to the multiplication of any whole number by a one digit number. Study the following examples and explanations to understand how that is done.

Example 1: 2 x 34 = ?

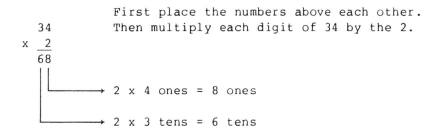

```
    34      First place the numbers above each other.
x    2      Then multiply each digit of 34 by the 2.
    68
```

→ 2 x 4 ones = 8 ones

→ 2 x 3 tens = 6 tens

Example 2: 4 x 316 = ?

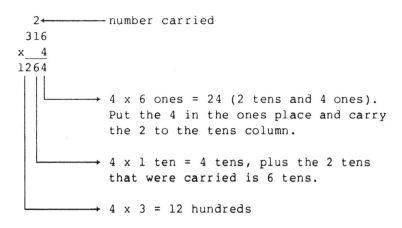

```
    2  ←——————— number carried
  316
x   4
 1264
```

→ 4 x 6 ones = 24 (2 tens and 4 ones). Put the 4 in the ones place and carry the 2 to the tens column.

→ 4 x 1 ten = 4 tens, plus the 2 tens that were carried is 6 tens.

→ 4 x 3 = 12 hundreds

Example 3: 7 x 6142 = ?

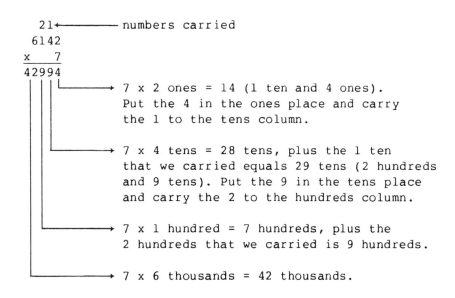

```
   21  ←——————— numbers carried
 6142
x   7
42994
```

→ 7 x 2 ones = 14 (1 ten and 4 ones). Put the 4 in the ones place and carry the 1 to the tens column.

→ 7 x 4 tens = 28 tens, plus the 1 ten that we carried equals 29 tens (2 hundreds and 9 tens). Put the 9 in the tens place and carry the 2 to the hundreds column.

→ 7 x 1 hundred = 7 hundreds, plus the 2 hundreds that we carried is 9 hundreds.

→ 7 x 6 thousands = 42 thousands.

By studying the previous examples you can see that if you multiply by a one digit number, just multiply each digit of the number by the one digit number. You must, however, remember to carry whenever a product has more than one digit.

Try to apply that principle by doing the following problems.

PROBLEM 1:   6 x 371 = ?   Answer: 2226

```
    4
   371
 x   6
  2226
```

PROBLEM 2:   8 x 1503 = ?   Answer: 12,024

```
   4 2
  1503
 x   8
 12024
```

PROBLEM 3:   5 x 28,746 = ?   Answer: 143,730

```
   4323
  28746
 x    5
 143730
```

PROBLEM 4:   7 x 900,707 = ?   Answer: 6,304,949

```
   4 4
  900707
 x    7
  6304949
```

PROBLEM 5:   3 x 67127 x 9 = ?   Answer: 1,812,429

```
   2 2              137
  67127           201381
 x    3          x     9
 201381          1812429
```

22

EXERCISE 1-6   SET A

NAME _____   DATE _____

MULTIPLY THE FOLLOWING: ANSWERS

1.   53   2.   87   1. _____
     x 4   x 6
                                                       2. _____

3.   78   4.   97   3. _____
     x 9   x 8
                                                       4. _____

5.   542   6.   715   5. _____
     x   3   x   4
                                                       6. _____

7.   193   8.   526   7. _____
     x   5   x   7
                                                       8. _____

9.   5706   10.   3904   9. _____
     x   2   x   6
                                                      10. _____

11. 15326   12. 24185   11. _____
     x   7   x   8
                                                      12. _____

13. 4 x 2576   13. _____

14. 5 x 195   14. _____

15. 6 x 23,400   15. _____

16. 7 x 75,674   16. _____

17. 8 x 1,238,475   17. _____

18. 9 x 15,768,207   18. _____

19. 3 x 4768 x 2   19. _____

EXERCISE 1-6   SET B

NAME _____   DATE _____

MULTIPLY THE FOLLOWING: ANSWERS

1.   81   2.   74   1. _____
      x 3   x 2
                                                        2. _____

3.   67   4.   87   3. _____
      x 9   x 8
                                                        4. _____

5.   423   6.   614   5. _____
      x 4   x 4
                                                        6. _____

7.   294   8.   876   7. _____
      x 5   x 7
                                                        8. _____

9. 5608   10. 8706   9. _____
      x 2   x 6
                                                       10. _____

11. 24376   12. 75149   11. _____
      x 7   x 8
                                                       12. _____

13. 4 x 1938   13. _____

14. 5 x 20,756   14. _____

15. 6 x 153,282   15. _____

16. 7 x 4,000,758   16. _____

17. 8 x 56,798   17. _____

18. 9 x 4,239,476   18. _____

19. 2 x 47,628 x 3   19. _____

1-7 Multiplying by Numbers Having More than One Digit

Using what you have learned about multiplying by one digit numbers, you can now move on to multiplying by larger whole numbers.

Let me show you how this is done by again working out and explaining some examples.

Example 1: 21 x 32 = ?

```
     tens
     ones
      21
    x 32
      42  ──→ 2 ones x 21 = 42 ones, so the last digit of 42
                is placed in the ones column.
      63  ──→ 3 tens x 21 = 63 tens, so the last digit of 63
     672       is placed in the tens column.
```

Now, add those products, making sure the 42 and the 63 are placed under their respective multipliers: the 42 ends under the 2; the 63 ends under the 3.

Example 2: 516 x 48 = ?

```
        2  ←── number carried when multiplying by the 4
       14  ←── numbers carried when multiplying by the 8

      516
    x  48
     4128  ──→ 8 ones x 516 = 4128 ones
     2064  ──→ 4 tens x 516 = 2064 tens
    24768  ──→ adding those products
```

Example 3: 327 x 402 = ?

```
       12  ←── numbers carried when multiplying by the 4
        1  ←── number carried when multiplying by the 2

      327
    x 402
      654  ──→ 2 ones x 327 = 654 ones
      000  ──→ 0 tens x 327 = 000 tens
     1308  ──→ 4 hundreds x 327 = 1308 hundreds
   131454  ──→ adding those products
```

Notice that the products 654, 000, and 1308 were placed under their multipliers. The 654 ended under its multiplier 2, the 000 ended under its multiplier 0, and the 1308 ended under its multiplier 4.

To check your answer, multiply the numbers in the reverse order.

PROBLEM 1: 571 x 25 = ?   Answer: 14,275   Check:

```
                                              1                       2
                                              3                       3
                                            571                      25
                                          x  25                   x 571
                                           2855                      25
                                           1142                     175
                                          14275                     125
                                                                  14275
```

PROBLEM 2: 382 x 560 = ?   Answer: 213,920   Check:
```
                                                                      1
                                             41                       4
                                             41                       1
                                            382                     560
                                          x 560                   x 382
                                            000                    1120
                                           2292                    4480
                                           1910                    1680
                                         213920                  213920
```

PROBLEM 3: 15,218 x 37 = ?   Answer: 563,066

```
                                            1  2
                                            3115
                                           15218
                                          x    37
                                          106526
                                           45654
                                          563066
```

PROBLEM 4: 5036 x 727 = ?   Answer: 3,661,172

```
                                             24
                                              1
                                             24
                                           5036
                                          x 727
                                          35252
                                          10072
                                          35252
                                        3661172
```

EXERCISE 1-7   SET A

NAME _____   DATE _____

MULTIPLY THE FOLLOWING: ANSWERS

1.   41   2.   34   1. _____
   x 23   x 12
                                                        2. _____

3.   984   4.   573   3. _____
   x 67   x 89
                                                        4. _____

5.   407   6.   503   5. _____
   x 53   x 84
                                                        6. _____

7.   518   8.   473   7. _____
   x 280   x 910
                                                        8. _____

9.   14256   10. 23819   9. _____
   x   568   x   476
                                                       10. _____

11.   5016 x 58   11. _____

12.   653 x 356   12. _____

13.   74,113 x 142   13. _____

14.   51,763 x 605   14. _____

15.   823 x 6515   15. _____

16.   41 x 816 x 7556   16. _____

27

EXERCISE 1-7   SET B

NAME _____   DATE _____

MULTIPLY THE FOLLOWING: ANSWERS

1.   43   2.   43   1. _____
    x 22   x 31
                                                 2. _____

3.   652   4.   539   3. _____
   x 84   x 76
                                                 4. _____

5.   701   6.   806   5. _____
   x 85   x 49
                                                 6. _____

7.   624   8.   752   7. _____
   x 180   x 920
                                                 8. _____

9. 17628   10. 34786   9. _____
   x 476   x 983
                                                10. _____

11. 6207 x 47   11. _____

12. 749 x 394   12. _____

13. 65,211 x 185   13. _____

14. 51,487 x 506   14. _____

15. 945 x 5165   15. _____

16. 45 x 184 x 6443   16. _____

1-8 Handling Zeros in Multiplying Whole Numbers

Zeros at the End of a Multiplier

If you were to use the process explained in the last section to multiply 21000 by 3000, it would look like this:

```
      21000
   x  3000
      00000
      00000
      00000
      63000
   63000000
```

Multiplying by all those zeros seems to just take space and waste time. There is a more efficient way to do that problem. Since any number times zero is equal to zero, and since the only nonzero answers occurred when you multiplied by the 3, why bother writing all those zeros down? Here is what you can do:

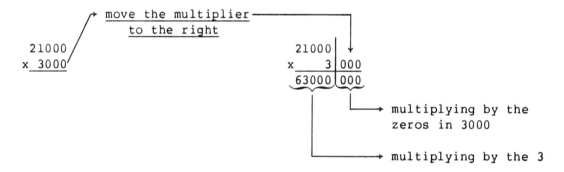

In either case you get 21000 x 3000 = 63,000,000.

Observe the similar procedure in the next example.

Example 1: 129 x 730,000,000 = ?

```
     26
      2
    129
  x  73|0,000,000
    387|0,000,000      Notice:
    903                    zeros placed to the right
   9417|0,000,000   → multiplying by the zeros
                    → multiplying 129 x 73
```

PROBLEM 1: 54 x 780,000 = ?   Answer: 42,120,000

```
      2
      3
     54
     78|0000
    432|0000
    378
   4212|0000
```

29

Zeros in the Middle of a Multiplier

If you likewise multiply 1312 by 2003 using the process explained in Section 1-7, it would look like this:

```
      1312
    x 2003
      3936
      0000
     0000
    2624
    2627936
```

$1312 \times 2003 = 2{,}627{,}936$

Those two rows of zeros again take space and waste time. Here is a faster way to do a problem like that:

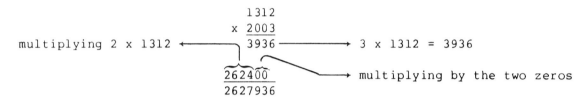

multiplying 2 x 1312 ←
```
      1312
    x 2003
      3936
    262400
    2627936
```
→ 3 x 1312 = 3936

→ multiplying by the two zeros

Note: You must be careful to place the products under their respective multipliers. The 3936 was placed under its multiplier 3, the two zeros were placed under their 0 multipliers, and the 2624 was placed under its multiplier 2.

Example 2:   572 x 60,004 = ?

```
         41
          2
        572
    x 60004
       2288  ───────→ 4 x 572 = 2288
    3432000  ───────→ 6 x 572 = 3432, and three zeros x 572 = 000
   34322288
```

PROBLEM 2:   4075 x 2008 = ?   Answer: 8,182,600

```
      11
      64
    4075
  x 2008
   32600
  815000
 8182600
```

PROBLEM 3:   25 x 100,003   Answer: 2,500,075

```
       1
      25
 x 100003
      75
  250000
 2500075
```

EXERCISE 1-8   SET A

NAME _____   DATE _____

MULTIPLY THE FOLLOWING: ANSWERS

1.   61   2.   87   1. _____
     x 70   x 50
                                                                           2. _____

3.   64   4.   75   3. _____
     x 509   x 807
                                                                           4. _____

5.   427   6.   634   5. _____
     x 23000   x 19000
                                                                           6. _____

7.   700   8.   600   7. _____
     x 400   x 500
                                                                           8. _____

9.   74000   10.   53000   9. _____
     x 8000   x 6000
                                                                          10. _____

11.   42   12.   63   11. _____
     x 3005   x 3006
                                                                          12. _____

13.   2504   14.   6305   13. _____
     x 4006   x 3009
                                                                          14. _____

15.   185 x 53,000   15. _____

16.   6,400 x 510,000   16. _____

17.   30,004 x 2,195   17. _____

18.   2,009 x 4,000,000   18. _____

31

EXERCISE 1-8   SET B

NAME _____   DATE _____

MULTIPLY THE FOLLOWING: ANSWERS

1.   52   2.   49   1. _____
   x 70   x 50
                                                      2. _____

3.   71   4.   95   3. _____
   x 309   x 708
                                                      4. _____

5.   398   6.   463   5. _____
   x 17000   x 43000
                                                      6. _____

7.   500   8.   900   7. _____
   x 600   x 700
                                                      8. _____

9. 45000   10. 76000   9. _____
   x 3000   x 7000
                                                     10. _____

11.   23   12.   36   11. _____
   x 4005   x 6003
                                                     12. _____

13. 1708   14. 2604   13. _____
   x 8002   x 9003
                                                     14. _____

15. 752 x 41,000   15. _____

16. 8,700 x 230,000   16. _____

17. 40,003 x 1,921   17. _____

18. 2,007 x 7,000,000   18. _____

1-9 Powers, Roots, Comparisons

IF YOU GET $2^3=6$, YOU HAVE EXPERIENCED <u>POWER</u> FAILURE.

There are a few other concepts that should also be learned as you continue to improve your math skills.

Powers

There is a short way to express a multiplication problem in which a number is repeatedly multiplied by itself, such as 3x3x3x3x3. Instead of writing all those 3's, we write 3^5. The raised number, 5, is called the **exponent** or **power** and the 3 is called the **base**. 3^5 is read "3 raised to the 5th power."

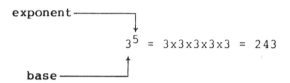

$$3^5 = 3\text{x}3\text{x}3\text{x}3\text{x}3 = 243$$

The **exponent** tells you how many times to multiply the **base** number by itself.

For example:

$7^2 = 7\text{x}7 = 49$
$2^3 = 2\text{x}2\text{x}2 = 8$
$1^8 = 1\text{x}1\text{x}1\text{x}1\text{x}1\text{x}1\text{x}1\text{x}1 = 1$

PROBLEM 1: $4^2 = ?$ Answer: 16 (4x4)

PROBLEM 2: $10^5 = ?$ Answer: 100,000 (10x10x10x10x10)

PROBLEM 3: $0^6 = ?$ Answer: 0 (0x0x0x0x0x0)

Square Roots

Raising a number to the second power is called **squaring** a number. For example, five squared means:

$5^2 = 5\text{x}5 = 25$ ⟶ the square of 5

The reverse of squaring a number is called taking the **square root** of a number. In the above example, the number multiplied by itself to get that answer of 25 is the square root of 25, written $\sqrt{25}$. That is:

$5^2 = 5\text{x}5 = 25$ ⟶ the square root of 25 ($\sqrt{25} = 5$)

> To find the square root of a number you must find what multiplied by itself results in the given number.

$\sqrt{4}$ = 2, since the number multiplied by itself that gives 4 is 2. (2x2 = 4)

$\sqrt{49}$ = 7, since the number multiplied by itself that gives 49 is 7. (7x7 = 49)

$\sqrt{169}$ = 13, since the number multiplied by itself that gives 169 is 13. (13x13 = 169)

PROBLEM 4: $\sqrt{64}$ = ?   Answer: 8, since 8x8 = 64

PROBLEM 5: $\sqrt{121}$ = ?   Answer: 11, since 11x11 = 121

PROBLEM 6: $\sqrt{0}$ = ?   Answer: 0, since 0x0 = 0

PROBLEM 7: $\sqrt{10,000}$ = ?   Answer: 100, since 100x100 = 10,000

Comparing Whole Numbers

There are three basic symbols that are used to compare the size of numbers or answers to problems in arithmetic. They are the "equals" sign (=), the "greater than" sign (>), and the "less than" sign (<).

The "equals" sign is used whenever two numbers or two answers are the same. The statement 7=7 is read "seven equals seven."

For example: 105 = 105,   4+2 = 2x3, $\sqrt{100}$ = 10,   8x7 = 7x8

The "greater than" sign is used whenever the first number is larger than the second number being compared. The statement 7>4 is read "seven is greater than four."

For example: 26 > 25,   3+10 > 5x1, 7^2 > 7,   49 > $\sqrt{49}$

The "less than" sign is used whenever the first number is smaller than the second number being compared. The statement 0<5 is read "zero is less than five."

For example: 456 < 500,   8x2 < 7+11, $\sqrt{1}$ < 2,   12 < 12+1

An inequality sign points to the smaller of the two numbers involved.

PROBLEMS: Replace the ? with =, >, or <:   Answers:

8. 8x0 ? 8x1   8. <, since 0<8

9. 15+7+6 ? 7+15+6   9. =, since 28=28

10. 9^2 ? $\sqrt{9}$ 10. >, since 81>3

EXERCISE 1-9   SET A

NAME _____   DATE _____

FIND THE FOLLOWING POWERS: ANSWERS

1. 3^2 2. 7^2 1. _____   2. _____

3. 12^2 4. 15^2 3. _____   4. _____

5. 4^3 6. 2^3 5. _____   6. _____

7. 0^5 8. 0^7 7. _____   8. _____

9. 1^4 10. 1^6 9. _____   10. _____

11. 2^5 12. 3^5 11. _____   12. _____

13. 10^6 14. 10^4 13. _____   14. _____

15. 34^2 16. 45^2 15. _____   16. _____

FIND THE FOLLOWING SQUARE ROOTS:

17. $\sqrt{4}$ 18. $\sqrt{9}$ 17. _____   18. _____

19. $\sqrt{25}$ 20. $\sqrt{16}$ 19. _____   20. _____

21. $\sqrt{36}$ 22. $\sqrt{49}$ 21. _____   22. _____

23. $\sqrt{1}$ 24. $\sqrt{64}$ 23. _____   24. _____

25. $\sqrt{0}$ 26. $\sqrt{225}$ 25. _____   26. _____

27. $\sqrt{100}$ 28. $\sqrt{81}$ 27. _____   28. _____

REPLACE THE ? WITH =, >, OR <:

29. 6x4 ? 4x6   29. _____

30. 0 x 156 ? 156 x 1   30. _____

31. 757 + 893 ? 893 + 757   31. _____

32. 3^2 ? 2^3 32. _____

33. $\sqrt{25}$? 5^2 33. _____

34. 10^4 ? 40   34. _____

35. 2x0 ? 2+0   35. _____

EXERCISE 1-9   SET B

NAME _____ DATE _____

FIND THE FOLLOWING POWERS: ANSWERS

1. 4^2 2. 6^2 1. _____   2. _____

3. 13^2 4. 18^2 3. _____   4. _____

5. 5^3 6. 3^3 5. _____   6. _____

7. 0^4 8. 0^6 7. _____   8. _____

9. 1^5 10. 1^7 9. _____ 10. _____

11. 3^5 12. 2^5 11. _____ 12. _____

13. 10^5 14. 10^3 13. _____ 14. _____

15. 43^2 16. 24^2 15. _____ 16. _____

FIND THE FOLLOWING SQUARE ROOTS:

17. $\sqrt{16}$ 18. $\sqrt{1}$ 17. _____ 18. _____

19. $\sqrt{36}$ 20. $\sqrt{25}$ 19. _____ 20. _____

21. $\sqrt{49}$ 22. $\sqrt{9}$ 21. _____ 22. _____

23. $\sqrt{0}$ 24. $\sqrt{100}$ 23. _____ 24. _____

25. $\sqrt{196}$ 26. $\sqrt{121}$ 25. _____ 26. _____

27. $\sqrt{81}$ 28. $\sqrt{4}$ 27. _____ 28. _____

REPLACE THE ? WITH =, >, OR <:

29. 12 x 9 ? 9 x 12   29. _____

30. 0 x 986 ? 986 x 1   30. _____

31. 793 + 927 ? 927 + 793   31. _____

32. 3^4 ? 4^3 32. _____

33. $\sqrt{100}$? 10^2 33. _____

34. 5^2 ? 10   34. _____

35. 8x0 ? 8+0   35. _____

1-10 Applications Involving Whole Numbers

When math problems are expressed in words, you may have some difficulty deciding what steps should be taken to solve the problem. There is no easy way to tell you what to do in a particular word problem. However, you should be able to apply your knowledge of arithmetic to word problems by thinking carefully and following the steps listed below.

1. **Read:** Read the problem slowly and carefully. Determine what is being asked for and what are the facts of the problem.

2. **Analyze:** Determine what operations are needed to answer the question presented in the problem.

3. **Solve:** Do the necessary computation to obtain the answer to the given problem.

Example 1: If one tire costs $63, how much would four tires cost?

Analyze: 1 tire costs $63, so 4 tires would cost 4 times as much. You must multiply 4 times $63.

Solve: $63
      x   4
       $252

Example 2: You spend $69 on a sweatsuit, $29 for tennis shoes, and $3 for socks. What is your total bill before sales tax?

Analyze: You want the total of the prices, so add them together.

Solve: $69
         29
       + 3
       $101

Example 3: At a local circus, tickets were priced $5 for adults and $3 for children. How much would a group, consisting of 7 adults and 26 children, pay to go to the circus?

Analyze: To find the total cost, you need to add the cost for the adults (7 x $5) and the cost for the children (26 x $3).

Solve: adults: 7 x $5 = $35
       children: 26 x $3 = $78
       total: = $113

37

Example 4:   In October your heating bill was $27. However, your November bill was double your October bill and your December bill was triple your November bill. How much did you pay for heating in those first three months?

Analyze: Double means to multiply by 2; triple means to multiply by 3. To find the November bill multiply the October bill by 2. To find the December bill multiply the November bill by 3. Add the amount from each month to get the total.

Solve:
October: = $ 27
November: 2 x $27   = $ 54
December: 3 x $54   = $162
Total: = $243

Example 5:   A truck is loaded with 7 sacks of cement weighing 90 pounds each, 20 bags of sand weighing 75 pounds each, and 12 posts weighing 24 pounds each. If the truck alone weighs 4500 pounds and the driver weighs 176 pounds, what is the total weight of the loaded truck?

Analyze: To find the total weight, add the weight of the cement, sand, posts, truck, and driver.

Solve:
cement: 7 x 90 = 630 lb
sand: 20 x 75 = 1500 lb   (Note: lb is an
posts: 12 x 24 = 288 lb   abbreviation for pounds)
truck: = 4500 lb
driver: = 176 lb
total: = 7094 lb

PROBLEM 1:   If an average double-spaced page of typing contains 250 words, how many words would an average 15 page term paper contain?

Answer: 3750 words

250 x 15 = 3750

PROBLEM 2:   In a football game, 9 touchdowns (6 pts. each), 6 field goals (3 pts. each), and 7 extra points (1 pt each) were scored. How many total points were scored in the football game?

Answer: 79 pts.

9 x 6 = 54
6 x 3 = 18
7 x 1 = 7
total = 79

PROBLEM 3:   If you now earn $975 per month, and are guaranteed a $35 raise each month you remain with the company, what will be your monthly salary after one year?

Answer: $1395

12 x $35 = $420
           +$975
salary: $1395

EXERCISE 1-10   SET A

NAME _____   DATE _____

                                                                    ANSWERS

1. At the bookstore you spend $15 on a math book, $28 on art
   books, and $9 on pens. What is your bill before sales tax?   1. _____

2. In four golf tournaments, Nancy Lopez won $12,000, $15,000,
   $16,500, and $15,000. What was her total earnings for the
   four tournaments?   2. _____

3. Your compact car gets 39 miles per gallon of gasoline. How
   many miles can you drive on a full tank of 15 gallons?   3. _____

4. A ream of paper contains 500 sheets of paper. How many
   sheets of paper are contained in a box of 24 reams?   4. _____

5. If you drink 3 cups of coffee a day and use 2 cubes of
   sugar in each cup, how many cubes of sugar would you use
   in one year (365 days)?   5. _____

6. A box of candy contains 28 pieces of chocolate. How many
   pieces of chocolate are contained in a dozen boxes?   6. _____

7. A house contains three bedrooms with 144 sq ft of floor
   space each, a master bedroom with 300 sq ft, two bathrooms
   with 56 sq ft each, a kitchen with 240 sq ft, a living room
   with 460 sq ft, and a hallway with 105 sq ft. What is the
   total floor space of the house?   7. _____

8. During a season, a basketball team made 623 free throws
   (1 pt. each), 1742 field goals (2 pts. each), and 47 three
   point baskets (3 pts. each). How many total points did the
   team score that season?   8. _____

9. Each piece of pipe is 16 ft long and each joint that connects
   two pieces of pipe adds 2 ft to the total length. The pieces
   of pipe are joined together and laid end to end in a straight
   line. If 24 sections of pipe and 23 joints are laid in such
   a manner, how far would the pipe reach?   9. _____

10. A small truck is carrying 25 crates of eggs. If each crate
    has 18 dozen eggs, how many eggs is the truck carrying?   10. _____

11. Tickets for a school production sold for $5 for general
    admission and $8 for reserved seats. If 575 general
    admission and 250 reserved tickets were sold, how much
    money was collected?   11. _____

EXERCISE 1-10   SET B

NAME _____   DATE _____

ANSWERS

1. During a shopping spree you wrote checks for $13, $27, $47, $78, and $9. What is the total of the checks you wrote?

   1. _____

2. On a special diet, you ate a lunch consisting of 1 apple (80 cal), 8 oz of yogurt (230 cal), 1 hard boiled egg (80 cal), and a glass of tomato juice (45 cal). What was your calorie intake for that lunch?

   2. _____

3. A computer prints invoices at a rate of 1256 per hour. How many invoices does it print in 8 hours?

   3. _____

4. Due to the stronger force of gravity, objects on earth weigh 6 times what they weigh on the moon. If an object weighs 28 lb on the moon, what is its weight on earth?

   4. _____

5. If a leaking faucet drips water at a rate of two gallons every hour, how many gallons of water is lost in a week?

   5. _____

6. If a bag of peanuts contains 48 peanuts, how many peanuts are contained in two dozen bags?

   6. _____

7. If you had three $50 bills, seven $20 bills, nine $10 bills, six $5 bills, and seventeen $1 bills, how much cash would you have?

   7. _____

8. During a season, a football team scored 39 touchdowns (6 pts. each), 23 field goals (3 pts. each), and 35 extra points (1 pt. each). How many total points did the team score that season?

   8. _____

9. If you started work earning $325 a week, then received two raises of $35 a week and one raise of $75 a week, how much would you now be earning per week?

   9. _____

10. A box of pens contains one dozen pens. How many pens would you have if you purchased a dozen boxes?

    10. _____

11. The local movie theater charges $5 for adults and $3 for children. How much would a family of 2 adults and 5 children pay to see a movie?

    11. _____

Chapter 1 Summary

Concepts

You may refer to the sections listed below to review how to:

1. read and write whole numbers using the place values of our number system. (1-1)

2. round off whole numbers by changing them to nearest multiples. (1-2)

3. add whole numbers by lining up the digits with corresponding place values. (1-3), (1-4)

4. multiply whole numbers by placing them above each other and using the carrying process. (1-5), (1-6), (1-7)

5. use short cuts for handling zeros in multiplying whole numbers. (1-8)

6. find powers and square roots of whole numbers. (1-9)

7. use the "equals" (=), "greater than" (>), and "less than" (<) signs. (1-9)

8. solve word problems involving whole numbers using the "read, analyze, solve" procedure. (1-10)

Terminology

This chapter's important terms and their corresponding page numbers are:

addend: name given to the numbers that are added. (13)

base: the number being raised to a power. (33)

digit: any of the ten numerals 1, 1, 2, 3, 4, 5, 6, 7, 8, 9. (5)

exponent or **power:** raised number that tells how many times to use a number as a factor. (33)

factors: name given to the numbers that are multiplied. (19)

place value: the value assigned to the position of a digit in a number. (5)

product: the answer from multiplication. (19)

square root: a number which, when multiplied by itself, results in the given number. (33)

squaring: the process of raising a number to the second power (multiplying it by itself). (33)

sum: the answer from addition. (13)

Student Notes

Chapter 1 Practice Test A

ANSWERS

1. Write using numbers and commas:
   two hundred fifty-four million, thirty thousand,
   one hundred seven.

   1. _____

2. Write in words: 3,407,123

   2. _____

Round off 753,250 to the nearest:

3. thousand   4. hundred thousand

   3. _____

   4. _____

5. 234 + 78 + 3804

   5. _____

6. 3456 + 907 + 78,954 + 3012 + 56

   6. _____

7. 769 x 6

   7. _____

8. 5382 x 879

   8. _____

9. 952 x 41,000,000

   9. _____

10. 365 x 700,006

    10. _____

11. 3^4 12. $\sqrt{9}$

    11. _____

    12. _____

Replace the ? with =, >, or <:

13. 5 x 10 ? 10^5 14. $\sqrt{16}$? 2^2

    13. _____

    14. _____

15. A delivery van is carrying 12 TV sets weighing 54 pounds each and 25 VCRs weighing 16 pounds each. What is the total weight of the merchandise in the van?

    15. _____

Chapter 1 Practice Test B

ANSWERS

1. Write using numbers and commas:
   one billion, twenty-five million, three hundred
   six thousand, five hundred seventeen   1. _____

2. Write in words: 678,309   2. _____

Round off 1,257,381 to the nearest:

3. ten   4. ten thousand   3. _____

                                                   4. _____

5. 347 + 23 + 1067   5. _____

6. 47 + 89,099 + 235 + 9 + 523 + 4128   6. _____

7. 478 x 8   7. _____

8. 2785 x 596   8. _____

9. 56 x 340,000   9. _____

10. 906 x 6,000,001   10. _____

11. 4^3 12. $\sqrt{4}$ 11. _____

                                                   12. _____

Replace the ? with =, >, or <:

13. 345 x 67 ? 67 x 345   14. 5^2 ? $\sqrt{25}$ 13. _____

                                                   14. _____

15. The low temperature on Monday was 67°. If there was a
    3° increase in the low temperature each day for six
    consecutive days, what was the low temperature on Sunday? 15. _____

Chapter 1 Supplementary Exercises

Section 1-1

Write in words:

1. 506
2. 1317
3. 260,000
4. 45,200
5. 107,070
6. 1,325,006
7. 82,000,090

Write using numbers:

8. four hundred seven
9. seventeen thousand
10. five thousand two
11. twenty-one thousand, six hundred
12. sixty thousand, two hundred eight
13. one million, thirty thousand
14. fourteen million, five hundred

Section 1-2

Round off to the nearest thousand:

1. 3706
2. 8397
3. 6542
4. 9765
5. 49,500
6. 126,499

Round off 6,984,155 to the nearest:

7. ten
8. hundred
9. thousand
10. ten thousand
11. hundred thousand
12. million

Section 1-4

1. 142 + 706
2. 7638 + 297
3. 3876 + 1849
4. 7500 + 9700
5. 517 + 6380
6. 276 + 4302
7. 8976 + 10,385
8. 46,529 + 9813
9. 19 + 126 + 4973
10. 26 + 384 + 4198
11. 9000 + 287 + 63,509
12. 400 + 19,876 + 2914

Section 1-6

1. 486 x 2
2. 4320 x 3
3. 487 x 5
4. 1906 x 7
5. 5006 x 8
6. 6008 x 9
7. 8966 x 6
8. 7699 x 6
9. 6 x 27,800
10. 7 x 43,900
11. 5 x 49,457
12. 4 x 76,899

Section 1-7

1. 45
   × 31
2. 728
   × 92
3. 96
   × 87
4. 788
   × 96

5. 835 × 496
6. 765 × 537
7. 257 × 4282
8. 536 × 4776
9. 826 × 1418
10. 195 × 6888
11. 43,216 × 128
12. 54,617 × 216
13. 16,736 × 608
14. 14,384 × 907
15. 8769 × 270
16. 7836 × 460
17. 12,548 × 96
18. 3280 × 875

Section 1-8

1. 96
   × 50
2. 85
   × 600
3. 72
   × 309
4. 52
   × 7001

5. 794 × 8000
6. 5000 × 658
7. 283 × 30,000
8. 918 × 60,000
9. 5019 × 370,000
10. 2097 × 240,000
11. 826 × 5007
12. 766 × 4008
13. 6008 × 436
14. 3009 × 983
15. 40,002 × 78
16. 67 × 30,007
17. 5000 × 420
18. 930 × 6800

Section 1-9

1. 8^2
2. 7^2
3. 6^3
4. 5^4
5. 3^3
6. 11^2
7. 0^6
8. 1^7
9. 4^3
10. 3^4
11. 4^2
12. 2^4
13. $\sqrt{9}$
14. $\sqrt{25}$
15. $\sqrt{36}$
16. $\sqrt{16}$
17. $\sqrt{64}$
18. $\sqrt{49}$
19. $\sqrt{81}$
20. $\sqrt{121}$
21. $\sqrt{1}$
22. $\sqrt{0}$
23. $\sqrt{4}$
24. $\sqrt{100}$

Replace the ? with =, >, or <:

25. 5+7 ? 5×7
26. 6+9 ? 9+6
27. 0×89 ? 0+89
28. 96×1 ? 96+1
29. 14+9 ? 19+4
30. 17×37 ? 37×17
31. 1^5 ? 0^5
32. $\sqrt{49}$? 7^2
33. 78×6 ? 86×7
34. 10^2 ? $\sqrt{100}$
35. 4^2 ? 2^4
36. 3^5 ? 5^2
37. 4^2 ? $\sqrt{4}$
38. $\sqrt{9}$? 3^2
39. 1^4 ? 1^5

Section 1-10

1. The attendance at six performances of a local production of "Chorus Line" was 314, 299, 324, 223, 198, and 327. What was the total attendance for the performances?

2. Alice scored 17 points more on the second test of the semester than she did on the first test. If her score on the first test was 76, what was her score on the second test?

3. What is the total cost for a dozen potted plants that sell for $6 each?

4. If a necklace requires 76 pearls, how many pearls would be needed to make 15 necklaces?

5. You rent a car for a week, paying $19 a day and buying your own gas. If the gas costs you $12 a day, how much does it cost to rent the car for the week?

6. If you buy seven cubic yards of gravel at a cost of $39 a cubic yard and pay an additional fee of $25 for delivery, what is the total amount you pay for the gravel?

7. If your regular wage is $6 an hour and you earn $9 an hour for overtime, how much do you earn for working 40 regular hours and 7 overtime hours?

8. If you had one $50 bill, nine $20 bills, and seven $5 bills, how much cash would you have?

9. A history test consists of 10 true-false questions worth 2 points each, 14 multiple choice questions worth 5 points each, and 4 essay questions worth 15 points each. How many total points does the test contain?

10. In a track meet, first place finishers score 5 points, second place finishers score 3 points, and third place finishers score 1 point. If your team took 7 firsts, 12 seconds, and 9 thirds, how many points did your team score?

Chapter 2

Whole Numbers: Subtraction, Division, and the Order of Operations

After finishing this chapter you should be able to do the following:

1. subtract whole numbers.
2. divide whole numbers.
3. perform computations that involve more than one operation.
4. factor whole numbers into primes.
5. solve word problems involving operations with whole numbers.

On the next page you will find a pretest for this chapter. The purpose of the pretest is to help you determine which sections in this chapter you need to study in detail and which sections you can review quickly. By taking and correcting the pretest according to the instructions on the next page you can better plan your pace through this chapter.

Chapter 2 Pretest

Take and correct this test using the answer section at the end of the book. Those problems that give you difficulty indicate which sections need extra attention. Section numbers are in parentheses before each problem.

ANSWERS

(2-2) 1. 683 − 352

1. _____

(2-2) 2. 77,003 − 10,399

2. _____

(2-2) 3. 8000 − 2947

3. _____

(2-3) 4. 54 ÷ 9   (2-4) 5. $7\overline{)41}$

4. _____

5. _____

(2-4) 6. 36 ÷ 0   (2-5) 7. 318 ÷ 6

6. _____

7. _____

(2-6) 8. $8\overline{)16507}$ (2-6) 9. $692\overline{)204140}$

8. _____

9. _____

(2-7) 10. $57\overline{)34562}$ (2-7) 11. $85\overline{)51000}$

10. _____

11. _____

(2-8) 12. $\sqrt{64} + 9 \times 10^3$

12. _____

(2-8) 13. 5 × (15 − 3 + 6) + 42 ÷ 6

13. _____

(2-9) 14. Factor 140 into primes.

14. _____

(2-10) 15. In your last five math tests you received scores of 95, 82, 86, 94, and 98. What is the average of your test scores?

15. _____

2-1 Subtraction Facts

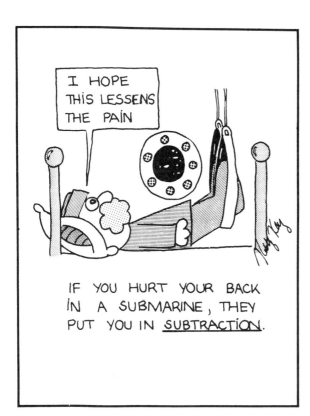

If you had $25 and spent $7 on lunch, how much would you have left? To solve the problm you would have to take $7 away from the $25 and get the answer, $18. You would have done the opposite of addition. This process is called subtraction. We could write that problem in the following ways:

25 subtract 7 is 18

25 minus 7 equals 18

25 - 7 = 18

The name given to each part of a subtraction problem is as follows:

minuend   subtrahend   difference
↓   ↓   ↓
25   -   7   =   18

Below is a chart containing the basic subtraction facts that you should know.

−	0	1	2	3	4	5	6	7	8	9
0	0	-	-	-	-	-	-	-	-	-
1	1	0	-	-	-	-	-	-	-	-
2	2	1	0	-	-	-	-	-	-	-
3	3	2	1	0	-	-	-	-	-	-
4	4	3	2	1	0	-	-	-	-	-
5	5	4	3	2	1	0	-	-	-	-
6	6	5	4	3	2	1	0	-	-	-
7	7	6	5	4	3	2	1	0	-	-
8	8	7	6	5	4	3	2	1	0	-
9	9	8	7	6	5	4	3	2	1	0
10	10	9	8	7	6	5	4	3	2	1
11	11	10	9	8	7	6	5	4	3	2
12	12	11	10	9	8	7	6	5	4	3
13	13	12	11	10	9	8	7	6	5	4
14	14	13	12	11	10	9	8	7	6	5
15	15	14	13	12	11	10	9	8	7	6
16	16	15	14	13	12	11	10	9	8	7
17	17	16	15	14	13	12	11	10	9	8
18	18	17	16	15	14	13	12	11	10	9

Note: These dashes are here since at this point you can not subtract a larger number from a smaller number.

To use the chart to find an answer, such as 9 - 6, find the 9 on the left hand side and go along that row until you reach the 6 at the top of the chart. That number (3) will be the answer to 9 - 6.

To help you master basic subtraction facts, on the next page you will find a self drill in subtraction. You might also find the use of "flash cards" helpful in learning the basic subtraction facts.

Self Drill in Subtraction

Directions

1. Cover the answers for a row.

2. Quickly figure out and write down your answer to each problem.

3. Check your work.

4. Repeat this for each row until you can do each one quickly and correctly.

2	5	3	9	9	6	2	3	1	10
−0	−1	−3	−2	−0	−3	−1	−2	−0	−3
2	4	0	7	9	3	1	1	1	7

10	7	4	3	5	8	4	8	2	9
−2	−0	−3	−1	−3	−2	−0	−1	−2	−1
8	7	1	2	2	6	4	7	0	8

8	4	9	3	11	1	8	4	7	6
−0	−2	−3	−0	−3	−1	−3	−1	−3	−2
8	2	6	3	8	0	5	3	4	4

6	5	6	7	10	0	11	7	12	5
−1	−2	−0	−1	−1	−0	−2	−2	−3	−0
5	3	6	6	9	0	9	5	9	5

4	6	10	8	13	11	10	7	7	10
−4	−5	−6	−4	−6	−5	−4	−5	−6	−5
0	1	4	4	7	6	6	2	1	5

15	14	6	5	12	13	13	9	11	14
−6	−6	−4	−5	−6	−4	−5	−6	−4	−5
9	8	2	0	6	9	8	3	7	9

12	5	11	9	8	7	9	8	6	12
−5	−4	−6	−5	−6	−4	−4	−5	−6	−4
7	1	5	4	2	3	5	3	0	8

9	11	8	11	7	12	16	14	8	14
−7	−9	−8	−8	−7	−9	−9	−7	−7	−9
2	2	0	3	0	3	7	7	1	5

9	15	9	10	13	10	13	11	17	12
−9	−8	−8	−7	−7	−8	−9	−7	−9	−8
0	7	1	3	6	2	4	4	8	4

18	16	12	16	14	13	15	15	17	10
−9	−8	−7	−7	−8	−8	−9	−7	−8	−9
9	8	5	9	6	5	6	8	9	1

2-2 Subtracting Whole Numbers

Using the knowledge of the basic subtraction facts, you can now subtract larger whole numbers. The basic procedure for subtracting is as follows:

> Place the numbers above each other so that the digits with the same place values are lined up. Subtract the digits in the ones, tens, hundreds, etc. places.

By following through the examples, this method will become clear.

Example 1: 7953 - 641 = ?

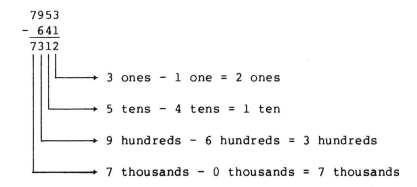

```
  7953
-  641
  7312
```

→ 3 ones - 1 one = 2 ones
→ 5 tens - 4 tens = 1 ten
→ 9 hundreds - 6 hundreds = 3 hundreds
→ 7 thousands - 0 thousands = 7 thousands

If you try to find the answer to $\begin{array}{r}52\\-17\end{array}$ the process in Example 1 fails to work out, since in the ones place you can not subtract a larger number (7) from a smaller number (2).

What do you do to solve that problem? Just as we have carrying in addition, there is a process called borrowing in subtraction. Let me show you what I mean.

Example 2: 52 - 17 = ?

— We borrow 1 ten from the 5 tens, leaving 4 tens.

```
  4 12
  5̷ 2̷
- 1 7     Now subtract.
  3 5
```

← Adding the 1 ten we borrowed to the 2 ones, we get 12 ones.

→ 12 ones - 7 ones = 5 ones
→ 4 tens - 1 ten = 3 tens

Example 3:   4628 - 593 = ?

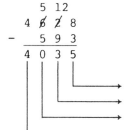

Since the 9 is larger than the 2 in the tens column, we borrow 1 hundred from the 6 hundreds, leaving 5 hundreds.

Adding the 1 hundred to the 2 tens we get 12 tens.

→ 8 ones - 3 ones = 5 ones
→ 12 tens - 9 tens = 3 tens
→ 5 hundreds - 5 hundreds = 0 hundreds
→ 4 thousands - 0 thousands = 4 thousands

Example 4:   603 - 178 = ?

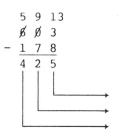

Since the 8 is larger than the 3 in the ones column, we must borrow. But there is a zero in the tens place, so we borrow 1 from the 6 hundreds (60 tens) leaving 59 tens. We then have 13 in the ones column.

→ 13 ones - 8 ones = 5 ones
→ 9 tens - 7 tens = 2 tens
→ 5 hundreds - 1 hundred = 4 hundreds

To check your work, add the difference to the subtrahend (the two bottom numbers), and you should get the minuend (the top number).

PROBLEM 1:   5713 - 4975 = ?   Answer: 738   Check:

```
        16 10                4975
      4 6 0 13              + 738
      5 7 1 3               5713
    - 4 9 7 5
        7 3 8
```

PROBLEM 2:   6003 - 527 = ?   Answer: 5476   Check:

```
    5 9 9 13                 527
    6 0 0 3                + 5476
    -   5 2 7               6003
    5 4 7 6
```

PROBLEM 3:   What's wrong with this subtraction problem?

```
      126
    -  48
       82
```

Answer:
Subtraction in the ones column was done in the reverse order.

EXERCISE 2-2 SET A

NAME _____   DATE _____

SUBTRACT THE FOLLOWING: ANSWERS

1.   96   2.   87   1. _____
   - 32   - 53
                                                 2. _____

3. 514   4. 816   3. _____
   - 13   - 14
                                                 4. _____

5. 9876   6. 5665   5. _____
  - 1034   - 1201
                                                 6. _____

7.   83   8.   94   7. _____
   - 27   - 66
                                                 8. _____

9. 329   10. 438   9. _____
   - 95   - 63
                                                 10. _____

11. 4563   12. 3762   11. _____
   - 2985   - 1974
                                                 12. _____

13. 607   14. 804   13. _____
   - 59   - 75
                                                 14. _____

15. 4003   16. 5004   15. _____
   - 758   - 639
                                                 16. _____

17. 5000   18. 7000   17. _____
   - 4638   - 5297
                                                 18. _____

19. 14,625 - 5,073   19. _____

20. 75,075 - 67,067   20. _____

21. 1,596,385 - 847,128   21. _____

EXERCISE 2-2 SET B

NAME _____   DATE _____

SUBTRACT THE FOLLOWING: ANSWERS

1.   57   2.   65   1. _____
   - 24   - 42
                                                   2. _____

3.   819   4.   715   3. _____
   - 14   - 13
                                                   4. _____

5. 6789   6. 5775   5. _____
   - 2025   - 3013
                                                   6. _____

7.   92   8.   83   7. _____
   - 57   - 65
                                                   8. _____

9.   435   10.   548   9. _____
   - 66   - 79
                                                  10. _____

11. 3762   12. 5418   11. _____
   - 1375   - 2329
                                                  12. _____

13. 507   14.   903   13. _____
   - 69   - 75
                                                  14. _____

15. 3006   16. 5005   15. _____
   - 629   - 738
                                                  16. _____

17. 7000   18. 9000   17. _____
   - 2561   - 3275
                                                  18. _____

19. 13,264 - 4,073   19. _____

20. 67,067 - 58,058   20. _____

21. 2,563,871 - 452,076   21. _____

2-3 Division Facts

SUCH A REPORT CARD GIVES ME DIVISIONS.

The last operation to cover is the division of whole numbers. Division occurs in problems such as this:

> If you had $28 and you wanted to buy concert tickets that cost $7 each, how many could you buy?

One way to do the problem is to determine how many 7's there are in 28 by subtracting $7 for each ticket, until the money runs out.

$$\begin{array}{cccc} 28 & 21 & 14 & 7 \\ \underline{-7} & \underline{-7} & \underline{-7} & \underline{-7} \\ 21 & 14 & 7 & 0 \end{array}$$ There are four 7's in 28.

Thus we say 28 divided by 7 is 4, and we write it:

$$28 \div 7 = 4$$

Or we say 7 divides into 28 four times and we write it:

$$7\overline{)28}^{\,4}$$

Notice the reversal of the order of the 28 and the 7 in these two ways to express division.

Each part of the division problem has a name:

$$\text{divisor} \rightarrow 7\overline{)28}^{\,4 \leftarrow \text{quotient}} \leftarrow \text{dividend}$$

If you had to use repeat subtraction whenever you divided numbers, a problem such as 13,676 ÷ 52 would take much work and many steps. There is luckily another way to consider division. Let's look at that example again.

$$7\overline{)28}^{\,4}$$

Notice: 4 x 7 = 28

$$7\overline{)28}^{\,4} \\ \underline{28}$$

That points out the basic fact about division. If you take the answer of a division problem (the quotient) and multiply it by the divisor, you get the number that you are dividing into (the dividend). Let me show you what this means by doing some examples.

Example 1: 18 ÷ 6 = ?

$$6\overline{)18}^{?} \qquad \text{Think:} \qquad ? \times 6 = 18 \qquad 6\overline{)18}^{\leftarrow ?}_{\hookrightarrow 18}$$

Since 3 × 6 = 18, we get 18 ÷ 6 = 3.

$$6\overline{)18}^{3}_{18}$$

Example 2: 72 ÷ 9 = ?

$$9\overline{)72}^{?} \qquad \text{Think:} \qquad ? \times 9 = 72 \qquad 9\overline{)72}^{\leftarrow ?}_{\hookrightarrow 72}$$

Since 8 × 9 = 72, we get 72 ÷ 9 = 8.

$$9\overline{)72}^{8}_{72}$$

Using the fact that the quotient times the divisor equals the dividend will make division quicker and easier to perform than using repeat subtraction.

PROBLEM 1:   Find 32 ÷ 8 using repeat subtraction.

Answer: 4

$$\begin{array}{cccc} 32 & 24 & 16 & 8 \\ \underline{-8} & \underline{-8} & \underline{-8} & \underline{-8} \\ 24 & 16 & 8 & 0 \end{array}$$

There are 4 eights in 32.

PROBLEM 2:   Find 45 ÷ 5 using the multiplication method.

Answer: 9

$$5\overline{)45}^{\leftarrow 9}_{\hookrightarrow 45}$$

PROBLEM 3:   Find 63 ÷ 9 using the multiplication method.

Answer: 7

$$9\overline{)63}^{\leftarrow 7}_{\hookrightarrow 63}$$

EXERCISE 2-3   SET A

NAME _____   DATE _____

DIVIDE THE FOLLOWING: ANSWERS

1. 8 ÷ 8   2. 4 ÷ 4   3. 7 ÷ 7   1. _____   2. _____   3. _____

4. 2)10   5. 6)24   6. 7)28   4. _____   5. _____   6. _____

7. 9 ÷ 3   8. 18 ÷ 6   9. 36 ÷ 9   7. _____   8. _____   9. _____

10. 8)40   11. 4)20   12. 3)21   10. _____ 11. _____ 12. _____

13. 5 ÷ 1   14. 12 ÷ 6   15. 56 ÷ 7   13. _____ 14. _____ 15. _____

16. 9)45   17. 5)10   18. 3)27   16. _____ 17. _____ 18. _____

19. 16 ÷ 2   20. 30 ÷ 5   21. 32 ÷ 8   19. _____ 20. _____ 21. _____

22. 8)48   23. 5)45   24. 3)15   22. _____ 23. _____ 24. _____

25. 9)72   26. 7)49   27. 8)48   25. _____ 26. _____ 27. _____

28. 64 ÷ 8   29. 63 ÷ 7   30. 81 ÷ 9   28. _____ 29. _____ 30. _____

31. 8)32   32. 7)35   33. 8)24   31. _____ 32. _____ 33. _____

34. 30 ÷ 6   35. 28 ÷ 4   36. 30 ÷ 5   34. _____ 35. _____ 36. _____

37. 8)56   38. 9)54   39. 7)21   37. _____ 38. _____ 39. _____

USE REPEAT SUBTRACTION TO SHOW THAT:

40. 20 ÷ 4 = 5   40. _____

    6
41. 3)18   41. _____

EXERCISE 2-3 SET B

NAME _____ DATE _____

DIVIDE THE FOLLOWING   ANSWERS

1. 8 ÷ 2   2. 10 ÷ 5   3. 15 ÷ 3   1. _____   2. _____   3. _____

4. 6)‾24   5. 4)‾36   6. 5)‾30   4. _____   5. _____   6. _____

7. 72 ÷ 8   8. 14 ÷ 7   9. 54 ÷ 9   7. _____   8. _____   9. _____

10. 2)‾10   11. 3)‾18   12. 2)‾12   10. _____ 11. _____ 12. _____

13. 25 ÷ 5   14. 16 ÷ 4   15. 48 ÷ 6   13. _____ 14. _____ 15. _____

16. 9)‾27   17. 7)‾56   18. 8)‾32   16. _____ 17. _____ 18. _____

19. 6)‾42   20. 8)‾16   21. 9)‾72   19. _____ 20. _____ 21. _____

22. 24 ÷ 3   23. 10 ÷ 5   24. 42 ÷ 7   22. _____ 23. _____ 24. _____

25. 3)‾21   26. 5)‾45   27. 7)‾35   25. _____ 26. _____ 27. _____

28. 18 ÷ 3   29. 40 ÷ 5   30. 49 ÷ 7   28. _____ 29. _____ 30. _____

31. 9)‾81   32. 8)‾56   33. 6)‾42   31. _____ 32. _____ 33. _____

34. 40 ÷ 8   33. 63 ÷ 9   36. 6 ÷ 6   34. _____ 35. _____ 36. _____

37. 8)‾48   38. 54 ÷ 6   39. 9)‾54   37. _____ 38. _____ 39. _____

USE REPEAT SUBTRACTION TO SHOW THAT:

40. 20 ÷ 5 = 4   40. _____

41. $\overset{5}{6\overline{)30}}$ 41. _____

2-4 Remainders, Dividing by Zero

MISSION IMPOSSIBLE: DIVIDING BY A ZERO.

Now that you have reviewed the basic division facts you can proceed onto more involved division problems.

For example, what is 26 ÷ 8?

Let's do it by repeat subtraction.

$$\begin{array}{r}26\\-8\\\hline18\end{array} \nearrow \begin{array}{r}18\\-8\\\hline10\end{array} \nearrow \begin{array}{r}10\\-8\\\hline2\end{array}$$

There are 3 eights in 26 with 2 ones left over. That left over amount is called the **remainder**.

Thus 26 ÷ 8 = 3 remainder 2, or 3 R2.

If you do that problem using the multiplication method, you would wirte:

$$\begin{array}{r}3\\8\overline{)26}\\24\\\hline 2\end{array} \longleftarrow \text{remainder}$$

Example 1: 57 ÷ 6 = ?

$6\overline{)57}^{?}$ There is no value for the ? x 6 that gives exactly 57.

$6\overline{)57}^{9}$ The closest we can get to 57, without going over it, is 9x6 = 54.
54
3 Subtracting the 54 from the 57 we get a remainder of 3.

Example 2: 36 ÷ 5 = ?

$5\overline{)36}^{?}$ There is no value for the ? x 5 that gives exactly 36.

$5\overline{)36}^{7}$ The closest we can get to 36, without going over it, is 7 x 5 = 35.
35
1

Subtracting the 35 from the 36 we get a remainder of 1.

61

We can summarize those examples by saying if a divisor does not divide exactly into a number, find the quotient that gets closest to the number without going over it and subtract to find the remainder.

PROBLEM 1: 23 ÷ 4 = ?   Answer: 5 R3

$$\begin{array}{r} 5 \\ 4\overline{)23} \\ \underline{20} \\ 3 \end{array}$$

PROBLEM 2: 74 ÷ 8 = ?   Answer: 9 R2

$$\begin{array}{r} 9 \\ 8\overline{)74} \\ \underline{72} \\ 2 \end{array}$$

Dividing by Zero

As you have seen, some division problems give exact answers while others have remainders. There are division problems, however, that do not have answers at all. Any division problem that has zero as the divisor can not be done. YOU CAN NOT DIVIDE BY ZERO.

Problems such as 7 ÷ 0 = ?, $0\overline{)56}$, or 340 ÷ 0 = ? are impossible!

Let me explain why this is so.

Consider 7 ÷ 0 = ?

1. If you used the first method of division, repeat subtraction, you will never get an answer. Look below:

    $$\begin{array}{cccc} 7 & 7 & 7 & 7 \\ \underline{-0} & \underline{-0} & \underline{-0} & \underline{-0} \\ 7 & 7 & 7 & 7 \end{array}$$ . . . You could subtract zeros forever and still be in the same place.

    The repeat subtraction method will not give you an answer for division by zero.

2. If you used the multiplication process for division, no answer can be obtained either. Look below:

    $$0\overline{)7}$$ with ? above and 7 below   . . . You simply can not find a value for the ?, since no number times zero equals 7.

    The multiplication process for division will also not give an answer for division by zero.

EXERCISE 2-4 SET A

NAME _____ DATE _____

ANSWERS

1. 25 ÷ 4   2. 32 ÷ 6   1. _____

                                            2. _____

3. 5)‾47‾   4. 4)‾29‾   3. _____

                                            4. _____

5. 68 ÷ 0   6. 75 ÷ 9   5. _____

                                            6. _____

7. 2)‾13‾   8. 0)‾16‾   7. _____

                                            8. _____

9. 49 ÷ 7   10. 81 ÷ 9   9. _____

                                            10. _____

11. 8)‾30‾   12. 9)‾50‾   11. _____

                                            12. _____

13. 5)‾33‾   14. 9)‾80‾   13. _____

                                            14. _____

15. 17 ÷ 6   16. 33 ÷ 4   15. _____

                                            16. _____

17. 7)‾58‾   18. 4)‾29‾   17. _____

                                            18. _____

19. 53 ÷ 7   20. 19 ÷ 8   19. _____

                                            20. _____

21. 75 ÷ 9   22. 46 ÷ 6   21. _____

                                            22. _____

23. 9)‾36‾   24. 8)‾64‾   23. _____

                                            24. _____

EXERCISE 2-4 SET B

NAME _____ DATE _____

ANSWERS

1. 17 ÷ 4   2. 31 ÷ 5   1. _____

2. _____

3. 4)‾38   4. 5)‾39   3. _____

4. _____

5. 15 ÷ 0   6. 79 ÷ 9   5. _____

6. _____

7. 2)‾17   8. 0)‾72   7. _____

8. _____

9. 36 ÷ 6   10. 64 ÷ 4   9. _____

10. _____

11. 7)‾52   12. 9)‾60   11. _____

12. _____

13. 6)‾57   14. 9)‾70   13. _____

14. _____

15. 19 ÷ 4   16. 38 ÷ 7   15. _____

16. _____

17. 7)‾66   18. 4)‾35   17. _____

18. _____

19. 29 ÷ 3   20. 15 ÷ 2   19. _____

20. _____

21. 8)‾70   22. 6)‾55   21. _____

22. _____

23. 9)‾45   24. 6)‾48   23. _____

24. _____

2-5 Dividing by One Digit Numbers

In the previous sections we covered the basic concepts about division. That knowledge will make division of whole numbers much easier. Let's study some examples and their explanations to see how division by a one digit number is accomplished.

Example 1:   693 ÷ 3 = ?

   If you try to use the concepts of the previous sections to find the answer, you might have some difficulty.

```
   ┌─ ?
 3 │ 693  . . .      It is not obvious what number times 3 gives 693.
   └→693
```

I will show you how to break a division problem into steps so that the answer can be obtained.

First: Instead of dividing the 3 into the entire number, divide the 3 into the 6 hundreds, getting:

```
   ┌─ 2
 3 │ 693     Notice:  The 2 is put in the hundreds place
   └→6                above the 6.
```

Second: Divide the 3 into the 9 in tens place.

   (We bring down the 9 as pictured to make it clear that the 3 is dividing into the 9 tens).

```
    ┌ 23 ┐
  3 │ 693      Notice:  The answer to the division (3) is
    │ 6↓               put in the tens place above the 9.
    │ ─
    │  9
    └→ 9
```

Third: Divide the 3 into the 3 in the ones place.

   (We bring down the 3 to make it clear that we are dividing into the 3).

```
    ┌ 231 ┐
  3 │ 693       Notice:  The answer to that division (1)
    │ 6  │              is put into the ones place above
    │ ─  │              the 3.
    │ 9  │
    │ 9↓ │
    │ ─  │
    │  3
    └→  3
```

Thus, 693 ÷ 3 = 231.

65

Example 2: 387 ÷ 4 = ?

```
    9
  ┌───
4 │ 387
  │ 36
    ──
     2
```
The divisor (4) is larger than the first digit of the dividend (3), so divide the 4 into the 38 tens, getting 9 tens and a remainder of 2 tens.

```
    9
  ┌───
4 │ 387
    36↓
    ──
    27
```
Bring down the next digit, the 7 ones, getting 27.

```
   96
  ┌───
4 │ 387
    36
    ──
    27
    24
    ──
     3
```
Divide the 4 into the 27, getting 6 with a remainder of 3.

PROBLEM 1: 6284 ÷ 2 = ?   Answer: 3142

```
     3142
  ┌──────
2 │ 6284
    6
    ─
    2
    2
    ─
     8
     8
     ─
      4
      4
      ─
```

PROBLEM 2: 897 ÷ 7 = ?   Answer: 128 R1

```
     128
  ┌─────
7 │ 897
    7
    ─
    19
    14
    ──
     57
     56
     ──
      1
```

PROBLEM 3: 2057 ÷ 6 = ?   Answer: 342 R5

```
     342
  ┌──────
6 │ 2057
    18
    ──
    25
    24
    ──
     17
     12
     ──
      5
```

EXERCISE 2-5   SET A

NAME _____   DATE _____

DIVIDE THE FOLLOWING   ANSWERS

1. 486 ÷ 2   2. 996 ÷ 3   1. _____

                                           2. _____

3. 4)96   4. 3)72   3. _____

                                           4. _____

5. 441 ÷ 7   6. 342 ÷ 6   5. _____

                                           6. _____

7. 5)433   8. 7)591   7. _____

                                           8. _____

9. 9)2313   10. 8)5224   9. _____

                                          10. _____

11. 3)7372   12. 6)1049   11. _____

                                          12. _____

13. 4)19089   14. 4)15638   13. _____

                                          14. _____

15. 7)47971   16. 8)19656   15. _____

                                          16. _____

17. 9)461241   18. 6)412752   17. _____

                                          18. _____

EXERCISE 2-5   SET B

NAME _____   DATE _____

ANSWERS

1. 682 ÷ 2   2. 696 ÷ 3   1. _____

2. _____

3. 5)65   3. 4)92   3. _____

4. _____

5. 378 ÷ 7   6. 672 ÷ 8   5. _____

6. _____

7. 9)551   8. 6)441   7. _____

8. _____

9. 8)4338   10. 9)1775   9. _____

10. _____

11. 6)1460   12. 3)1972   11. _____

12. _____

13. 5)16081   14. 4)11133   13. _____

14. _____

15. 8)19968   16. 7)17276   15. _____

16. _____

17. 6)147498   18. 9)520578   17. _____

18. _____

2-6 Dividing by Numbers Having More than One Digit

The process for division that has been explained in the last two sections will enable you to divide even larger numbers. Since those same principles as shown in Section 2-5 are involved in the next examples, I will concentrate on the division method rather than giving detailed explanations.

Example 1: 9065 ÷ 37 = ?

```
      ↓
37 ⟌ 9065
```

Step 1: Decide above which digit the quotient will start:
Does the 37 divide into the 9? No.
Does the 37 divide into the 90? Yes.
So the quotient starts above the 0.

```
       2
37 ⟌ 9065
     →74
       16
```

Step 2: Divide 37 into 90.
37 divides into 90 about 2 times.
The 2 is placed above the 0.
2 × 37 = 74.
The 74 is subtracted from the 90 to get a remainder of 16.

```
        24
37 ⟌ 9065
      74↓
      166
     →148
        18
```

Step 3: Bring down the next digit, the 6.
Divide the 37 into the 166.
37 divides into 166 about 4 times.
4 × 37 = 148.
The 148 is subtracted from the 166 to get a remainder of 18.

```
         245
37 ⟌ 9065
      74
      166
      148↓
        185
      →185
```

Step 4: Bring down the next digit, the 5.
Divide the 37 into the 185.
37 divides into 185 exactly 5 times.

Example 2: 13678 ÷ 243 = ?

```
        ↓
243 ⟌ 13678
```

Step 1: Decide where the quotient should start:
243 does not divide into 1 or 13 or 136.
243 divides into the 1367, so the quotient should start above the 7.

```
           5
243 ⟌ 13678
     →1215
        152
```

Step 2: Divide 243 into 1367.
243 divides into 1367 about 5 times.
5 × 243 = 1215.
The 1215 is subtracted from the 1367 to get a remainder of 152.

```
           56
243 ⟌ 13678
       1215↓
         1528
       →1458
           70
```

Step 3: Bring down the next digit, the 8.
Divide 243 into 1528.
243 divides into 1528 about 6 times.
6 × 243 = 1458.
The 1458 is subtracted from the 1528 to get a remainder of 70.

The process of division as described in those two examples is quite involved. There are a few suggestions that I can give to help you arrive at correct answers when doing division problems.

1. Make sure you start the quotient above the correct digit.

   For example:
   $$5287 \overline{)365425}$$
   with an arrow pointing to the 2.

   5287 will not divide into 3 since 5287 > 3.
   5287 will not divide into 36 since 5287 > 36.
   5287 will not divide into 365 since 5287 > 365.
   5287 will not divide into 3654 since 5287 > 3654.

   5287 will divide into 36542 since that is the first part of the dividend that 5287 is not larger than.

   Thus the quotient in this example will start above the 2.

2. Every remainder in the division process should be less than the divisor. If you get a remainder that is not less than the divisor, the digit used in the quotient was too small.

   For example:

   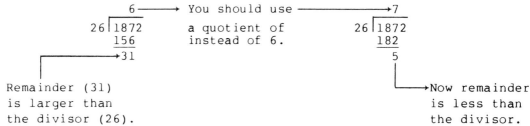

   Remainder (31) is larger than the divisor (26).

   Now remainder is less than the divisor.

3. How do you determine approximately how many times one number will divide into another number? If you round off your divisor, you should be better able to make a good guess about how many times one number divides into another.

   For example:
   $$86 \overline{)63124}$$

   The divisor 86 rounds off to 90.
   Now the 9 in 90 divides into the 63 of 631 about 7 times.
   So 86 should divide into 631 about 7 times.

   $$\begin{array}{r} 7 \\ 86 \overline{)63124} \\ 602 \\ \hline 29 \end{array}$$

   The remainder is less than the divisor so the guess of 7 was correct.

70

4. How do you check to see if your answer is correct? You could check your work by doing the following:

quotient x divisor + remainder = dividend

Example:

```
        237              Check:    quotient ——→   237
    15 ) 3559                      x divisor ——→   15
         30                                      ─────
         ──                                       1185
         55                                        237
         45                                      ─────
         ──                                       3555
         109            + remainder ——→           + 4
         105                                     ─────
         ───           = dividend ——→             3559
           4
```

Since the quotient times the divisor plus the remainder does equal the dividend (3559), I am sure that the answer, 237 R4, is correct.

PROBLEM 1: In the problem 419) 254670 above which digit should the quotient start?

Answer: Above the 6, since 419 is larger than 2, 25, and 254, but not larger than 2546.

PROBLEM 2: What is wrong with this division problem?

```
          3
    15 ) 67
         45
         ──
         22
```

Answer: The remainder (22) is larger than the divisor (15). You should use a quotient of 4.

PROBLEM 3: Use rounding off to find about how many times 77 divides into 5035.

Answer: 6 times. 77 rounds off to 80. The 8 in 80 divides into the 50 in 5035 about 6 times.

PROBLEM 4: Use the check method above to determine which are correct:

Answers:

a.
```
          57
    306 ) 17442
```

a. Correct, since

    quotient x divisor = dividend
       57   x   306   = 17442

b.
```
         465 R7
    82 ) 38133
```

b. Not correct, since

    quotient x divisor + remainder = 38137
       465   x   82   +   7   = 38137
    and that does not equal the dividend, which is 38133.

71

PROBLEM 5:   1482 ÷ 26 = ?   Answer: 57

```
        57
26 )1482
    130
    182
    182
```

PROBLEM 6:   30,378 ÷ 416 = ?   Answer: 73 R10

```
            73
416 )30378
     2912
     1258
     1248
       10
```

PROBLEM 7:   5406 ÷ 17 = ?   Answer: 318

```
       318
17 )5406
    51
    30
    17
    136
    136
```

PROBLEM 8:   445,588 ÷ 5,003 = ?   Answer: 89 R321

```
              89
5003 )445588
      40024
      45348
      45027
        321
```

PROBLEM 9:   1,894,077 ÷ 783 = ?   Answer: 2419

```
            2419
783 )1894077
     1566
     3280
     3132
     1487
      783
     7047
     7047
```

EXERCISE 2-6 SET A

NAME _____   DATE _____

DIVIDE THE FOLLOWING   ANSWERS

1. 23)391   2. 41)943   1. _____

                                              2. _____

3. 47)2491   4. 58)2088   3. _____

                                              4. _____

5. 69)3249   6. 75)4595   5. _____

                                              6. _____

7. 232)15080   8. 314)20410   7. _____

                                              8. _____

9. 247)33357   10. 643)55399   9. _____

                                             10. _____

11. 706)319818   12. 508)181356   11. _____

                                             12. _____

13. 4125)354968   14. 3176)149395   13. _____

                                             14. _____

15. 5806)1016050   16. 6904)1346280   15. _____

                                             16. _____

EXERCISE 2-6 SET B

NAME _____ DATE _____

DIVIDE THE FOLLOWING   ANSWERS

1. 32)832   2. 14)938   1. _____

                                                  2. _____

3. 57)1995   4. 48)2976   3. _____

                                                  4. _____

5. 86)4569   6. 79)2096   5. _____

                                                  6. _____

7. 322)20930   8. 414)31050   7. _____

                                                  8. _____

9. 536)25397   10. 463)34485   9. _____

                                                  10. _____

11. 506)175582   12. 706)302168   11. _____

                                                  12. _____

13. 5314)255396   14. 4236)305817   13. _____

                                                  14. _____

15. 6908)1761540   16. 8506)2339150   15. _____

                                                  16. _____

2-7 Zeros in the Quotient

The number zero quite frequently causes difficulty when it should be part of the quotient in a division problem. Students have a tendency to just leave it out of their answers. If you remember the next suggestion about division, you will be less prone to make errors with zeros in the quotient.

<u>After you place the first digit in the quotient, you should have a digit above each remaining digit in the dividend.</u>

Example 1:
```
      608
   3)1824
     18
      2
      0
      24
      24
```
Notice after we started the quotient above the 8, there is a digit above each digit in the dividend.

Example 2:
```
       32
   2)640
     6
      4
      4
```
Notice there is no number above the 0 in the dividend. Something must be wrong. The actual answer is 320. The zero at the end of the quotient was left out.

Example 3:
```
       37
   4)1228
     12
      28
      28
```
Something is wrong again. There is no number above the 8 in the dividend. The actual answer is 307. Again a zero was omitted from the quotient.

Example 4:
```
       3 ─────→
   7)2100
     21
```
7 divides into the 21, three times.

```
       30
   7)2100
     21↓
      0 ─────→
      0
```
Bring the first zero down. 7 divides into zero, 0 times. Thus a 0 was placed in the tens place of the quotient.

```
       300
   7)2100
     21
      0
      0↓
       0 ─────→
       0
```
Bring down the second zero. 7 divides into zero, 0 times. Thus a 0 was placed in the ones place of the quotient.

Example 5:

```
        4
    12 ) 4836
         48
```
→ 12 divides into 48, four times

```
        40
    12 ) 4836
         48↓
          3
          0
          3
```
→ Bring down the 3. 12 is larger than 3, so 12 divides into 3, zero times. Thus 0 is placed above the 3 in the tens place of the quotient.

```
        403
    12 ) 4836
         48
          3
          0↓
          36
          36
```
→ Bring down the 6. 12 divides into 36, three times. Thus the 3 is placed above the 6 in the ones place.

Example 6:

```
           9
    207 ) 1864256
          1863
             1
```
→ 207 divides into 1864, nine times with a remainder of one.

```
           90
    207 ) 1864256
          1863↓
            12
             0
            12
```
→ Bring down the 2. 207 is larger than 12, so 207 divides into 12, zero times. Thus a 0 is placed above the 2.

```
           900
    207 ) 1864256
          1863
            12
             0↓
           125
             0
           125
```
→ Bring down the 5. 207 is larger than 125, so 207 divides into 125, zero times. Thus a 0 is placed above the 5.

```
           9006
    207 ) 1864256
          1863
            12
             0
           125
             0↓
          1256
          1242
            14
```
→ Bring down the 6. Now 207 divides into 1256, 6 times with a remainder of 14.

EXERCISE 2-7 SET A

NAME _____ DATE _____

DIVIDE THE FOLLOWING: ANSWERS

1. 6 ⟌ 2400   2. 4 ⟌ 2800   1. _____

                                                  2. _____

3. 24 ⟌ 6000   4. 36 ⟌ 9000   3. _____

                                                  4. _____

5. 70 ⟌ 28000   6. 90 ⟌ 45000   5. _____

                                                  6. _____

7. 43 ⟌ 21629   8. 57 ⟌ 11742   7. _____

                                                  8. _____

9. 68 ⟌ 27648   10. 79 ⟌ 24066   9. _____

                                                  10. _____

11. 23 ⟌ 23046   12. 34 ⟌ 34102   11. _____

                                                  12. _____

13. 503 ⟌ 302856   14. 604 ⟌ 424652   13. _____

                                                  14. _____

15. 345 ⟌ 700350   16. 425 ⟌ 1283500   15. _____

                                                  16. _____

EXERCISE 2-7   SET B

NAME _____   DATE _____

DIVIDE THE FOLLOWING: ANSWERS

1. 7)3500   2. 3)2400   1. _____

                                                     2. _____

3. 32)8000   4. 28)7000   3. _____

                                                     4. _____

5. 80)48000   6. 60)54000   5. _____

                                                     6. _____

7. 53)16271   8. 47)19176   7. _____

                                                     8. _____

9. 49)14835   10. 58)23573   9. _____

                                                    10. _____

11. 34)34238   12. 23)23207   11. _____

                                                    12. _____

13. 602)184834   14. 405)83045   13. _____

                                                    14. _____

15. 275)551100   16. 385)1157310   15. _____

                                                    16. _____

2-8 The Order of Operations

Now that you have reviewed the four basic operations, powers, and square roots, you will encounter problems that have combined these operations.

    For example: 3 + 7 x 5 = ?

Some of you may get an answer of 50 by figuring 3 + 7 = 10 and 10 x 5 = 50. Others may get an answer of 38 by figuring 7 x 5 = 35 and 3 + 35 = 38. Which answer is correct: 50 or 38?

Both answers can not be correct. Actually 38 is the correct answer since it is the result of the proper order in which the operations are done.

Computation in mathematics must be done in a standardized order so that everyone could get the same answer for the same problem. The order that is agreed upon for doing computation in mathematics is as follows:

The Order of Operations

First: Do operations inside parentheses.
Second: Do powers or square roots.
Third: Do multiplications or divisions from left to right.
Fourth: Do additions or subtractions from left to right.

By "left to right" we mean whichever of the operations you come across first as you move from left to right through the problem.

    Example 1:   3 x 5 - (5 + 6) = ?

        = 3 x 5 - 11   Do the parentheses.
        = 15 - 11   Do the multiplication.
        = 4   Do the subtraction.

Note: To help assure accuracy, do one step at a time and write down the intermediate results below the computation.

    Example 2:   36 ÷ (3 + 1) x 6 = ?

        = 36 ÷ 4 x 6   Do the parentheses.
        = 9 x 6   Do the multiplications or divisions
        = 54   left to right. (Note: We did the division first since we came across a division first, moving left to right.)

Example 3: $16 + \sqrt{49} \times 3^2 = ?$

     = $16 + 7 \times 9$ Do the powers and square roots.

     = $16 + 63$ Do the multiplication.

     = 79 Do the addition.

Example 4: $(20 + 25) \div 5 - 2^3 + 1 = ?$

     = $45 \div 5 - 2^3 + 1$ Do the parentheses

     = $45 \div 5 - 8 + 1$ Do the powers.

     = $9 - 8 + 1$ Do the division.

     = 2 Do the additions or subtractions from left to right. (Note: We did the subtraction first since we came across it first as we moved left to right.)

PROBLEM 1: $65 - 4 \times 8 = ?$ Answer: 33

$$65 - 4 \times 8 = 65 - 32$$
$$= 33$$

PROBLEM 2: $6 \times (10 - 4) + \sqrt{100} = ?$ Answer: 46

$$6 \times (10-4) + \sqrt{100} = 6 \times 6 + \sqrt{100}$$
$$= 6 \times 6 + 10$$
$$= 36 + 10$$
$$= 46$$

PROBLEM 3: $5 \times 3^2 - 9 \div 3 \times 2 = ?$ Answer: 39

$$5 \times 3^2 - 9 \div 3 \times 2 = 5 \times 9 - 9 \div 3 \times 2$$
$$= 45 - 6$$
$$= 39$$

PROBLEM 4: $(7 + 3)^4 \times (2 + 3 \times 4) = ?$ Answer: 140,000

$$(7+3)^4 \times (2+3 \times 4) = 10^4 \times 14$$
$$= 10000 \times 14$$
$$= 140,000$$

EXERCISE 2-8   SET A

NAME _____   DATE _____

DO THE FOLLOWING COMPUTATIONS: ANSWERS

1. 5 + 3 x 4   2. 7 + 2 x 6   1. _____

                                                              2. _____

3. (5 + 3) x 4   4. (7 + 2) x 6   3. _____

                                                              4. _____

5. 24 x 4 ÷ 6   6. 20 x 4 ÷ 5   5. _____

                                                              6. _____

7. 24 ÷ 4 x 6   8. 20 ÷ 4 x 5   7. _____

                                                              8. _____

9. $4 \times 3^2 - 7 \times 5$ 10. $3 \times 2^3 - 2 \times 6$ 9. _____

                                                              10. _____

11. $(2 + 1)^4 \times (12 - 2 \times 5)$ 12. $(3 + 2)^2 \times (15 - 3 \times 4)$ 11. _____

                                                              12. _____

13. $\sqrt{100} + 3 \times \sqrt{49}$ 14. $\sqrt{81} + 2 \times \sqrt{64}$ 13. _____

                                                              14. _____

15. (12 + 18) ÷ 5 + 10   16. (15 + 18) ÷ 3 + 8   15. _____

                                                              16. _____

17. $\sqrt{36} \times 10^5 - 3 \times 10^3$ 18. $\sqrt{25} \times 10^6 - 3 \times 10^4$ 17. _____

                                                              18. _____

19. 16 - 12 ÷ 4 x 3 + 8 x 2   20. 28 - 36 ÷ 9 x 4 + 6 x 3   19. _____

                                                              20. _____

EXERCISE 2-8 SET B

NAME _____   DATE _____

DO THE FOLLOWING COMPUTATIONS: ANSWERS

1. 6 + 2 x 7   2. 4 + 5 x 8   1. _____

                                                          2. _____

3. (6 + 2) x 7   4. (4 + 5) x 8   3. _____

                                                          4. _____

5. 36 x 4 ÷ 9   6. 56 x 7 ÷ 8   5. _____

                                                          6. _____

7. 36 ÷ 4 x 9   8. 56 ÷ 7 x 8   7. _____

                                                          8. _____

9. $5 \times 4^2 - 3 \times 6$ 10. $4 \times 3^3 - 5 \times 6$ 9. _____

                                                          10. _____

11. $(3 + 2)^2 \times (17 - 3 \times 5)$ 12. $(1 + 2)^4 \times (22 - 5 \times 3)$ 11. _____

                                                          12. _____

13. $\sqrt{121} + 2 \times \sqrt{16}$ 14. $\sqrt{49} + 3 \times \sqrt{9}$ 13. _____

                                                          14. _____

15. (18 + 14) ÷ 8 + 10   16. (18 + 18) ÷ 4 + 8   15. _____

                                                          16. _____

17. $\sqrt{16} \times 10^4 - 2 \times 10^2$ 18. $\sqrt{64} \times 10^6 - 5 \times 10^3$ 17. _____

                                                          18. _____

19. 29 - 15 ÷ 3 x 5 + 6 x 4   20. 34 - 16 ÷ 8 x 2 + 8 x 4   19. _____

                                                          20. _____

2-9 Primes, Divisibility, and Factoring into Primes

Primes

A **prime number** is a whole number greater than 1 that is evenly divisible by only 1 and itself. If a number is evenly divisible by other numbers besides 1 and itself, it is a **composite number**.

2, 3, 5, 7, 11, 13, 17, 19, 23, 29 . . . are prime numbers while

4, 6, 8, 9, 10, 12, 14, 15, 16, 18 . . . are composite numbers.

Examples: Determine whether the following are prime or composite:

31   prime   (divisible by only 1 and 31)

22   composite (divisible by 1 and 22, but also by 2 and 11)

51   composite (divisible by 1 and 51, but also by 3 and 17)

Divisibility

In determining if a given number is prime or composite, you need to decide if any numbers besides 1 and itself divide evenly into the given number. Below are some facts about divisibility that will help you figure out what numbers divide evenly into a given number.

Divisibility by 2: If a number ends in 0, 2, 4, 6, or 8, it is an even number. Even numbers are divisible by 2.

96 is divisible by 2, since it is an even number.

Divisibility by 3: Add up the digits of the number. If you can divide the sum evenly by 3, the original number is divisible by 3.

441 is divisible by 3. The sum of its digits is 9 (4+4+1 = 9), and 9 can be divided by 3.

Divisibility by 5: If a number ends in either 5 or 0, it is divisible by 5.

935 is divisible by 5, since it ends in 5.

PROBLEMS: is 510 divisible by:
1. 2 ?

2. 3 ?

3. 5 ?

4. 7 ?

Answers:
1. Yes, it is an even number.

2. Yes, sum of its digits is 6, which is divisible by 3.

3. Yes, it ends in 0.

4. No, dividing by 7 leaves a remainder of 6.

Factoring into Primes

Since the numbers used in multiplication are called factors, **factoring** is the process of writing a number as the product of other numbers. The number 12 can be factored as:

$$12 = 1 \times 12 \qquad 12 = 2 \times 6 \qquad 12 = 3 \times 4 \qquad 12 = 2 \times 2 \times 3$$

If all the factors in this process are prime numbers, you have written the number as a product of primes. In the above example, $12 = 2 \times 2 \times 3$ represents 12 factored into primes. An effective way to factor a composite number into primes is to find the prime divisors of the number.

Example 1:   Factor 30 into primes.   Note: $2\overline{)30}$ is being written $2\underline{|30}$
$\phantom{Example 1:   Factor 30 into primes.   Note: 2\overline{)30} is being written}$ $\overline{15}$

prime divisors
↓

$2\underline{|30}$ ←——— (2 is a prime divisor; 30 is even.)
$3\underline{|15}$ ←——— (3 is a prime divisor, sum of digits is 6.)
$5\underline{|5}$ ←——— (5 is a prime divisor.)
$\overline{1}$ ←——— (1, so stop.)

The numbers on the left are the prime factors. $30 = 2 \times 3 \times 5$
The order you pick or write the prime divisors does not make a difference. You could have begun with the 3 or the 5.

Example 2:   Factor 570 into primes.

prime divisors
↓

$2\underline{|570}$ ←——— (2 is a prime divisor; 570 is even.)
$5\underline{|285}$ ←——— (5 is a prime divisor; 285 ends in 5.)
$3\underline{|57}$ ←——— (3 is a prime divisor, sum of digits is 12.)
$19\underline{|19}$ ←——— (19 is a prime divisor.)
$\overline{1}$ ←——— (1, so stop.)

$570 = 2 \times 5 \times 3 \times 19$

Example 3:   Factor 539 into primes.

(539 is not divisible by 2, 3, or 5, so try larger prime divisors.)

$7\underline{|539}$ ←——— (7 is a prime divisor.)
$7\underline{|77}$ ←——— (7 is a prime divisor.)
$11\underline{|11}$ ←——— (11 is a prime divisor.)
$\overline{1}$ ←——— (1, so stop.)

$539 = 7 \times 7 \times 11$

PROBLEM 5:   Factor 150 into primes.   Answer: $2 \times 3 \times 5 \times 5$

$2\underline{|150}$
$3\underline{|75}$
$5\underline{|25}$
$5\underline{|5}$
$\overline{1}$

EXERCISE 2-9 SET A

NAME _____   DATE _____

DETERMINE IF THE FOLLOWING ARE PRIME OR COMPOSITE: ANSWERS

1. 17   2. 23   3. 21   1. _____

                                                 2. _____

                                                 3. _____

4. 49   5. 315   6. 441   4. _____

                                                 5. _____

                                                 6. _____

FACTOR THE FOLLOWING INTO THE PRODUCT OF PRIMES:

7. 18   8. 24   7. _____

                                                 8. _____

9. 46   10. 34   9. _____

                                                 10. _____

11. 42   12. 66   11. _____

                                                 12. _____

13. 55   14. 35   13. _____

                                                 14. _____

15. 75   16. 245   15. _____

                                                 16. _____

17. 1375   18. 1625   17. _____

                                                 18. _____

19. 660   20. 650   19. _____

                                                 20. _____

21. 222   22. 387   21. _____

                                                 22. _____

23. 133   24. 143   23. _____

                                                 24. _____

EXERCISE 2-9 SET B

NAME _____   DATE _____

DETERMINE IF THE FOLLOWING ARE PRIME OR COMPOSITE: ANSWERS

1. 35   2. 19   3. 31   1. _____

                                                            2. _____

                                                            3. _____

4. 501   5. 121   6. 775   4. _____

                                                            5. _____

                                                            6. _____

FACTOR THE FOLLOWING INTO THE PRODUCT OF PRIMES:

7. 16   8. 27   7. _____

                                                            8. _____

9. 58   10. 62   9. _____

                                                            10. _____

11. 36   12. 40   11. _____

                                                            12. _____

13. 50   14. 125   13. _____

                                                            14. _____

15. 98   16. 54   15. _____

                                                            16. _____

17. 1925   18. 525   17. _____

                                                            18. _____

19. 520   20. 420   19. _____

                                                            20. _____

21. 234   22. 1722   21. _____

                                                            22. _____

23. 209   24. 161   23. _____

                                                            24. _____

2-10 More Applications Involving Whole Numbers

In Section 1-10 the basic steps for solving word problems were covered. You must remember to **read** the problem carefully, **analyze** the problem to determine what operations to use, and **solve** to find the answer to the question asked in the problem. In some problems you may have difficulty analyzing what operations should be used to calculate the answer. If that happens, you might try substituting very simple numbers for the original numbers in the problem. Decide what should be done with the simpler numbers, then solve the original problem, using the same operations. Study the next example to see what I mean.

Example 1: You drove 637 miles in 13 hours. How many miles did you travel per hour?

Analyze: If you are not sure what to do, try substituting simple numbers for the 637 and the 13. For instance, change the problem to "You drove 10 miles in 2 hours. How many miles did you travel per hour?" The answer (5 miles per hour) is fairly obvious, and was obtained by dividing 2 into 10. So divide the original numbers, 13 into 637.

Solve:
```
       49 miles per hour
   13)637
       52
       117
       117
```

Example 2: The scores on your last four math tests were 96, 88, 83, and 93. What is the average of your scores?

Analyze: To find the **average** of a list of numbers, add the numbers and divide by how many numbers are in the list.

Solve:
```
   96         90 is the average
   88      4)360
   83         36
 + 93          0
  360          0
```

Example 3: The area of the largest state, Alaska, is 589,757 square miles. The smallest state, Rhode Island, has an area of 1,214 square miles. What is the difference in their areas?

Analyze: To find the difference between two numbers you must subtract the two numbers.

Solve:
```
   589,757
 -   1,214
   588,543 square miles
```

Example 4: Out of a batch of 450 tennis balls, 21 balls were found to be defective and were discarded. The remaining balls were packed in cans containing 3 balls each. How many cans of tennis balls were packed?

Analyze: Find the number of good balls by subtracting those that were discarded. Since each can contain 3 balls, divide the number of good balls by 3 to find the number of cans.

Solve: good balls: 450 − 21 = 429

number of cans: 429 ÷ 3 = 143

Example 5: The total amount of an auto loan is $11,376. If you pay off the loan with equal monthly payments over four years, how much would you pay each month?

Analyze: Find the number of months over which the payments will be made by multiplying 4 times 12. Divide the total loan by the number of months to get the monthly payment.

Solve: number of months: 4 x 12 = 48

monthly payment: $11,376 ÷ 48 = $237

PROBLEM 1: The high temperatures during seven consecutive days in May were 75, 68, 80, 83, 77, 72, and 70. What was the average temperature for the week?

Answer: 75

sum of temperatures:
75 + 68 + 80 + 83 + 77 + 72 + 70 = 525

average:
525 ÷ 7 = 75

PROBLEM 2: If you type at a rate of 45 words per minute, how long would you take to type a nine page report containing 250 words per page?

Answer: 50 min

number of words:
9 x 250 = 2250

amount of time:
2250 ÷ 45 = 50 min.

PROBLEM 3: An amusement park has a family rate of $25 for a family of four. If the regular price is $9 for adults and $6 for children, how much does a family of 2 adults and 2 children save with the family rate?

Answer: $5

regular:
2 x $9 = $18
2 x $6 = $12
total: = $30

savings:
$30 − $25 = $5

EXERCISE 2-10   SET A

NAME _____   DATE _____

ANSWERS

1. How much would a $1199 freezer cost after a $250 discount?   1. _____

2. At the start of your vacation, your car's odometer read
   55,743. When you returned, it read 56,891. How many
   miles did you drive on your vacation?   2. _____

3. You want to save $3600 for a trip two years from now. How
   much should you save each month to have $3600 by then?   3. _____

4. At a canning factory a machine seals cans at a rate of 75
   cans per minute. At that rate, how long will it take to
   seal 3000 cans?   4. _____

5. In your last three games of bowling, you bowled 171, 149
   and 184. What is your average for those three games?   5. _____

6. The players on a fifth grade basketball team weigh 59 lb,
   66 lb, 75 lb, 71 lb, and 84 lb. What is the average weight
   of the members of the team?   6. _____

7. If you earn $4 per hour and $6 for each hour worked over a
   regular 40 hour week, how much do you earn if you work
   52 hours in a week?   7. _____

8. You borrowed $260 from a friend. After gving him $25
   each month for 7 months, how much do you still owe him?   8. _____

9. If you earn $18,192 per year, how much do you earn per
   month?   9. _____

10. Jill is twice as old as her sister Mary. She is also 5
    years older than her brother Jason. If Jill is 18, how
    old is Mary? How old is Jason?   10. _____

11. At a garage sale held by four families, $787 was collected.
    If the expenses for advertising the garage sale were $35 and
    the four families divide the profit equally, how much money
    does each family receive?   11. _____

EXERCISE 2-10   SET B

NAME _____   DATE _____

ANSWERS

1. In 1972, a Holstein-Friesian cow sold for $122,000. If the
   cow weighed 2000 lb, how much was it worth per pound?   1. _____

2. How many $7 cconcert tickets can you buy with $133?   2. _____

3. In 1979, the Chrysler Corporation had sales of $16,136,100,000
   and expenses of $16,340,700,000. How much of a loss was that? 3. _____

4. If you earn $678 per week but $89 is taken out for income tax
   and social security, how much is your take home pay?   4. _____

5. In five games, the Renegades football team scored 21 points,
   35 points, 17 points, 29 points, and 13 points. What was
   the average points scored per game?   5. _____

6. The numbers of students enrolled in six basic math classes
   were 34, 29, 37, 23, 30, and 21. What was the average
   enrollment in these classes?   6. _____

7. If you earn $5 per hour and $7 for each hour worked over a
   regular 8 hour day, how much would you earn working 13 hours
   in one day?   7. _____

8. If a realtor sells $665,400 worth of real estate in a year
   what is his average per month?   8. _____

9. If a plane trip of 2215 miles takes 5 hours, how many miles
   is traveled per hour?   9. _____

10. What do you pay for a $679 stereo after a $99 discount is
    given and $24 sales tax is added?   10. _____

11. For a field trip, a school orders three buses, which can
    seat 55 people each. If 192 people show up for the field
    trip, how many can not be seated on the buses?   11. _____

Chapter 2 Summary

Concepts

You may refer to the sections listed below to review how to:

1. subtract whole numbers using the borrowing process. (2-1). (2-2)

2. divide whole numbers using the long division process. (2-3), (2-4), (2-5), (2-6)

3. handle zeros in the quotient of division problems. (2-7)

4. solve problems involving a variety of operations by using the following order of operations:

    a. operations inside parentheses
    b. powers and square roots
    c. multiplication or division from left to right
    d. addition or subtraction from left to right. (2-8)

5. factor composite numbers into a product of primes. (2-9)

6. solve word problems involving whole numbers. (2-10)

7. find the average of a list of numbers by adding the numbers and dividing by how many are in the list. (2-10)

Terminology

This chapter's important terms and their corresponding page numbers are:

average: an estimate obtained by dividing the sum of quantities by the number of quantities. (87)

composite number: a number that is evenly divisible by other numbers besides 1 and itself. (83)

difference: the answer from subtraction. (51)

dividend: the number in a division problem that is being divided. (57)

divisor: the number by which the dividend is divided. (57)

factoring: the process of writing a number as the product of other numbers. (84)

minuend: the number from which another number is subtracted. (51)

order of operations: the order in which operations of arithmetic are performed. (79)

prime number: a whole number greater than 1 that is evenly divisible by only 1 and itself. (83)

quotient: the answer from division. (57)

remainder: whole number that is left over in a division problem whose quotient is not exact. (61)

subtrahend: the number that is subtracted from another number. (51)

Student Notes

Chapter 2 Practice Test A

ANSWERS

1. 768 − 543   2. 83,002 − 54,178   1. _____

2. _____

3. 37,000 − 6,584   4. 72 ÷ 9   3. _____

4. _____

5. 67 ÷ 8   6. 22 ÷ 0   5. _____

6. _____

7. 488 ÷ 7   8. 6⟌17832   7. _____

8. _____

9. 384⟌252288   10. 48⟌14837   9. _____

10. _____

11. 45⟌36000   12. $5 \times \sqrt{36} + 3 \times 10^4$ 11. _____

12. _____

13. 56 ÷ (20 − 14 + 2) × 7   13. _____

14. Factor 750 into primes.   14. _____

15. In your last four games of bowling, you scored 136, 168, 192, and 156. What is your average bowling score for those games?   15. _____

Chapter 2 Practice Test B

ANSWERS

1. 957 − 436   2. 74,003 − 6,348   1. _____

2. _____

3. 6000 − 786   4. 45 ÷ 9   3. _____

4. _____

5. 57 ÷ 7   6. $0\overline{)18}$ 5. _____

6. _____

7. 469 ÷ 6   8. $8\overline{)16504}$ 7. _____

8. _____

9. $654\overline{)232824}$ 10. $53\overline{)21697}$ 9. _____

10. _____

11. $85\overline{)68000}$ 12. $4 \times 10^2 - 3 \times \sqrt{100}$ 11. _____

12. _____

13. 7 × (15 − 4 + 5) ÷ 2   13. _____

14. Factor 660 into primes.   14. _____

15. Out of a batch of 500 light bulbs, 76 were found to be defective and were discarded. The remaining bulbs were shipped in boxes containing four bulbs each. How many boxes of light bulbs were shipped?   15. _____

Chapter 2 Supplementary Exercises

Section 2-2

1. 86
   − 44

2. 95
   − 37

3. 367
   − 180

4. 5683
   − 1392

5. 503 − 185
6. 605 − 387
7. 5683 − 399
8. 4382 − 799
9. 6000 − 5936
10. 4000 − 3761
11. 90,000 − 329
12. 60,000 − 476
13. 70,063 − 19,806
14. 80,019 − 29,671

Section 2-3

1. 63 ÷ 9
2. 24 ÷ 4
3. 56 ÷ 8
4. 72 ÷ 9
5. 72 ÷ 8
6. 54 ÷ 6
7. 56 ÷ 7
8. 32 ÷ 8

9. $5\overline{)45}$
10. $7\overline{)49}$
11. $2\overline{)18}$
12. $9\overline{)36}$

Section 2-4

1. 32 ÷ 5
2. 19 ÷ 2
3. 85 ÷ 9
4. 76 ÷ 7
5. 19 ÷ 7
6. 38 ÷ 5
7. 56 ÷ 9
8. 16 ÷ 6

9. $6\overline{)41}$
10. $8\overline{)61}$
11. $8\overline{)43}$
12. $9\overline{)89}$

Section 2-5

1. 1848 ÷ 4
2. 145 ÷ 5
3. 326 ÷ 2
4. 4113 ÷ 3
5. 2826 ÷ 9
6. 1927 ÷ 7

7. $8\overline{)52715}$
8. $6\overline{)59247}$
9. $6\overline{)64511}$
10. $9\overline{)83801}$
11. $5\overline{)396380}$
12. $7\overline{)41328}$
13. $8\overline{)30123}$
14. $5\overline{)289765}$
15. $7\overline{)34365}$
16. $6\overline{)73694}$

Section 2-6

1. 645 ÷ 43
2. 234 ÷ 39
3. 351 ÷ 27
4. 344 ÷ 43
5. 2763 ÷ 87
6. 3142 ÷ 66
7. 51,255 ÷ 75
8. 64,155 ÷ 65
9. 16,180 ÷ 59
10. 27,080 ÷ 78
11. 401 ⟌ 17638
12. 305 ⟌ 17690
13. 516 ⟌ 19608
14. 727 ⟌ 18102
15. 368 ⟌ 458641
16. 298 ⟌ 361552
17. 432 ⟌ 235473
18. 365 ⟌ 239809
19. 1853 ⟌ 237453
20. 5216 ⟌ 281664

Section 2-7

1. 4200 ÷ 7
2. 3600 ÷ 4
3. 40,000 ÷ 8
4. 30,000 ÷ 6
5. 584,000 ÷ 73
6. 667,000 ÷ 29
7. 100,000 ÷ 25
8. 300,000 ÷ 75
9. 5481 ÷ 27
10. 7839 ÷ 39
11. 43 ⟌ 4386
12. 56 ⟌ 5836
13. 68 ⟌ 34612
14. 38 ⟌ 7752
15. 32 ⟌ 160128
16. 76 ⟌ 228532
17. 701 ⟌ 216699
18. 603 ⟌ 487296
19. 457 ⟌ 917856
20. 238 ⟌ 714576

Section 2-8

1. 6 + 3 x 2
2. 3 + 8 x 2
3. 17 − 5 x 3
4. 28 − 4 x 5
5. 36 ÷ 4 + 8
6. 50 ÷ 2 + 8
7. 18 x (9 ÷ 3)
8. 16 ÷ (8 x 2)
9. 5 x (3 + 8)
10. 4 x (8 − 1)
11. $3^3 - 4 \times \sqrt{25}$
12. $2^3 + 3 \times \sqrt{9}$
13. $6^2 - 28 \div 4 + 8$
14. $5^2 - 36 \div 9 + 6$
15. $19 + 3 \times (\sqrt{16} - 1)$
16. $20 + 2 \times (\sqrt{49} - 2)$
17. 14 ÷ 2 x 7 + 14 x 2 ÷ 7
18. 32 ÷ 2 x 8 + 32 x 2 ÷ 8
19. (3 + 4 x 2) x (4 x 2 − 3)
20. (16 − 2 x 3) x (2 x 3 + 16)

Section 2-9

Determine if the following are prime or composite numbers.

1. 9
2. 19
3. 31
4. 18
5. 47
6. 49
7. 23
8. 73

Factor the following into primes:

9. 14
10. 30
11. 32
12. 65
13. 39
14. 95
15. 57
16. 81
17. 152
18. 184
19. 216
20. 111
21. 312
22. 342
23. 1000
24. 1500

Section 2-10

1. If the cost of seven balcony tickets for an opera is $91, what is the cost of each ticket?

2. The tallest student in a fifth grade class is 62 inches tall. The shortest student is 37 inches tall. What is the difference in their heights?

3. For five performances of "The Christmas Carol," the attendance was 203, 147, 191, 162 and 242. What was the average attendance?

4. The ages of the starting nine for the "Masters" baseball team are 42, 46, 36, 50, 41, 40, 52, 38, and 42. What is the average age of that team?

5. A part-time cook at "McB's Burgers" works a 5 hour shift four days a week. If he frys 92 hamburgers in an hour, how many working days would it take him to fry 13,800 hamburgers?

6. If your pay for the year is $13,500 before deductions, and deductions for the year are $2,400, what is your average monthly salary?

7. If 12 cubic yards of sand cost $324, what is the cost for one cubic yard?

8. The total amount of a loan for a motor home is $23,640. If you pay off the loan with equal monthly payments over five years, how much would you pay each month?

9. The length of Joan's stride for the last 600 feet of a 3 mile race was 6 feet. If the length of her stride was 5 feet for the rest of the race, how many steps did Joan take in the race? (1 mi = 5280 ft)

10. Suppose a person watches television an average of 2 hours a day, 360 days a year. This is equivalent to how many full days of television watching?

Unit I Exam

Chapters 1 and 2   ANSWERS

1. Write 5,000,012,576 in words.   1. _____
                                                                                       _____
                                                                                       _____

2. 37 + 456 + 3215 + 9   3. 957 x 87   2. _____

                                                                                    3. _____

4. 875 x 4003   5. 2179 x 80,000   4. _____

                                                                                    5. _____

6. $8^2 + \sqrt{81}$ 7. 5209 - 573   6. _____

                                                                                    7. _____

8. 60,000 - 14,357   9. 2786 ÷ 7   8. _____

                                                                                    9. _____

10. 46,989 ÷ 573   11. 111,222 ÷ 37   10. _____

                                                                                    11. _____

12. 5 x (8 + 24 ÷ 2) - 8 x 3   12. _____

13. Factor 680 into a product of primes.   13. _____

14. A waiter at the end of his shift finds that he has two
    $20 bills, five $10 bills, seven $5 bills, and thirty-
    two $1 bills from tips. If he gives the bus boy $17,
    how much cash does the waiter have left?   14. _____

15. During the last 5 weeks your grocery bills were $180,
    $176, $154, $161, and $149. What was the average amount
    spent on groceries per week?   15. _____

98

Unit II

Fractions and Decimals

In Unit I we worked with whole numbers. Even though whole numbers occur very frequently in our daily lives, we also come across other kinds of numbers. Your bill at the bookstore is usually not exactly $10 or $20. It would probably be more like $10.52 or $23.17. The course you jogged was more than 3 miles; it was 3 1/4 miles.

In the grocery store the sign below the canned peaches reads "5.6¢ per ounce." You hear that 4/5 of all dentists surveyed recommend "Best" toothpaste. Your grade point average was 2.95. All of these give examples of fractional or decimal numbers.

The ability to work with these types of numbers is a necessary step in the development of your math skills. In this unit you will learn to operate with fractional and decimal numbers. When you have finished this unit you should be able to do the following:

1. interpret fractions and find equivalent fractions.
2. read and round off decimal numbers
3. perform the basic operations of fractional and decimal numbers.
4. compare the size of fractions and decimals.
5. do computations involving a combination of fractions and decimals.
6. solve word problems involving operations with fractions and decimals.

Unit II

Brain Buster

If you had a billion dollar bills, how long would it take you to count them, if you count a dollar each second, 24 hours a day?

(Hint: The answer is in years.)

Chapter 3

Fractions

After finishing this chapter you should be able to do the following:

1. interpret and find equivalent fractions.
2. multiply fractions.
3. divide fractions.
4. add fractions.
5. subtract fractions.
6. operate with mixed numbers.
7. simplify complex fractions.
8. compare the size of fractions.
9. solve word problems involving fractions.

On the next page you will find a pretest for this chapter. The purpose of the pretest is to help you determine which sections in this chapter you need to study in detail and which sections you can review quickly. By taking and correcting the pretest according to the instructions on the next page you can better plan your pace through this chapter.

Chapter 3 Pretest

Take and correct this test using the answer section at the end of the book. Those problems that give you difficulty indicate which sections need extra attention. Section numbers are in parentheses before each problem.

ANSWERS

(3-2)   1. Find the number represented by the ?: $\dfrac{7}{8} = \dfrac{?}{56}$ 1. _____

(3-3)   2. Express $\dfrac{33}{9}$ as a reduced mixed number   2. _____

(3-4)   3. $\dfrac{9}{10} \times \dfrac{3}{5}$ (3-5)   4. $\dfrac{6}{10} \div \dfrac{6}{7}$

3. _____

4. _____

(3-6)   5. $3\dfrac{2}{3} \times 1\dfrac{4}{7}$ 5. _____

(3-7)   6. A manufacturing company employs 222 people. During the first week of July, one-sixth of the employees are on vacation. How many employees are not on vacation in the first week of July?   6. _____

(3-8)   7. $\dfrac{1}{9} + \dfrac{5}{9}$ (3-9)   8. $\dfrac{3}{8} + \dfrac{1}{4} - \dfrac{3}{16}$

7. _____

8. _____

(3-10)   9. $\dfrac{4}{15} + \dfrac{7}{16} + \dfrac{5}{12}$ (3-10)   10. $\dfrac{5}{9} - \dfrac{11}{24}$

9. _____

10. _____

(3-11)   11. $2\dfrac{1}{3} - 1\dfrac{4}{9}$ (3-11)   12. $3\dfrac{2}{3} + 2\dfrac{4}{5}$

11. _____

12. _____

(3-12)   13. $\dfrac{3 + \dfrac{1}{2}}{3 - \dfrac{4}{5}}$ 13. _____

(3-13)   14. Arrange from largest to smallest: $\dfrac{8}{9}, \dfrac{7}{8}, \dfrac{2}{3}$ 14. _____

(3-14)   15. Find the cost per ounce of 6 1/3 ounces of juice that sells for 76¢.   15. _____

3-1 The Meaning of Fractions

I am sure you have seen numbers such as $\frac{1}{2}, \frac{5}{8}, \frac{1}{4}, \frac{3}{16}, \frac{2}{7}$, etc. These numbers are called **fractions**. They are not whole numbers since they represent a part or a portion of a whole.

HE OWES BEING ELECTED TO <u>DENOMINATOR</u>.

For example, what part of the figure below is shaded?

One out of the three equal sections is shaded. So we write $\frac{1}{3}$ (one-third) of the whole figure is shaded.

What part of the figure below is shaded?

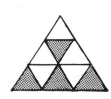

Four out of the nine equal sections are shaded. So we write $\frac{4}{9}$ (four-ninths) of the whole figure is shaded.

> Every fraction has three parts. The top number is the numerator, the bottom number is the denominator, and middle line is the fraction line.

fraction line ⟶ $\frac{5}{16}$ ⟵ numerator (tells how many parts you have)
⟵ denominator (tells the total number of parts)

PROBLEM 1: What part of the figure is:

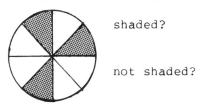

shaded?   Answer: 3/8, since 3 out of the 8 sections are shaded.

not shaded?   Answer: 5/8, since 5 out of the 8 sections are not shaded.

PROBLEM 2: If on a 50 question test you get 37 correct, what fraction did you get correct?   Answer: 37/50

Proper Fractions

All the fractions shown so far have displayed a portion of the whole object or a part of one unit. Each of those fractions has a value that is less than 1. We call such fractions **proper fractions**.

Proper fractions are easy to recognize since, besides having a value that is less than 1, their numerators are less than their denominators.

For example:

$$\frac{3}{5} \longleftarrow \text{numerator is less than the denominator}$$

Improper Fractions

Besides proper fractions that have a value that is less than 1, there are other fractions which have a value that is either equal to 1 or greater than 1. These are called **improper fractions**. Let me show you some examples:

Example 1: In the figure below, four out of the four sections are shaded.

$\frac{4}{4}$ (four-fourths) is shaded.

But that represents the whole object.

So $\frac{4}{4} = 1$.

Thus 4/4 is a fraction whose value is not <u>part</u> of a whole; it <u>is</u> the whole. It equals 1. Likewise 8/8 = 1, 10/10 = 1, 327/327 = 1, etc.

Example 2: In the figures below, we have 4/4 of figure A shaded and 3/4 of figure B shaded.

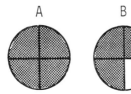

$\frac{4}{4}$ and $\frac{3}{4}$ is $\frac{7}{4}$.

We have one whole object and 3/4 of another that is shaded.

We have 1 and 3/4 (written $1\frac{3}{4}$) shaded.

So $\frac{7}{4} = 1\frac{3}{4}$. Thus 7/4 is a fraction whose value is greater than 1. Similarly 13/5, 110/17, 9/2, etc., have values that are larger than 1.

Improper fractions such as those are also easily recognized since, besides having a value that is greater than or equal to 1, their numerators are <u>not</u> less than their denominators.

For example: $\frac{16}{7} \longleftarrow$ numerators are <u>not</u> less $\longrightarrow \frac{14}{14}$
than the denominators

EXERCISE 3-1   SET A

NAME _____   DATE _____

WHAT PART OF THE FIGURE IS: ANSWERS

    1. shaded?   1. _____

    2. not shaded?   2. _____

WHAT PART OF THE FIGURE IS:

    3. shaded?   3. _____

    4. not shaded?   4. _____

WHAT PART OF THE FIGURE IS:

    5. shaded?   5. _____

    6. not shaded?   6. _____

IF, ON A TEST YOU GET 17 CORRECT OUT OF 20 QUESTIONS:

7. What fraction did you get correct?   7. _____

8. What fraction did you get wrong?   8. _____

IF, OUT OF 11 PEOPLE AT A PARTY, 3 ARE BLONDES:

9. What fraction are blondes?   9. _____

10. What fraction are not blondes?   10. _____

IF, IN A CLASS OF 30 STUDENTS, 17 ARE FEMALES:

11. What fraction are females?   11. _____

12. What fraction are males?   12. _____

CONSIDER THE FRACTIONS $\frac{1}{2}, \frac{3}{5}, \frac{7}{7}, \frac{9}{6}, \frac{8}{1}, \frac{15}{16}$:

13. List all the proper fractions.   13. _____

14. List all the improper fractions.   14. _____

CONSIDER THE FRACTIONS $\frac{12}{7}, \frac{7}{8}, \frac{3}{4}, \frac{19}{19}, \frac{7}{3}, \frac{10}{1}$:

15. List all the proper fractions.   15. _____

16. List all the improper fractions.   16. _____

EXERCISE 3-1   SET B

NAME _____   DATE _____

WHAT PART OF THE FIGURE IS: ANSWERS

    1. shaded?   1. _____

    2. not shaded?   2. _____

WHAT PART OF THE FIGURE IS:

    3. shaded?   3. _____

    4. not shaded?   4. _____

WHAT PART OF THE FIGURE IS:

    5. shaded?   5. _____

    6. not shaded?   6. _____

IF, ON A TEST YOU GET 19 CORRECT OUT OF 25 QUESTIONS:

7. What fraction did you get correct?   7. _____

8. What fraction did you get wrong?   8. _____

IF, OUT OF 12 DINNER GUESTS, 5 DRINK MILK:

9. What fraction drink milk?   9. _____

10. What fraction do not drink milk?   10. _____

IF, IN A CLASS OF 40 STUDENTS, 23 ARE MALES:

11. What fraction are males?   11. _____

12. What fraction are females?   12. _____

CONSIDER THE FRACTIONS $\frac{1}{5}, \frac{9}{4}, \frac{2}{3}, \frac{7}{1}, \frac{3}{3}, \frac{5}{8}$:

13. List all the proper fractions.   13. _____

14. List all the improper fractions.   14. _____

CONSIDER THE FRACTIONS $\frac{11}{8}, \frac{17}{32}, \frac{5}{1}, \frac{5}{5}, \frac{3}{5}, \frac{1}{10}$:

15. List all the proper fractions.   15. _____

16. List all the improper fractions.   16. _____

3-2 Equivalent Fractions

OIL REDUCES FRACTION.

Consider the two shaded blocks below:

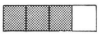 $\frac{3}{4}$ of the block is shaded.

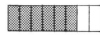 $\frac{6}{8}$ of the block is shaded.

The shaded sections represent the same amount of each block. So those two fractions, 3/4 and 6/8, represent the same amount of shaded area. They have the same value.

$\frac{3}{4}$ and $\frac{6}{8}$ are equivalent. $\frac{3}{4} = \frac{6}{8}$.

You can change 3/4 to 6/8 by multiplying its numerator and denominator by 2. You can change 6/8 to 3/4 by dividing its numerator and denominator by 2. Both procedures yield fractions that have the same value.

Raising Fractions to Higher Terms

By multiplying the numerator and denominator of a fraction by the same number, you **raise the fraction to higher terms**, getting an equivalent fraction.

$$\frac{3}{4} \xrightarrow{\times 2} \frac{6}{8} \qquad \frac{1}{3} \xrightarrow{\times 3} \frac{3}{9} \qquad \frac{5}{8} \xrightarrow{\times 10} \frac{50}{80}$$

Example 1: Change $\frac{3}{8}$ to a fraction with a numerator of 12.

$\frac{3}{8} \xrightarrow{\times 4} \frac{12}{?} = \frac{12}{32}$ To change the numerator from 3 to 12, you must multiply by 4. So you must also multiply the denominator by 4.

Example 2: Change $\frac{3}{2}$ to a fraction with a denominator of 18.

$\frac{3}{2} \xrightarrow{\times 9} \frac{?}{18} = \frac{27}{18}$ To change the denominator from 2 to 18, you must multiply by 9. So you must also multiply the numerator by 9.

PROBLEM 1: $\frac{5}{6} = \frac{35}{?}$ Answer: $\frac{35}{42}$ (Multiply numerator and denominator by 7.)

Reducing Fractions to Lowest Terms--Method 1

To **reduce a fraction to lower terms**, you must find a number other than 1 that divides evenly into both the numerator and denominator of the fraction. To **reduce to lowest terms**, you must continue to divide both the numerator and the denominator until no whole number except 1 divides evenly into each.

Example 3: Reduce $\frac{30}{42}$ to lowest terms.

2 divides evenly into both 30 and 42.

$$\frac{30}{42} \xrightarrow{\div 2} \frac{15}{21}$$

Now 3 divides evenly into both 15 and 21.

$$\frac{30}{42} = \frac{15}{21} \xrightarrow{\div 3} \frac{5}{7}$$

Example 4: Reduce $\frac{140}{910}$ to lowest terms.

Since numbers that end in zero are divisible by 10, 10 divides evenly into both 140 and 910.

$$\frac{140}{910} \xrightarrow{\div 10} \frac{14}{91}$$

Now 7 divides evenly into both 14 and 91

$$\frac{140}{910} = \frac{14}{91} \xrightarrow{\div 7} \frac{2}{13}$$

The key to reducing fractions by this method is to find numbers that divide evenly into both the numerator and the denominator of the fraction. Using the divisibility facts covered in Section 2-9 will be very helpful in this process.

PROBLEM 2:   Reduce $\frac{12}{24}$ to lowest terms.   Answer: $\frac{1}{2}$

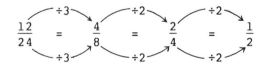

PROBLEM 3:   Reduce $\frac{135}{360}$ to lowest terms.   Answer: $\frac{3}{8}$

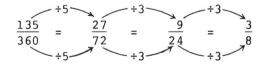

Reducing Fractions to Lowest Terms--Method 2

Finding numbers that divide evenly into both the numerator and the denominator of a fraction may be troublesome, even with the use of the divisibility facts learned previously. This second method allows you to work separately with the numerator and denominator, instead of searching for a divisor of both.

Example 5: Reduce $\frac{28}{210}$ to lowest terms.

1. Factor the numerator into primes: $28 = 2 \times 2 \times 7$

2. Factor the denominator into primes: $210 = 2 \times 3 \times 5 \times 7$

3. Cancel like factors:

$$\frac{28}{210} = \frac{\cancel{2} \times 2 \times \cancel{7}}{\cancel{2} \times 3 \times 5 \times \cancel{7}}$$

(If the numerator and denominator have the same factor, divide both by that number, giving the answer of 1 (1/1). This process is called **canceling**.)

4. Multiply the resulting factors: $\frac{28}{210} = \frac{1 \times 2 \times 1}{1 \times 3 \times 5 \times 1} = \frac{2}{15}$

By following the steps shown in the above example, you have a second method for reducing fractions to lowest terms. However, you must know how to factor numbers into primes to use this method. See Section 2-9, if you need to review that factoring process.

Example 6: Reduce $\frac{52}{78}$ to lowest terms.

$$\frac{52}{78} = \frac{\cancel{2} \times 2 \times \cancel{13}}{\cancel{2} \times 3 \times \cancel{13}} = \frac{2}{3}$$

Example 7: Reduce $\frac{102}{85}$ to lowest terms.

$$\frac{102}{85} = \frac{2 \times 3 \times \cancel{17}}{5 \times \cancel{17}} = \frac{6}{5}$$

We can summarize the methods for **reducing fractions** as follows:

> **Method 1**--Divide the numerator and denominator by a whole number (larger than 1) that divides evenly into each. Repeat until no more divisors can be found.
>
> **Method 2**--Factor the numerator and denominator into primes, cancel factors that are the same in both, and multiply the resulting factors.

Both methods work quite effectively in reducing fractions to lowest terms. Use the method that you find easier to work with.

PROBLEM 4:   Reduce $\frac{100}{150}$ to lowest terms.   Answer: $\frac{2}{3}$

Method 1:
$$\frac{100}{150} \xrightarrow{\div 10} = \frac{10}{15} \xrightarrow{\div 5} = \frac{2}{3}$$

Method 2:
$$\frac{100}{150} = \frac{\cancel{2} \times 2 \times \cancel{5} \times \cancel{5}}{\cancel{2} \times 3 \times \cancel{5} \times \cancel{5}} = \frac{2}{3}$$

PROBLEM 5:   Reduce $\frac{65}{143}$ to lowest terms.   Answer: $\frac{5}{11}$

$$\frac{65}{143} = \frac{5 \times \cancel{13}}{11 \times \cancel{13}} = \frac{5}{11}$$

PROBLEM 6:   Reduce $\frac{90}{36}$ to lowest terms.   Answer: $\frac{5}{2}$

$$\frac{90}{36} = \frac{\cancel{2} \times \cancel{3} \times \cancel{3} \times 5}{\cancel{2} \times 2 \times \cancel{3} \times \cancel{3}} = \frac{5}{2}$$

PROBLEM 7:   Reduce $\frac{21}{31}$ to lowest terms.   Answer: $\frac{21}{31}$

It is in lowest terms. No number, except 1, divides into both the numerator and denominator.

EXERCISE 3-2 SET A

NAME _____   DATE _____

RAISE EACH FRACTION TO HIGHER TERMS AS INDICATED: ANSWERS

1. $\frac{1}{2} = \frac{?}{6}$ 2. $\frac{1}{2} = \frac{?}{8}$ 1. _____ 2. _____

3. $\frac{2}{3} = \frac{?}{9}$ 4. $\frac{3}{5} = \frac{?}{15}$ 3. _____ 4. _____

5. $\frac{8}{7} = \frac{32}{?}$ 6. $\frac{9}{8} = \frac{36}{?}$ 5. _____ 6. _____

7. $\frac{6}{5} = \frac{72}{?}$ 8. $\frac{3}{4} = \frac{27}{?}$ 7. _____ 8. _____

9. $\frac{1}{6} = \frac{?}{180}$ 10. $\frac{1}{6} = \frac{?}{72}$ 9. _____ 10. _____

REDUCE THE FOLLOWING FRACTIONS TO LOWEST TERMS:

11. $\frac{3}{15}$ 12. $\frac{4}{12}$ 11. _____ 12. _____

13. $\frac{10}{18}$ 14. $\frac{6}{9}$ 13. _____ 14. _____

15. $\frac{21}{33}$ 16. $\frac{27}{39}$ 15. _____ 16. _____

17. $\frac{80}{120}$ 18. $\frac{90}{240}$ 17. _____ 18. _____

19. $\frac{75}{180}$ 20. $\frac{132}{180}$ 19. _____ 20. _____

21. $\frac{78}{441}$ 22. $\frac{84}{351}$ 21. _____ 22. _____

23. $\frac{144}{108}$ 24. $\frac{255}{195}$ 23. _____ 24. _____

25. $\frac{34}{85}$ 26. $\frac{51}{68}$ 25. _____ 26. _____

27. $\frac{52}{91}$ 28. $\frac{78}{91}$ 27. _____ 28. _____

EXERCISE 3-2   SET B

NAME _____   DATE _____

RAISE EACH FRACTION TO HIGHER TERMS AS INDICATED: ANSWERS

1. $\dfrac{1}{2} = \dfrac{?}{10}$ 2. $\dfrac{1}{2} = \dfrac{?}{4}$ 1. _____   2. _____

3. $\dfrac{4}{5} = \dfrac{?}{15}$ 4. $\dfrac{2}{3} = \dfrac{?}{12}$ 3. _____   4. _____

5. $\dfrac{9}{7} = \dfrac{36}{?}$ 6. $\dfrac{7}{6} = \dfrac{28}{?}$ 5. _____   6. _____

7. $\dfrac{3}{8} = \dfrac{?}{64}$ 8. $\dfrac{5}{8} = \dfrac{?}{56}$ 7. _____   8. _____

9. $\dfrac{7}{20} = \dfrac{?}{180}$ 10. $\dfrac{5}{12} = \dfrac{?}{144}$ 9. _____   10. _____

REDUCE THE FOLLOWING FRACTIONS TO LOWEST TERMS:

11. $\dfrac{3}{12}$ 12. $\dfrac{7}{14}$ 11. _____   12. _____

13. $\dfrac{8}{12}$ 14. $\dfrac{6}{10}$ 13. _____   14. _____

15. $\dfrac{27}{33}$ 16. $\dfrac{21}{39}$ 15. _____   16. _____

17. $\dfrac{90}{120}$ 18. $\dfrac{60}{150}$ 17. _____   18. _____

19. $\dfrac{125}{180}$ 20. $\dfrac{88}{180}$ 19. _____   20. _____

21. $\dfrac{96}{315}$ 22. $\dfrac{87}{414}$ 21. _____   22. _____

23. $\dfrac{531}{423}$ 24. $\dfrac{188}{148}$ 23. _____   24. _____

25. $\dfrac{95}{114}$ 26. $\dfrac{57}{76}$ 25. _____   26. _____

27. $\dfrac{78}{104}$ 28. $\dfrac{72}{58}$ 27. _____   28. _____

3-3 Mixed Numbers

Improper fractions have a value that is greater than or equal to 1. For example, in Section 3-1 we showed that 4/4 = 1 and 7/4 = 1 3/4. In fact, every improper fraction can be expressed as either a whole number or as the sum of a whole number and a proper fraction (a **mixed number**).

The process for doing this is as follows:

> The fraction line not only separates the numerator and the denominator, it also implies a division--the numerator divided by the denominator.

For that reason some improper fractions give whole number results:

NUMBERS THAT ARE CONFUSED ARE MIXED NUMBERS.

$$\frac{4}{4} = 4\overline{)4} = 1$$

$$\frac{15}{3} = 3\overline{)15} = 5$$

$$\frac{120}{12} = 12\overline{)120} = 10$$

Some improper fractions give mixed number results:

$\frac{7}{4} = 4\overline{)7} = 1\ R3 = 1\frac{3}{4}$ — Since we are dividing by 4, the remainder is 3 out of 4, or 3/4. So the mixed number is 1 and 3/4.

$\frac{19}{5} = 5\overline{)19} = 3\ R4 = 3\frac{4}{5}$ — Since we are dividing by 5, the remainder is 4 out of 5, or 4/5. So the mixed number is 3 and 4/5.

To simplify an improper fraction all you need to do is to divide the numerator by the denominator and represent any remainder as a proper fraction with the divisor as its denominator. If you get confused on which number to divide by, you can remember **D** for **d**enominator, **d**own number, and **d**ivisor.

PROBLEM 1: Simplify $\frac{18}{3}$

Answer: 6

$$\frac{18}{3} = 3\overline{)18} = 6$$

PROBLEM 2: Simplify $\frac{39}{7}$

Answer: $5\frac{4}{7}$

$$\frac{39}{7} = 7\overline{)39} = 5\ R4 = 5\frac{4}{7}$$

In working with fractions you will find it sometimes necessary to change whole or mixed numbers into improper fractions. This is the reverse of what we just did in this section.

For example: $\frac{2}{1} = 2 \div 1 = 2$

Thus the whole number 2 is the improper fraction $\frac{2}{1}$.

Similarly, any whole number can be converted into an improper fraction by putting the whole number over a denominator of 1.

That is, $7 = \frac{7}{1}$, $25 = \frac{25}{1}$, $156 = \frac{156}{1}$, and so on.

We can also obtain a method for converting mixed numbers into improper fractions by studying how we changed improper fractions into mixed numbers earlier in this section.

For example: $\frac{7}{4} = 1\frac{3}{4}$

dividend ─── quotient ─── remainder

$$\frac{7}{4} = 4\overline{)7} = 1 \text{ R}3 = 1\frac{3}{4}$$

── divisor ──

Thus if you have a mixed number, you have the quotient, the remainder, and the divisor of a division problem. All you need, to find the improper fraction, is the dividend.

But remember the dividend = divisor x quotient + remainder (from 2-6).

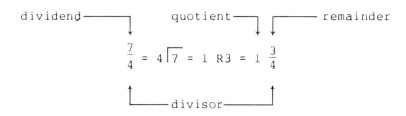

$$1\frac{3}{4} = \frac{4 \times 1 + 3}{4} = \frac{7}{4}$$

If you understand that process for changing a mixed number into a fraction, the short cut below will make those problems easier to do.

$5\frac{3}{4} = \frac{23}{4}$ $10\frac{1}{6} = \frac{61}{6}$ $122\frac{2}{3} = \frac{368}{3}$

───

PROBLEM 3:   Change $6\frac{2}{7}$ to an improper fraction.   Answer: 44/7

$$6\frac{2}{7} = \frac{7 \times 6 + 2}{7} = \frac{44}{7}$$

PROBLEM 4:   Change $140\frac{1}{2}$ to an improper fraction.   Answer: $\frac{281}{2}$

$$140\frac{1}{2} = \frac{281}{2}$$

EXERCISE 3-3   SET A

NAME _____   DATE _____

CHANGE THE FOLLOWING TO WHOLE OR MIXED NUMBERS: ANSWERS

1. $\dfrac{8}{1}$ 2. $\dfrac{7}{1}$ 1. _____   2. _____

3. $\dfrac{15}{3}$ 4. $\dfrac{24}{6}$ 3. _____   4. _____

5. $\dfrac{17}{3}$ 6. $\dfrac{25}{6}$ 5. _____   6. _____

7. $\dfrac{10}{7}$ 8. $\dfrac{11}{8}$ 7. _____   8. _____

9. $\dfrac{37}{5}$ 10. $\dfrac{47}{5}$ 9. _____   10. _____

11. $\dfrac{125}{16}$ 12. $\dfrac{140}{17}$ 11. _____   12. _____

13. $\dfrac{457}{7}$ 14. $\dfrac{396}{7}$ 13. _____   14. _____

CHANGE THE FOLLOWING TO IMPROPER FRACTIONS:

15. 9   16. 8   15. _____   16. _____

17. $1\dfrac{1}{2}$ 18. $1\dfrac{3}{5}$ 17. _____   18. _____

19. $2\dfrac{5}{8}$ 20. $2\dfrac{6}{7}$ 19. _____   20. _____

21. $9\dfrac{1}{4}$ 22. $8\dfrac{1}{3}$ 21. _____   22. _____

23. $37\dfrac{1}{2}$ 24. $52\dfrac{1}{6}$ 23. _____   24. _____

25. $10\dfrac{5}{6}$ 26. $20\dfrac{8}{9}$ 25. _____   26. _____

27. $124\dfrac{3}{7}$ 28. $209\dfrac{2}{3}$ 27. _____   28. _____

29. $45\dfrac{15}{17}$ 30. $19\dfrac{19}{32}$ 29. _____   30. _____

EXERCISE 3-3 SET B

NAME _____   DATE _____

CHANGE THE FOLLOWING TO WHOLE OR MIXED NUMBERS: ANSWERS

1. $\dfrac{6}{1}$ 2. $\dfrac{9}{1}$ 1. _____ 2. _____

3. $\dfrac{18}{3}$ 4. $\dfrac{24}{4}$ 3. _____ 4. _____

5. $\dfrac{19}{3}$ 6. $\dfrac{25}{4}$ 5. _____ 6. _____

7. $\dfrac{11}{6}$ 8. $\dfrac{12}{5}$ 7. _____ 8. _____

9. $\dfrac{37}{7}$ 10. $\dfrac{58}{7}$ 9. _____ 10. _____

11. $\dfrac{140}{19}$ 12. $\dfrac{150}{13}$ 11. _____ 12. _____

13. $\dfrac{547}{7}$ 14. $\dfrac{593}{7}$ 13. _____ 14. _____

CHANGE THE FOLLOWING TO IMPROPER FRACTIONS:

15. 7 16. 6   15. _____ 16. _____

17. $1\dfrac{2}{3}$ 18. $1\dfrac{3}{4}$ 17. _____ 18. _____

19. $2\dfrac{5}{6}$ 20. $2\dfrac{7}{8}$ 19. _____ 20. _____

21. $8\dfrac{1}{4}$ 22. $9\dfrac{1}{3}$ 21. _____ 22. _____

23. $24\dfrac{1}{2}$ 24. $46\dfrac{1}{5}$ 23. _____ 24. _____

25. $20\dfrac{5}{8}$ 26. $10\dfrac{7}{9}$ 25. _____ 26. _____

27. $142\dfrac{2}{7}$ 28. $107\dfrac{3}{8}$ 27. _____ 28. _____

29. $39\dfrac{17}{19}$ 30. $17\dfrac{17}{32}$ 29. _____ 30. _____

3-4 Multiplying Fractions

The easiest operation with fractions is multiplication, since to multiply fractions you just have to multiply their numerators and multiply their denominators to get the numerator and the denominator of the answer.

For example: $\frac{1}{2} \times \frac{3}{5} = \frac{1 \times 3}{2 \times 5} = \frac{3}{10}$

$\frac{4}{7} \times \frac{1}{5} \times \frac{6}{13} = \frac{4 \times 1 \times 6}{7 \times 5 \times 13} = \frac{24}{455}$

The difficulty, however, with working with fractions is that answers should be reduced to lowest terms.

For example: $\frac{6}{15} \times \frac{42}{63} = \frac{6 \times 42}{15 \times 63} = \frac{252}{945}$

We must now reduce that answer:

$$\frac{252}{945} \xrightarrow{\div 3} \frac{84}{315} \xrightarrow{\div 3} \frac{28}{105} \xrightarrow{\div 7} \frac{4}{15}$$

The reducing process in that example was quite involved since the numerator and the denominator of the fraction were large numbers. However, if we reduced the fractions before we multiplied, that problem would be a little easier.

Consider $\frac{6}{15} \times \frac{42}{63}$ again: $\frac{6}{15}$ reduces to $\frac{2}{5}$ (÷ by 3) $\frac{42}{63}$ reduces to $\frac{2}{3}$ (÷ by 21)

Thus $\frac{6}{15} \times \frac{42}{63} = \frac{2}{5} \times \frac{2}{3} = \frac{4}{15}$.

Note: To eliminate some of the writing involved in doing a problem such as that, we would write the problem as shown below:

$$\frac{\cancel{6}^{2}}{\cancel{15}_{5}} \times \frac{\cancel{42}^{2}}{\cancel{63}_{3}} = \frac{2 \times 2}{5 \times 3} = \frac{4}{15}$$

Notice that the canceling of numbers is similar to what was done on page 109.

There is another short cut that can be used when multiplying fractions. Let me show you by another example. Consider:

$$\frac{3}{70} \times \frac{14}{25} = \frac{3 \times 14}{70 \times 25} = \frac{42}{1750}$$

We must now reduce that answer:

$$\frac{42}{1750} \xrightarrow[\div 2]{\div 2} \frac{21}{875} \xrightarrow[\div 7]{\div 7} \frac{3}{125}$$

Notice that the original fractions 3/70 and 14/25 are not reducible. However, the product 42/1750 is reducible to lower terms. It would be nice if we could reduce before we multiplied, as we did in the previous example. What we could have done is reduce the numerator of the 2nd fraction with the denominator of the 1st fraction.

$$\frac{3}{\cancel{70}_{5}} \times \frac{\cancel{14}^{1}}{25} = \frac{3 \times 1}{5 \times 25} = \frac{3}{125}$$

This process is called cross canceling and can <u>only</u> be used when multiplying fractions. The process requires that you divide the same number evenly into any numerator and any denominator. You can do this canceling process in any order, and you should continue until all possible combinations are canceled.

$$2 \times \frac{18}{10} \times \frac{15}{26} = ?$$

$$= \frac{2}{1} \times \frac{\cancel{18}^{9}}{\cancel{10}_{5}} \times \frac{15}{26} \qquad \text{Change 2 to the fraction 2/1 and reduce 18/10 to 9/5.}$$

$$= \frac{\cancel{2}^{1}}{1} \times \frac{9}{\cancel{5}_{1}} \times \frac{\cancel{15}^{3}}{\cancel{26}_{13}} \qquad \text{Cross cancel the 2 and the 26; cross cancel the 5 and the 15.}$$

$$= \frac{27}{13} = 2\frac{1}{13} \qquad \text{Multiply the remaining fractions and change the improper fraction to a mixed number.}$$

PROBLEM 1: $\frac{3}{4} \times \frac{20}{7} = ?$ Answer: 2 1/7

$$\frac{3}{\cancel{4}_{1}} \times \frac{\cancel{20}^{5}}{7} = \frac{15}{7} = 2\frac{1}{7}$$

PROBLEM 2: $\frac{12}{15} \times \frac{8}{25} \times \frac{75}{100} = ?$ Answer: 24/125

$$\frac{\cancel{12}^{4}}{\cancel{15}_{5}} \times \frac{8}{25} \times \frac{\cancel{75}^{3}}{\cancel{100}_{4}} = \frac{4}{5} \times \frac{8}{25} \times \frac{3}{\cancel{4}_{1}}^{2} = \frac{24}{125}$$

EXERCISE 3-4   SET A

NAME _____   DATE _____

MULTIPLY AND REDUCE: ANSWERS

1. $\frac{1}{2} \times \frac{3}{4}$ 2. $\frac{1}{3} \times \frac{2}{5}$ 1. _____ 2. _____

3. $\frac{7}{8} \times \frac{5}{9}$ 4. $\frac{9}{10} \times \frac{7}{8}$ 3. _____ 4. _____

5. $3 \times \frac{1}{5}$ 6. $2 \times \frac{1}{3}$ 5. _____ 6. _____

7. $\frac{2}{3} \times 5$ 8. $\frac{3}{7} \times 6$ 7. _____ 8. _____

9. $\frac{3}{7} \times \frac{5}{9}$ 10. $\frac{2}{7} \times \frac{5}{6}$ 9. _____ 10. _____

11. $\frac{10}{12} \times \frac{11}{7}$ 12. $\frac{6}{9} \times \frac{11}{5}$ 11. _____ 12. _____

13. $\frac{12}{35} \times \frac{5}{16}$ 14. $\frac{10}{27} \times \frac{9}{20}$ 13. _____ 14. _____

15. $3 \times \frac{4}{25} \times \frac{5}{12}$ 16. $2 \times \frac{8}{49} \times \frac{7}{16}$ 15. _____ 16. _____

17. $\frac{13}{35} \times \frac{14}{21} \times 5$ 18. $\frac{11}{42} \times \frac{6}{8} \times 6$ 17. _____ 18. _____

19. $\frac{1}{3} \times \frac{4}{5} \times \frac{8}{7}$ 20. $\frac{1}{4} \times \frac{3}{5} \times \frac{7}{2}$ 19. _____ 20. _____

21. $\frac{1}{3} \times \frac{6}{13} \times \frac{39}{15}$ 22. $\frac{1}{4} \times \frac{12}{33} \times \frac{11}{9}$ 21. _____ 22. _____

23. $\frac{20}{21} \times \frac{19}{36} \times \frac{33}{38}$ 24. $\frac{22}{13} \times \frac{11}{36} \times \frac{39}{22}$ 23. _____ 24. _____

25. $\frac{150}{320} \times \frac{5}{6} \times \frac{28}{180}$ 26. $\frac{108}{320} \times \frac{7}{8} \times \frac{28}{144}$ 25. _____ 26. _____

EXERCISE 3-4 SET B

NAME _____ DATE _____

MULTIPLY AND REDUCE: ANSWERS

1. $\dfrac{1}{4} \times \dfrac{3}{4}$ 　　　　　 2. $\dfrac{1}{2} \times \dfrac{1}{3}$ 　　　　　 1. _____ 2. _____

3. $\dfrac{6}{7} \times \dfrac{5}{11}$ 　　　　 4. $\dfrac{7}{10} \times \dfrac{3}{4}$ 　　　　 3. _____ 4. _____

5. $4 \times \dfrac{1}{5}$ 　　　　　　　 6. $3 \times \dfrac{1}{4}$ 　　　　　　　 5. _____ 6. _____

7. $\dfrac{3}{4} \times 5$ 　　　　　　　 8. $\dfrac{2}{3} \times 7$ 　　　　　　　 7. _____ 8. _____

9. $\dfrac{4}{5} \times \dfrac{7}{12}$ 　　　　 10. $\dfrac{3}{7} \times \dfrac{5}{6}$ 　　　　 9. _____ 10. _____

11. $\dfrac{10}{15} \times \dfrac{13}{3}$ 　　　 12. $\dfrac{8}{12} \times \dfrac{13}{7}$ 　　　 11. _____ 12. _____

13. $\dfrac{24}{35} \times \dfrac{7}{18}$ 　　　 14. $\dfrac{12}{25} \times \dfrac{35}{36}$ 　　 13. _____ 14. _____

15. $4 \times \dfrac{6}{25} \times \dfrac{5}{24}$ 　　 16. $3 \times \dfrac{6}{49} \times \dfrac{7}{18}$ 　　 15. _____ 16. _____

17. $\dfrac{11}{45} \times \dfrac{8}{12} \times 5$ 　 18. $\dfrac{7}{30} \times \dfrac{6}{9} \times 6$ 　 17. _____ 18. _____

19. $\dfrac{1}{2} \times \dfrac{10}{18} \times \dfrac{54}{15}$ 　 20. $\dfrac{1}{5} \times \dfrac{12}{27} \times \dfrac{18}{30}$ 　 19. _____ 20. _____

21. $\dfrac{1}{3} \times \dfrac{2}{5} \times \dfrac{7}{9}$ 　　 22. $\dfrac{1}{2} \times \dfrac{3}{4} \times \dfrac{5}{7}$ 　　 21. _____ 22. _____

23. $\dfrac{14}{30} \times \dfrac{11}{36} \times \dfrac{15}{22}$ 　 24. $\dfrac{10}{49} \times \dfrac{19}{36} \times \dfrac{21}{38}$ 　 23. _____ 24. _____

25. $\dfrac{144}{180} \times \dfrac{7}{3} \times \dfrac{56}{108}$ 　 26. $\dfrac{54}{108} \times \dfrac{4}{9} \times \dfrac{56}{180}$ 　 25. _____ 26. _____

3-5 Dividing Fractions

An understanding of how to multiply fractions is essential in this section. You will learn how division of fractions can be changed into the multiplication of fractions.

Look at the problem 12 ÷ 3. We can consider dividing a number by 3 the same as taking 1/3 of it. Since "of" means "times" in math, we get:

$$12 \div 3 = 12 \times \frac{1}{3} = 4$$

Also in a problem such as 3/5 ÷ 2, we can consider dividing a number by 2 the same as taking a half of it.

$$\frac{3}{5} \div 2 = \frac{3}{5} \times \frac{1}{2} = \frac{3}{10}$$

Let's look closely at those two problems. We know that 3 = 3/1 and 2 = 2/1, so those two problems become:

A GOOD CAR SALESMAN <u>CANCEL</u>.

$$12 \div 3 = 12 \div \frac{3}{1} = 12 \times \frac{1}{3} \qquad \frac{3}{5} \div 2 = \frac{3}{5} \div \frac{2}{1} = \frac{3}{5} \times \frac{1}{2}$$

Those two examples point out a method for **dividing fractions**.

> Invert the divisor (find its **reciprocal**) and change the division into multiplication.

$$12 \div \frac{3}{1} = 12 \times \frac{1}{3} \qquad \frac{3}{5} \div \frac{2}{1} = \frac{3}{5} \times \frac{1}{2}$$

(÷ changed to ×) with (reciprocal) arrows shown on both.

Using that method, problems that involve dividing fractions become multiplication problems with the original divisor inverted. Let's try some examples.

Example 1: $\frac{2}{3} \div \frac{5}{7} = ?$

$$= \frac{2}{3} \times \frac{7}{5} \qquad \text{Invert the divisor and multiply.}$$

$$= \frac{14}{15}$$

Example 2: $\frac{10}{15} \div \frac{7}{27} = ?$

$= \frac{10}{15} \times \frac{27}{7}$ Invert the divisor and multiply.

$= \frac{\overset{2}{\cancel{10}}}{\underset{3}{\cancel{15}}} \times \frac{27}{7}$ Reduce.

$= \frac{2}{\underset{1}{\cancel{3}}} \times \frac{\overset{9}{\cancel{27}}}{7}$ Cross cancel.

$= \frac{18}{7} = 2\frac{4}{7}$ Multiply and change the improper fraction to a mixed number.

Example 3: $35 \div \frac{28}{6} = ?$

$= 35 \times \frac{6}{28}$ Invert the divisor and multiply.

$= \frac{35}{1} \times \frac{3}{14}$ Change 35 to 35/1 and reduce 6/28 to 3/14.

$= \frac{\overset{5}{\cancel{35}}}{1} \times \frac{3}{\underset{2}{\cancel{14}}}$ Cross cancel.

$= \frac{15}{2} = 7\frac{1}{2}$ Multiply and change the improper fraction to a mixed number.

We can summarize **division of fractions** as follows:

1. Invert the divisor and change to multiplication.
2. Reduce and cross cancel, if possible.
3. Multiply the resulting fractions.
4. Change any improper fraction into a mixed number.

PROBLEM 1: $\frac{10}{3} \div 45 = ?$

Answer: 2/27

$$\frac{10}{3} \div \frac{45}{1} = \frac{\overset{2}{\cancel{10}}}{3} \times \frac{1}{\underset{9}{\cancel{45}}} = \frac{2}{27}$$

PROBLEM 2: $\frac{28}{16} \div \frac{21}{10} = ?$

Answer: 5/6

$$= \frac{\overset{7}{\cancel{28}}}{\underset{4}{\cancel{16}}} \times \frac{10}{21} = \frac{\overset{1}{\cancel{7}}}{4} \times \frac{10}{\underset{3}{\cancel{21}}} = \frac{1}{\underset{2}{\cancel{4}}} \times \frac{\overset{5}{\cancel{10}}}{3} = \frac{5}{6}$$

EXERCISE 3-5 SET A

NAME _____ DATE _____

DIVIDE AND REDUCE: ANSWERS

1. $\dfrac{1}{3} \div \dfrac{1}{2}$ 2. $\dfrac{1}{4} \div \dfrac{1}{3}$ 1. _____

                                                                2. _____

3. $\dfrac{2}{3} \div \dfrac{3}{5}$ 4. $\dfrac{5}{6} \div \dfrac{6}{7}$ 3. _____

                                                                4. _____

5. $\dfrac{6}{9} \div \dfrac{7}{5}$ 6. $\dfrac{8}{10} \div \dfrac{9}{7}$ 5. _____

                                                                6. _____

7. $\dfrac{5}{6} \div \dfrac{7}{6}$ 8. $\dfrac{4}{7} \div \dfrac{3}{7}$ 7. _____

                                                                8. _____

9. $\dfrac{2}{3} \div 4$ 10. $\dfrac{3}{5} \div 6$ 9. _____

                                                               10. _____

11. $\dfrac{22}{7} \div \dfrac{22}{21}$ 12. $\dfrac{19}{16} \div \dfrac{19}{4}$ 11. _____

                                                               12. _____

13. $18 \div \dfrac{8}{9}$ 14. $40 \div \dfrac{25}{7}$ 13. _____

                                                               14. _____

15. $\dfrac{5}{18} \div \dfrac{35}{33}$ 16. $\dfrac{4}{21} \div \dfrac{24}{35}$ 15. _____

                                                               16. _____

17. $\dfrac{20}{42} \div \dfrac{15}{28}$ 18. $\dfrac{24}{50} \div \dfrac{16}{15}$ 17. _____

                                                               18. _____

19. $\dfrac{32}{30} \div \dfrac{24}{70}$ 20. $\dfrac{36}{28} \div \dfrac{30}{35}$ 19. _____

                                                               20. _____

EXERCISE 3-5 SET B

NAME _____   DATE _____

DIVIDE AND REDUCE: ANSWERS

1. $\dfrac{1}{6} \div \dfrac{1}{5}$ 2. $\dfrac{1}{5} \div \dfrac{1}{3}$ 1. _____

                                                    2. _____

3. $\dfrac{4}{5} \div \dfrac{5}{7}$ 4. $\dfrac{3}{4} \div \dfrac{4}{5}$ 3. _____

                                                    4. _____

5. $\dfrac{9}{12} \div \dfrac{7}{5}$ 6. $\dfrac{6}{8} \div \dfrac{11}{7}$ 5. _____

                                                    6. _____

7. $\dfrac{5}{9} \div \dfrac{4}{9}$ 8. $\dfrac{3}{8} \div \dfrac{5}{8}$ 7. _____

                                                    8. _____

9. $\dfrac{3}{4} \div 9$ 10. $\dfrac{5}{8} \div 10$ 9. _____

                                                    10. _____

11. $\dfrac{18}{7} \div \dfrac{18}{35}$ 12. $\dfrac{20}{7} \div \dfrac{20}{21}$ 11. _____

                                                    12. _____

13. $24 \div \dfrac{9}{5}$ 14. $36 \div \dfrac{20}{7}$ 13. _____

                                                    14. _____

15. $\dfrac{7}{40} \div \dfrac{21}{25}$ 16. $\dfrac{11}{50} \div \dfrac{33}{35}$ 15. _____

                                                    16. _____

17. $\dfrac{28}{30} \div \dfrac{21}{55}$ 18. $\dfrac{14}{40} \div \dfrac{21}{25}$ 17. _____

                                                    18. _____

19. $\dfrac{36}{54} \div \dfrac{42}{72}$ 20. $\dfrac{32}{40} \div \dfrac{56}{35}$ 19. _____

                                                    20. _____

3-6 Multiplying and Dividing Mixed Numbers

In Sections 3-4 and 3-5 you learned how to multiply and divide fractions. If you look over the examples and problems in those sections, you will notice that you operated with both proper and improper fractions. The methods shown worked for both types of fractions. Therefore, the logical way to multiply and divide mixed numbers is to first change them into improper fractions and then use those same methods. Let me show you what I mean.

THE MAGICIAN HAD A GREAT TRICK: FROM ONE CROSS HE MADE MANY OTHERS APPEAR. HE MULTIPLIED ACROSS.

Multiplying Mixed Numbers

Example 1: $11 \times 5\frac{2}{9} = ?$

$= \frac{11}{1} \times \frac{47}{9}$ Change to improper fractions.

$= \frac{517}{9} = 57\frac{4}{9}$ Multiply across and simplify.

Example 2: $3\frac{3}{4} \times 5\frac{1}{3} = ?$

$= \frac{15}{4} \times \frac{16}{3}$ Change the mixed numbers to improper fractions.

$= \frac{\cancel{15}^{5}}{\cancel{4}_{1}} \times \frac{\cancel{16}^{4}}{\cancel{3}_{1}}$ Cross cancel the 4 and the 16; cross cancel the 15 and the 3.

$= \frac{20}{1} = 20$ Multiply across and simplify.

PROBLEM 1: $7\frac{2}{3} \times 10\frac{1}{2} = ?$

Answer: 80 1/2

$7\frac{2}{3} \times 10\frac{1}{2} = \frac{23}{\cancel{3}_1} \times \frac{\cancel{21}^7}{2}$

$= \frac{161}{2} = 80\frac{1}{2}$

Dividing Mixed Numbers

When dividing mixed numbers you will again change the mixed numbers into improper fractions. Remember, however, that when you divide fractions you must invert the divisor and change the division to multiplication.

Example 3: $10\frac{2}{15} \div 6\frac{2}{5} = ?$

$= \frac{152}{15} \div \frac{32}{5}$ Change the mixed numbers to improper fractions.

$= \frac{\cancel{152}^{19}}{\cancel{15}_{3}} \times \frac{\cancel{5}^{1}}{\cancel{32}_{4}}$ Invert the divisor; change to multiplication; cross cancel the 15 and the 5, and the 152 and the 32.

$= \frac{19}{12} = 1\frac{7}{12}$ Multiply across and simplify.

Example 4: $9\frac{5}{6} \div 12 = ?$

$= \frac{59}{6} \div \frac{12}{1}$ Change the mixed numbers to improper fractions.

$= \frac{59}{6} \times \frac{1}{12}$ Invert the divisor and multiply.

$= \frac{59}{72}$

PROBLEM 2: $8\frac{1}{3} \div 9\frac{4}{9} = ?$ Answer: 15/17

$8\frac{1}{3} \div 9\frac{4}{9} = \frac{25}{3} \div \frac{85}{9}$

$= \frac{\cancel{25}^{5}}{\cancel{3}_{1}} \times \frac{\cancel{9}^{3}}{\cancel{85}_{17}} = \frac{15}{17}$

PROBLEM 3: $76 \div 2\frac{5}{8} = ?$ Answer: 28 20/21

$76 \div 2\frac{5}{8} = \frac{76}{1} \div \frac{21}{8}$

$= \frac{76}{1} \times \frac{8}{21} = \frac{608}{21} = 28\frac{20}{21}$

EXERCISE 3-6 SET A

NAME _____   DATE _____

MULTIPLY OR DIVIDE AND REDUCE: ANSWERS

1. $2\frac{3}{4} \times 1\frac{2}{3}$ 2. $5\frac{1}{4} \times 1\frac{2}{7}$ 1. _____

                                                                                                       2. _____

3. $5\frac{3}{8} \times 16$ 4. $12 \times 7\frac{3}{5}$ 3. _____

                                                                                                       4. _____

5. $26\frac{2}{3} \times 4\frac{4}{35}$ 6. $67\frac{1}{5} \times 7\frac{8}{21}$ 5. _____

                                                                                                       6. _____

7. $7\frac{3}{5} \div 6\frac{3}{10}$ 8. $10\frac{5}{6} \div 12\frac{2}{3}$ 7. _____

                                                                                                       8. _____

9. $24\frac{5}{8} \div 4$ 10. $12\frac{15}{16} \div 6$ 9. _____

                                                                                                       10. _____

11. $5 \div 7\frac{1}{2}$ 12. $22 \div 3\frac{2}{3}$ 11. _____

                                                                                                       12. _____

EXERCISE 3-6 SET B

NAME _____   DATE _____

MULTIPLY OR DIVIDE AND REDUCE: ANSWERS

1. $3\frac{2}{3} \times 1\frac{3}{4}$ 2. $4\frac{1}{2} \times 2\frac{5}{6}$ 1. _____

                                                                        2. _____

3. $4\frac{3}{8} \times 24$ 4. $20 \times 3\frac{5}{16}$ 3. _____

                                                                        4. _____

5. $12\frac{6}{7} \times 4\frac{8}{33}$ 6. $8\frac{4}{7} \times 11\frac{2}{3}$ 5. _____

                                                                        6. _____

7. $5\frac{4}{5} \div 3\frac{7}{10}$ 8. $3\frac{3}{4} \div 5\frac{2}{5}$ 7. _____

                                                                        8. _____

9. $35\frac{3}{8} \div 4$ 10. $40\frac{5}{8} \div 5$ 9. _____

                                                                        10. _____

11. $9 \div 10\frac{1}{2}$ 12. $33 \div 2\frac{3}{4}$ 11. _____

                                                                        12. _____

3-7 Fractional Parts of Numbers

Fractions are often used to find a part of a number or quantity. You might encounter phrases such as "4/5 of those surveyed", "1/3 off the retail price", or "3/4 of the class". In statements such as these, you are finding a part of a total amount. The word "of" in this usage indicates multiplying the fraction and the number.

For example: $\frac{4}{5}$ of 4725 means $\frac{4}{5} \times 4725$

$\frac{1}{3}$ of $630 means $\frac{1}{3} \times \$630$

$\frac{3}{4}$ of 36 means $\frac{3}{4} \times 36$

When you take a fraction of a number, you simply multiply the fraction and the number and reduce your answer to lowest terms.

Example 1: $\frac{5}{6}$ of 4200 = ?

$$\frac{5}{6} \text{ of } 4200 = \frac{5}{6} \times 4200$$

$$= \frac{5}{\cancel{6}_1} \times \frac{\cancel{4200}^{700}}{1} = 3500$$

Example 2: $\frac{1}{4}$ of $25\frac{3}{8}$ = ?

$$\frac{1}{4} \text{ of } 25\frac{3}{8} = \frac{1}{4} \times 25\frac{3}{8}$$

$$= \frac{1}{4} \times \frac{203}{8} = \frac{203}{32} = 6\frac{11}{32}$$

PROBLEM 1: $\frac{2}{3}$ of 951 = ?   Answer: 634

$$\frac{2}{\cancel{3}_1} \times \frac{\cancel{951}^{317}}{1} = 634$$

PROBLEM 2: $\frac{5}{16}$ of 9 = ?   Answer: 2 13/16

$$\frac{5}{16} \times \frac{9}{1} = \frac{45}{16} = 2\frac{13}{16}$$

Finding the fractional part of a quantity as shown on the previous page is useful in many applications involving fractions. Let's examine some of them.

Example 3: 3/5 of those surveyed preferred "Best" toothpaste. If 100,000 people were surveyed, how many preferred "Best" toothpaste?

Analyze: You need to take a fraction (3/5) of the total number of people surveyed (100,000). Multiply 3/5 times 100,000.

Solve: $\dfrac{3}{5}$ of $100,000 = \dfrac{3}{\cancel{5}} \times \dfrac{\cancel{100,000}^{\,20,000}}{1} = 60,000$ people

Example 4: A store is advertising a discount of 1/4 off all retail prices. What would be the cost of a $304 item after such a discount?

Analyze:
1. Find the amount of the discount (1/4 of $304).
2. Subtract the discount from the retail price.

Solve:
1. $\dfrac{1}{\cancel{4}} \times \dfrac{\cancel{\$304}^{\,76}}{1} = \$76$

2. $304 - $76 = $228

PROBLEM 3: 1/12 of the 100,800 runners at a Bay to Breakers Race in San Francisco did not officially register for the race. If 2/3 of the unregistered runners were males, how many females were not officially registered for the race?

Answer: 2800 females

unregistered:
1/12 x 100,800 = 8400

unregistered males:
2/3 x 8400 = 5600

unregistered females:
8400 - 5600 = 2800

PROBLEM 4: What amounts would you use to make half of the amount of Nut Bread given in the recipe below?

Nut Bread

1/2 cup sugar   3 1/4 cups biscuit mix
2 eggs   1 1/2 cups chopped nuts
1/8 tsp salt   1 1/4 cups milk

Answers: (Multiply each amount by 1/2)

sugar: 1/4 cup
eggs: 1
salt: 1/16 tsp
mix: 1 5/8 cups
nuts: 3/4 cup
milk: 5/8 cup

EXERCISE 3-7 SET A

NAME _____ DATE _____

FIND THE REDUCED ANSWERS: ANSWERS

1. $\frac{1}{2}$ of 537   2. $\frac{1}{3}$ of 441   1. _____

                                                              2. _____

3. $\frac{2}{5}$ of 4270   4. $\frac{3}{4}$ of 2105   3. _____

                                                              4. _____

5. $\frac{3}{8}$ of $7\frac{1}{2}$ 6. $\frac{5}{7}$ of $4\frac{2}{3}$ 5. _____

                                                              6. _____

7. $\frac{1}{4}$ of $36\frac{5}{8}$ 8. $\frac{1}{5}$ of $83\frac{1}{2}$ 7. _____

                                                              8. _____

9. 3/4 of the club members voted for Jane Adams.   9. _____
   If 2452 members voted, how many voted for Jane?

10. 5/8 of the light bulbs produced at a factory are   10. _____
    50-watt bulbs. If the factory produces 10,000 light
    bulbs a day, how many of them are 50-watt bulbs?

11. A shoe store is having a 1/4 off sale. How much   11. _____
    would a $52 pair of shoes cost after the discount?

12. During a flu epidemic, 1/3 of the students were   12. _____
    absent on Monday. If the school has an enrollment
    of 942, how many were in school on that Monday?

13. 3/4 of the coins in a bank are pennies. 2/3 of those   13. _____
    pennies are dated 1983 or later. If the bank has 876
    coins, how many are pennies dated before 1983?

14. An analysis of the 96 member Spartan football team   14. _____
    reveals that 2/3 of them weigh over 200 pounds and
    that 1/4 of those over 200 pounds are also over 6'2"
    tall. How many Spartan football players are both
    over 200 pounds and over 6'2" tall?

EXERCISE 3-7   SET B

NAME _____   DATE _____

FIND THE REDUCED ANSWERS: ANSWERS

1. $\frac{1}{3}$ of 39   2. $\frac{1}{2}$ of 57   1. _____

                                                                   2. _____

3. $\frac{2}{3}$ of 571   4. $\frac{4}{5}$ of 610   3. _____

                                                                   4. _____

5. $\frac{3}{4}$ of $25\frac{1}{8}$ 6. $\frac{5}{8}$ of $16\frac{1}{2}$ 5. _____

                                                                   6. _____

7. $\frac{1}{2}$ of $107\frac{5}{16}$ 8. $\frac{1}{4}$ of $326\frac{1}{4}$ 7. _____

                                                                   8. _____

9. 3/4 of all entering freshmen at a college live on   9. _____
    campus. If the freshman enrollment is 1240, how
    many live on campus?

10. A 64 ounce jug is 2/3 full of water. How many   10. _____
    ounces of water does the jug contain?

11. 1/3 of the 5343 students registering for classes   11. _____
    paid their fees with cash. How many students did
    not pay with cash?

12. At a fire sale, items are being sold at 3/4 off the   12. _____
    marked price. What is the sale price of an item
    which has a marked price of $156?

13. Of the 2310 people polled, 2/3 favored Proposition A.   13. _____
    However, only 1/5 of those who favored Proposition A
    also favored Proposition B. How many favored both
    propositions?

14. A factory has 1710 workers. 3/5 of the workers are   14. _____
    women, and 1/6 of the women have a college degree.
    How many of the workers are women with a college degree?

3-8 Adding and Subtracting Like Fractions

The next step in developing your math skills is adding and subtracting fractions. We will start by adding and subtracting **like fractions,** that is, fractions with the same denominators.

For example: $\frac{2}{5} + \frac{1}{5} = ?$

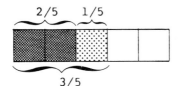

Thus $\frac{2}{5} + \frac{1}{5} = \frac{3}{5}$.

And $\frac{6}{8} - \frac{1}{8} = ?$

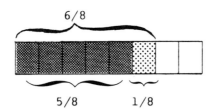

Thus $\frac{6}{8} - \frac{1}{8} = \frac{5}{8}$.

It seems very simple then to add or subtract like fractions. The numerator of the answer is the sum of the numerators if you are adding, or the difference between the numerators if you are subtracting. In either case, the denominator of the answer is the same as the denominators in the problem. Notice we did <u>not</u> add or subtract the denominators. As with all computations with fractions, you must remember to reduce your final answer.

Example 1: $\frac{3}{7} + \frac{2}{7} = ?$

$$= \frac{3 + 2}{7}$$

$$= \frac{5}{7}$$

Example 2: $\frac{5}{6} - \frac{2}{6} = ?$

$$= \frac{5 - 2}{6}$$

$$= \frac{3}{6}$$

$$= \frac{1}{2} \qquad \text{Reduce the answer.}$$

Example 3: $\frac{5}{8} + \frac{2}{8} + \frac{3}{8} = ?$

$$= \frac{5 + 2 + 3}{8}$$

$$= \frac{10}{8}$$

$$= \frac{5}{4} \quad \text{Reduce the answer.}$$

$$= 1\frac{1}{4} \quad \text{Change the improper fraction to a mixed number.}$$

Example 4: $\frac{63}{25} - \frac{11}{25} + \frac{8}{25} = ?$

$$= \frac{63 - 11 + 8}{25}$$

$$= \frac{60}{25}$$

$$= \frac{12}{5} \quad \text{Reduce the answer.}$$

$$= 2\frac{2}{5} \quad \text{Change the improper fraction to a mixed number.}$$

PROBLEM 1: $\frac{7}{18} + \frac{3}{18} = ?$ Answer: 5/9

$$\frac{7}{18} + \frac{3}{18} = \frac{10}{18} = \frac{5}{9}$$

PROBLEM 2: $\frac{11}{30} - \frac{5}{30} = ?$ Answer: 1/5

$$\frac{11}{30} - \frac{5}{30} = \frac{6}{30} = \frac{1}{5}$$

PROBLEM 3: $\frac{2}{9} + \frac{5}{9} + \frac{7}{9} - \frac{1}{9} = ?$ Answer: 1 4/9

$$\frac{2}{9} + \frac{5}{9} + \frac{7}{9} - \frac{1}{9} = \frac{2+5+7-1}{9}$$

$$= \frac{13}{9}$$

$$= 1\frac{4}{9}$$

EXERCISE 3-8   SET A

NAME _____   DATE _____

ADD OR SUBTRACT AND REDUCE: ANSWERS

1. $\dfrac{1}{7} + \dfrac{1}{7}$ 2. $\dfrac{1}{3} + \dfrac{1}{3}$ 1. _____

                                                          2. _____

3. $\dfrac{3}{9} - \dfrac{2}{9}$ 4. $\dfrac{4}{7} - \dfrac{2}{7}$ 3. _____

                                                          4. _____

5. $\dfrac{3}{8} + \dfrac{1}{8}$ 6. $\dfrac{5}{12} + \dfrac{1}{12}$ 5. _____

                                                          6. _____

7. $\dfrac{7}{18} - \dfrac{5}{18}$ 8. $\dfrac{8}{15} - \dfrac{2}{15}$ 7. _____

                                                          8. _____

9. $\dfrac{4}{3} + \dfrac{5}{3}$ 10. $\dfrac{5}{2} + \dfrac{7}{2}$ 9. _____

                                                          10. _____

11. $\dfrac{19}{5} - \dfrac{4}{5}$ 12. $\dfrac{17}{6} + \dfrac{5}{6}$ 11. _____

                                                          12. _____

13. $\dfrac{3}{8} + \dfrac{2}{8} + \dfrac{3}{8}$ 14. $\dfrac{3}{9} + \dfrac{4}{9} + \dfrac{2}{9}$ 13. _____

                                                          14. _____

15. $\dfrac{11}{32} + \dfrac{5}{32} - \dfrac{2}{32}$ 16. $\dfrac{13}{32} + \dfrac{8}{32} - \dfrac{3}{32}$ 15. _____

                                                          16. _____

17. $\dfrac{25}{54} + \dfrac{21}{54} + \dfrac{17}{54}$ 18. $\dfrac{27}{64} + \dfrac{15}{64} + \dfrac{18}{64}$ 17. _____

                                                          18. _____

EXERCISE 3-8 SET B

NAME _____ DATE _____

ADD OR SUBTRACT AND REDUCE: ANSWERS

1. $\frac{1}{9} + \frac{1}{9}$ 2. $\frac{1}{5} + \frac{1}{5}$ 1. _____

                                                                         2. _____

3. $\frac{3}{7} - \frac{2}{7}$ 4. $\frac{4}{9} - \frac{2}{9}$ 3. _____

                                                                         4. _____

5. $\frac{4}{10} + \frac{1}{10}$ 6. $\frac{2}{6} + \frac{1}{6}$ 5. _____

                                                                         6. _____

7. $\frac{7}{15} - \frac{2}{15}$ 8. $\frac{7}{18} - \frac{3}{18}$ 7. _____

                                                                         8. _____

9. $\frac{7}{2} + \frac{3}{2}$ 10. $\frac{7}{3} + \frac{5}{3}$ 9. _____

                                                                         10. _____

11. $\frac{17}{6} - \frac{5}{6}$ 12. $\frac{12}{5} - \frac{2}{5}$ 11. _____

                                                                         12. _____

13. $\frac{4}{7} + \frac{2}{7} + \frac{1}{7}$ 14. $\frac{3}{8} + \frac{4}{8} + \frac{1}{8}$ 13. _____

                                                                         14. _____

15. $\frac{15}{32} + \frac{7}{32} - \frac{4}{32}$ 16. $\frac{17}{32} + \frac{10}{32} - \frac{3}{32}$ 15. _____

                                                                         16. _____

17. $\frac{19}{64} + \frac{16}{64} + \frac{27}{64}$ 18. $\frac{14}{54} + \frac{25}{54} + \frac{18}{54}$ 17. _____

                                                                         18. _____

3-9 Adding and Subtracting Unlike Fractions

The fractions in the last section were like fractions since the fractions in each problem had the same denominators. We added and subtracted them very easily. In fact, in order to add and subtract any fractions, they must be like fractions. In the problem

$$\frac{3}{4} + \frac{1}{5} = ?$$

We must change each fraction to equivalent fractions that have the same denominators before we can add them. The denominator that we change each fraction into should be the smallest number that is divisible by each denominator. We call it the *least* **common denominator (LCD)**.

In the above problem, the LCD is 20, since 20 is the smallest number that is divisible by both 4 and 5. That means we have to change 3/4 and 1/5 into 20ths before we can add them.

$$\frac{3}{4} = \frac{?}{20} \qquad \frac{3}{4} \xrightarrow{\times 5} \frac{15}{20}$$

$$\frac{1}{5} = \frac{?}{20} \qquad +\frac{1}{5} \xrightarrow{\times 4} \frac{4}{20}$$

$$= \frac{19}{20}$$

— Now add the like fractions.

Quite frequently you can determine the LCD by inspection. That is, you look at the denominators in an addition or subtraction problem and think, "What is the smallest number that can be evenly divided by each denominator?"

For example, in these problems:

$\frac{2}{3} - \frac{1}{2} = ?$ The LCD is 6, since 6 is the smallest number divisible by 3 and 2.

$\frac{1}{2} + \frac{2}{5} + \frac{9}{10} = ?$ The LCD is 10, since 10 is the smallest number divisible by 2, 5, and 10.

$\frac{3}{4} + \frac{5}{6} + \frac{3}{2} - \frac{2}{3} = ?$ The LCD is 12, since 12 is the smallest number divisible by 4, 6, 2, and 3.

The method, then, for **adding and subtracting fractions** is as follows:

> 1. Find the least common denominator.
> 2. Change each fraction to a fraction with the LCD as its denominator.
> 3. Add or subtract those like fractions.
> 4. Reduce your answer and change any improper fraction into a mixed number.

Let's use those steps in doing the three previous examples.

Example 1: $\frac{2}{3} - \frac{1}{2} = ?$

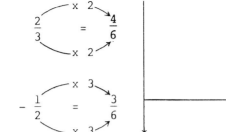

The LCD is 6.

Change each fraction to a fraction with a denominator of 6.

Subtract the like fractions.

Example 2: $\frac{1}{2} + \frac{2}{5} + \frac{9}{10} = ?$

$$\frac{1}{2} = \frac{5}{10}$$

$$+ \frac{2}{5} = \frac{4}{10}$$

$$+ \frac{9}{10} = \frac{9}{10}$$

$$= \frac{18}{10} = \frac{9}{5} = 1\frac{4}{5}$$

The LCD is 10.

Change each fraction to a fraction with a denominator of 10.

Add the like fractions.

Reduce the answer and change it to a mixed number.

Example 3: $\frac{3}{4} + \frac{5}{6} + \frac{3}{2} - \frac{2}{3} = ?$

$$\frac{3}{4} = \frac{9}{12}$$

$$+ \frac{5}{6} = \frac{10}{12}$$

$$+ \frac{3}{2} = \frac{18}{12}$$

$$- \frac{2}{3} = \frac{8}{12}$$

$$= \frac{29}{12} = 2\frac{5}{12}$$

The LCD is 12.

Change each fraction to a fraction with a denominator of 12.

Add and subtract the like fractions.

Change the improper fraction to a mixed number.

EXERCISE 3-9   SET A

NAME _____   DATE _____

ADD OR SUBTRACT AND REDUCE: ANSWERS

1. $\dfrac{1}{2} + \dfrac{1}{4}$ 2. $\dfrac{2}{3} + \dfrac{1}{6}$ 1. _____

                                                                  2. _____

3. $\dfrac{3}{4} - \dfrac{5}{8}$ 4. $\dfrac{5}{8} - \dfrac{1}{4}$ 3. _____

                                                                  4. _____

5. $\dfrac{3}{4} - \dfrac{1}{3}$ 6. $\dfrac{2}{3} - \dfrac{1}{4}$ 5. _____

                                                                  6. _____

7. $\dfrac{3}{20} + \dfrac{3}{4}$ 8. $\dfrac{13}{20} + \dfrac{1}{4}$ 7. _____

                                                                  8. _____

9. $\dfrac{5}{6} - \dfrac{4}{5}$ 10. $\dfrac{5}{8} - \dfrac{3}{5}$ 9. _____

                                                                  10. _____

11. $\dfrac{7}{10} + \dfrac{5}{2} + \dfrac{4}{5}$ 12. $\dfrac{9}{10} + \dfrac{7}{2} + \dfrac{3}{5}$ 11. _____

                                                                  12. _____

13. $\dfrac{3}{8} + \dfrac{15}{32} + \dfrac{7}{16}$ 14. $\dfrac{1}{8} + \dfrac{13}{32} + \dfrac{5}{16}$ 13. _____

                                                                  14. _____

15. $\dfrac{1}{4} + \dfrac{3}{6} + \dfrac{9}{2} - \dfrac{7}{3}$ 16. $\dfrac{1}{4} + \dfrac{2}{6} + \dfrac{7}{2} - \dfrac{5}{3}$ 15. _____

                                                                  16. _____

EXERCISE 3-9 SET B

NAME _____   DATE _____

ADD OR SUBTRACT AND REDUCE: ANSWERS

1. $\dfrac{1}{3} + \dfrac{1}{6}$ 2. $\dfrac{1}{4} + \dfrac{1}{2}$ 1. _____

                                                           2. _____

3. $\dfrac{3}{4} - \dfrac{3}{8}$ 4. $\dfrac{3}{4} - \dfrac{1}{8}$ 3. _____

                                                           4. _____

5. $\dfrac{1}{3} - \dfrac{1}{4}$ 6. $\dfrac{3}{4} - \dfrac{2}{3}$ 5. _____

                                                           6. _____

7. $\dfrac{7}{20} + \dfrac{2}{5}$ 8. $\dfrac{3}{20} + \dfrac{3}{5}$ 7. _____

                                                           8. _____

9. $\dfrac{4}{5} - \dfrac{3}{4}$ 10. $\dfrac{5}{6} - \dfrac{3}{5}$ 9. _____

                                                          10. _____

11. $\dfrac{9}{10} + \dfrac{3}{2} + \dfrac{3}{5}$ 12. $\dfrac{7}{10} + \dfrac{5}{2} + \dfrac{4}{5}$ 11. _____

                                                          12. _____

13. $\dfrac{1}{8} + \dfrac{11}{32} + \dfrac{7}{16}$ 14. $\dfrac{3}{8} + \dfrac{9}{32} + \dfrac{5}{16}$ 13. _____

                                                          14. _____

15. $\dfrac{1}{4} + \dfrac{5}{6} + \dfrac{3}{2} - \dfrac{4}{3}$ 16. $\dfrac{1}{4} + \dfrac{4}{6} + \dfrac{9}{2} - \dfrac{8}{3}$ 15. _____

                                                          16. _____

3-10 Finding the Least Common Denominator (LCD)

The major difficulty with adding and subtracting unlike fractions is finding the LCD. Sometimes it can not be found by mere inspection. If fractions have denominators of 12, 18, and 10, the LCD is not obvious. You know you must find a number that is evenly divisible by 12, 18, and 10, but how do you find that number?

Since each denominator divides evenly into the LCD, the LCD must contain the prime factors of each denominator. the method illustrated below will enable you to find the LCD when inspection fails you.

> **Finding the LCD**
>
> 1. Factor each denominator into primes.
> 2. Form the LCD by using each prime the <u>most</u> number of times it occurs as a factor in <u>any one</u> denominator.

Example 1: Find the LCD for $\frac{5}{6}$ and $\frac{3}{8}$.

1. Factor each denominator into primes. $6 = 2 \times 3$
   $8 = 2 \times 2 \times 2$

2. Determine the most number of times each prime occurs as a factor in any one denominator.

   2 occurs as a factor once in 6 and three times in 8, so use it three times in the LCD.
   3 occurs as a factor only once in 6, so use it once in the LCD.

Thus, the **LCD = 2x2x2x3 = 24.**

Example 2: Find the LCD for $\frac{1}{12}$, $\frac{5}{18}$, and $\frac{7}{10}$.

1. Factor each denominator into primes. $12 = 2 \times 2 \times 3$
   $18 = 2 \times 3 \times 3$
   $10 = 2 \times 5$

2. Determine the most number of times each prime occurs as a factor in any one denominator.

   2 occurs as a factor twice in 12, once in 18, and once in 10, so use it twice in the LCD.
   3 occurs as a factor once in 12 and twice in 18, so use it twice in the LCD.
   5 occurs as a factor only once in 10, so use it once in the LCD.

Thus, the **LCD = 2x2x3x3x5 = 180.**

Example 3: $\dfrac{11}{30} - \dfrac{7}{25} = ?$

1. Find the LCD.

   $30 = 2 \times 3 \times 5$
   $25 = 5 \times 5$

   $\text{LCD} = 2 \times 3 \times 5 \times 5$
   $= 150$

2. Convert each fraction to a fraction with a denominator of 150.

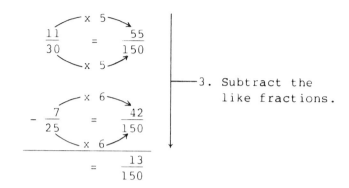

3. Subtract the like fractions.

Example 4: $\dfrac{5}{12} + \dfrac{11}{15} + \dfrac{1}{18} = ?$

1. Find the LCD.

   $12 = 2 \times 2 \times 3$
   $15 = 3 \times 5$
   $18 = 2 \times 3 \times 3$

   $\text{LCD} = 2 \times 2 \times 3 \times 3 \times 5$
   $= 180$

2. Convert each fraction to a fraction with a denominator of 180.

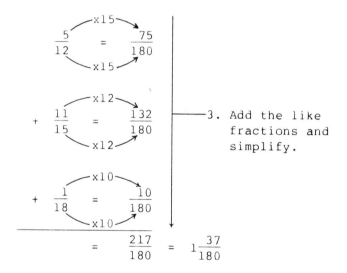

3. Add the like fractions and simplify.

PROBLEM 1: $\dfrac{5}{28} - \dfrac{1}{24} = ?$ Answer: $\dfrac{23}{168}$

$28 = 2 \times 2 \times 7$
$24 = 2 \times 2 \times 2 \times 3$

$\text{LCD} = 2 \times 2 \times 2 \times 3 \times 7$
$= 168$

$\dfrac{5}{28} = \dfrac{30}{168}$

$-\dfrac{1}{24} = \dfrac{7}{168}$

$= \dfrac{23}{168}$

EXERCISE 3-10   SET A

NAME _____   DATE _____

ADD OR SUBTRACT AND REDUCE: ANSWERS

1. $\dfrac{1}{6} + \dfrac{2}{9}$ 2. $\dfrac{1}{6} + \dfrac{3}{8}$ 1. _____

                                                                    2. _____

3. $\dfrac{11}{20} - \dfrac{7}{15}$ 4. $\dfrac{9}{10} - \dfrac{8}{15}$ 3. _____

                                                                    4. _____

5. $\dfrac{5}{18} + \dfrac{1}{12}$ 6. $\dfrac{7}{16} + \dfrac{1}{12}$ 5. _____

                                                                    6. _____

7. $\dfrac{13}{18} - \dfrac{11}{24}$ 8. $\dfrac{11}{16} - \dfrac{4}{15}$ 7. _____

                                                                    8. _____

9. $\dfrac{5}{42} + \dfrac{7}{36} + \dfrac{1}{21}$ 10. $\dfrac{9}{28} + \dfrac{11}{42} + \dfrac{1}{14}$ 9. _____

                                                                    10. _____

11. $\dfrac{7}{64} + \dfrac{11}{48} - \dfrac{5}{40}$ 12. $\dfrac{19}{72} + \dfrac{11}{63} - \dfrac{7}{56}$ 11. _____

                                                                    12. _____

EXERCISE 3-10 SET B

NAME _____   DATE _____

ADD OR SUBTRACT AND REDUCE: ANSWERS

1. $\dfrac{1}{6} + \dfrac{5}{8}$ 2. $\dfrac{1}{6} + \dfrac{4}{9}$ 1. _____

                                                                    2. _____

3. $\dfrac{7}{10} - \dfrac{4}{15}$ 4. $\dfrac{11}{20} - \dfrac{7}{15}$ 3. _____

                                                                    4. _____

5. $\dfrac{5}{12} + \dfrac{1}{16}$ 6. $\dfrac{7}{12} + \dfrac{1}{18}$ 5. _____

                                                                    6. _____

7. $\dfrac{13}{16} - \dfrac{7}{15}$ 8. $\dfrac{11}{18} - \dfrac{7}{24}$ 7. _____

                                                                    8. _____

9. $\dfrac{13}{28} + \dfrac{5}{42}$ 10. $\dfrac{13}{42} + \dfrac{7}{36}$ 9. _____

                                                                    10. _____

11. $\dfrac{5}{64} + \dfrac{11}{48} - \dfrac{7}{40}$ 12. $\dfrac{13}{72} + \dfrac{7}{63} - \dfrac{5}{56}$ 11. _____

                                                                    12. _____

3-11 Adding and Subtracting Mixed Numbers

Method 1: Change to Improper Fractions

An effective way to add and subtract mixed numbers is to first change them into improper fractions as you did in the previous section. Let me show you how this is done.

Example 1: $3\frac{2}{7} + 1\frac{8}{21} = ?$

$$3\frac{2}{7} = \frac{23}{7} = \frac{69}{21}$$

Change the mixed numbers to improper fractions.

$$+\ 1\frac{8}{21} = \frac{29}{21} = \frac{29}{21}$$

The LCD is 21. Change each fraction to a denominator of 21.

$$= \frac{98}{21} = 4\frac{14}{21} = 4\frac{2}{3}$$

Add the like fractions, simplify, and reduce.

Example 2: $40 - 36\frac{3}{5} = ?$

$$40 = \frac{40}{1} = \frac{200}{5}$$

Change the mixed numbers to improper fractions.

$$-\ 36\frac{3}{5} = \frac{183}{5} = \frac{183}{5}$$

The LCD is 5. Change each fraction to a denominator of 5.

$$= \frac{17}{5} = 3\frac{2}{5}$$

Subtract the like fractions and simplify.

PROBLEM 1: $75\frac{1}{2} - 56\frac{4}{5} = ?$

Answer: 18 7/10

$$75\frac{1}{2} = \frac{151}{2} = \frac{755}{10}$$

$$-\ 56\frac{4}{5} = \frac{284}{5} = \frac{568}{10}$$

$$= \frac{187}{10} = 18\frac{7}{10}$$

PROBLEM 2: $2\frac{1}{12} + 3\frac{2}{3} = ?$

Answer: 5 3/4

$$2\frac{1}{12} = \frac{25}{12} = \frac{25}{12}$$

$$+\ 3\frac{2}{3} = \frac{11}{3} = \frac{44}{12}$$

$$= \frac{69}{12} = 5\frac{9}{12} = 5\frac{3}{4}$$

Method 2: Leave as Mixed Numbers

The process described in Method 1 does not require you to learn anything new. However, it may cause you to work with some rather large numbers. You may find it easier to operate separately with the fractional and whole number parts of the mixed numbers.

Example 3: $182\frac{2}{3} + 48\frac{3}{4} = ?$

$$182\frac{2}{3} = 182\frac{8}{12}$$
$$+ 48\frac{3}{4} = 48\frac{9}{12}$$
$$= 230\frac{17}{12}$$

Change the fractional parts to an LCD of 12; leave the whole numbers alone.

Add the fractional parts.
Add the whole number parts.

$$= 230 + 1\frac{5}{12} = 231\frac{5}{12}$$

Convert the improper fraction to a mixed number and add the whole numbers together.

Example 4: $37\frac{4}{7} - 12\frac{6}{7} = ?$

Since the fractional part being subtracted (6/7) is larger than the 4/7, you must borrow 1 (in the form 7/7) from 37 and add it to the 4/7 (making 11/7).

$$37\frac{4}{7} = 36\frac{7}{7} + \frac{4}{7} = 36\frac{11}{7}$$
$$-12\frac{6}{7} = 12\frac{6}{7} = -12\frac{6}{7}$$
$$= 24\frac{5}{7}$$

Borrow $1 = \frac{7}{7}$ and add it to $\frac{4}{7}$.

Subtract the fractional parts.

Subtract the whole number parts.

PROBLEM 3: $26\frac{4}{9} - 17\frac{3}{4} = ?$ Answer: 8 25/36

$$26\frac{4}{9} = 26\frac{16}{36} = 25\frac{36}{36} + \frac{16}{36} = 25\frac{52}{36}$$
$$- 17\frac{3}{4} = 17\frac{27}{36} = 17\frac{27}{36} \qquad = 17\frac{27}{36}$$
$$= 8\frac{25}{36}$$

EXERCISE 3-11   SET A

NAME _____   DATE _____

ADD OR SUBTRACT AND REDUCE: ANSWERS

1. $4\frac{5}{6} + 2\frac{1}{6}$ 　　　　2. $8\frac{5}{32} + 3\frac{7}{32}$ 1. _____

　　　　　　　　　　　　　　　　　　　　　　　　　　　　　　　　2. _____

3. $9\frac{13}{16} + 6\frac{5}{8}$ 　　　4. $7\frac{11}{24} + 15\frac{7}{12}$ 3. _____

　　　　　　　　　　　　　　　　　　　　　　　　　　　　　　　　4. _____

5. $76\frac{3}{8} + 52\frac{7}{12}$ 　　6. $84\frac{7}{15} + 57\frac{5}{18}$ 5. _____

　　　　　　　　　　　　　　　　　　　　　　　　　　　　　　　　6. _____

7. $8\frac{5}{8} - 5\frac{7}{8}$ 　　　　8. $9\frac{1}{4} - 4\frac{3}{4}$ 7. _____

　　　　　　　　　　　　　　　　　　　　　　　　　　　　　　　　8. _____

9. $94\frac{5}{14} - 21\frac{11}{21}$ 　10. $86\frac{11}{15} - 32\frac{3}{10}$ 9. _____

　　　　　　　　　　　　　　　　　　　　　　　　　　　　　　　　10. _____

11. $9 - 4\frac{5}{9}$ 　　　　　　12. $13 - 7\frac{4}{7}$ 11. _____

　　　　　　　　　　　　　　　　　　　　　　　　　　　　　　　　12. _____

EXERCISE 3-11 SET B

NAME _____   DATE _____

ADD OR SUBTRACT AND REDUCE: ANSWERS

1. $5\frac{3}{5} + 2\frac{2}{5}$ 2. $9\frac{5}{8} + 4\frac{7}{8}$ 1. _____

                                                                                     2. _____

3. $7\frac{11}{16} + 8\frac{3}{4}$ 4. $4\frac{5}{18} + 9\frac{7}{9}$ 3. _____

                                                                                     4. _____

5. $62\frac{5}{8} + 43\frac{5}{12}$ 6. $75\frac{3}{14} + 61\frac{11}{21}$ 5. _____

                                                                                     6. _____

7. $9\frac{3}{8} - 6\frac{5}{8}$ 8. $8\frac{2}{7} - 5\frac{5}{7}$ 7. _____

                                                                                     8. _____

9. $95\frac{3}{10} - 32\frac{7}{15}$ 10. $132\frac{1}{18} - 17\frac{11}{27}$ 9. _____

                                                                                     10. _____

11. $8 - 5\frac{3}{9}$ 12. $10 - 6\frac{7}{11}$ 11. _____

                                                                                     12. _____

3-12 Complex Fractions

This section is concerned with complex fractions. A **complex fraction** is a fraction that contains more than one fraction line.

The objective for this section is to learn to simplify complex fractions and express them as a single fraction. Let me show you how this is done by simplifying the three examples below.

Example 1: $\dfrac{\frac{2}{3}}{\frac{3}{4}} = ?$ The fraction line indicates that you are to divide the fraction above the line by the fraction below the line.

Thus: $\dfrac{\frac{2}{3}}{\frac{3}{4}} = \frac{2}{3} \div \frac{3}{4}$

$= \frac{2}{3} \times \frac{4}{3} = \frac{8}{9}$ Invert the divisor and multiply.

Example 2: $\dfrac{\frac{5}{6} + \frac{2}{6}}{4 - \frac{2}{3}} = ?$ The procedure for simplifying a complex fraction is to get an answer for the fractions above the fraction line and divide it by the answer from the fractions below the line.

$\frac{5}{6} + \frac{2}{6} = \frac{7}{6}$ Add the fractions above the line.

$4 - \frac{2}{3} = \frac{4}{1} - \frac{2}{3} = \frac{12}{3} - \frac{2}{3} = \frac{10}{3}$ Subtract the fractions below the line.

$\dfrac{\frac{5}{6} + \frac{2}{6}}{4 - \frac{2}{3}} = \dfrac{\frac{7}{6}}{\frac{10}{3}} = \frac{7}{6} \div \frac{10}{3}$ Now divide those two results.

$= \frac{7}{\cancel{6}_2} \times \frac{\cancel{3}^1}{10} = \frac{7}{20}$ Invert the divisor and multiply.

149

Example 3: $\dfrac{2\frac{2}{3} + 1\frac{3}{8}}{\frac{7}{12} + \frac{11}{30}} = ?$

$$2\frac{2}{3} = \frac{8}{3} = \frac{64}{24}$$
$$+\, 1\frac{3}{8} = \frac{11}{8} = \frac{33}{24}$$
$$= \frac{97}{24}$$

— Add the fractions above the fraction line. (Change mixed numbers to improper fractions and use an LCD of 24.)

$$\frac{7}{12} = \frac{35}{60}$$
$$+\, \frac{11}{30} = \frac{22}{60}$$
$$= \frac{57}{60} = \frac{19}{20}$$

— Add the fractions below the fraction line. (Change to an LCD of 60.)

Reduce the answer (by dividing by 3).

$$\dfrac{2\frac{2}{3} + 1\frac{3}{8}}{\frac{7}{12} + \frac{11}{30}} = \dfrac{\frac{97}{24}}{\frac{19}{20}} = \frac{97}{24} \div \frac{19}{20}$$

Now divide those results.

$$= \frac{97}{\cancel{24}_{6}} \times \frac{\cancel{20}^{5}}{19}$$

Invert the divisor and multiply.

$$= \frac{485}{114} = 4\frac{29}{114}$$

PROBLEM 1: $\dfrac{\frac{4}{5}}{3\frac{1}{2}} = ?$

Answer: 8/35

$$\dfrac{\frac{4}{5}}{\frac{7}{2}} = \frac{4}{5} \div \frac{7}{2} = \frac{4}{5} \times \frac{2}{7} = \frac{8}{35}$$

PROBLEM 2: $\dfrac{\frac{2}{3} - \frac{5}{12}}{5 + \frac{1}{2}} = ?$

Answer: 1/22

$$\dfrac{\frac{8}{12} - \frac{5}{12}}{\frac{5}{1} + \frac{1}{2}} = \dfrac{\frac{3}{12}}{\frac{10}{2} + \frac{1}{2}} = \dfrac{\frac{1}{4}}{\frac{11}{2}}$$

$$= \frac{1}{4} \div \frac{11}{2} = \frac{1}{\cancel{4}_{2}} \times \frac{\cancel{2}^{1}}{11} = \frac{1}{22}$$

EXERCISE 3-12   SET A

NAME _____   DATE _____

SIMPLIFY EACH COMPLETELY: ANSWERS

1. $\dfrac{\frac{3}{4}}{\frac{2}{3}}$ 2. $\dfrac{\frac{5}{7}}{\frac{3}{4}}$ 1. _____

    2. _____

3. $\dfrac{\frac{7}{8}}{\frac{3}{4}}$ 4. $\dfrac{\frac{5}{6}}{\frac{2}{3}}$ 3. _____

    4. _____

5. $\dfrac{3\frac{7}{10}}{4\frac{1}{15}}$ 6. $\dfrac{2\frac{5}{12}}{3\frac{7}{30}}$ 5. _____

    6. _____

7. $\dfrac{5 + \frac{1}{2}}{4 - \frac{1}{2}}$ 8. $\dfrac{6 + \frac{2}{3}}{5 - \frac{2}{3}}$ 7. _____

    8. _____

9. $\dfrac{\frac{3}{8} + \frac{5}{8}}{\frac{2}{3} + \frac{3}{4}}$ 10. $\dfrac{\frac{7}{10} + \frac{3}{10}}{\frac{4}{5} + \frac{2}{3}}$ 9. _____

    10. _____

11. $\dfrac{1\frac{5}{16} + 2\frac{1}{2}}{24}$ 12. $\dfrac{5\frac{7}{12} - 2\frac{1}{6}}{24}$ 11. _____

    12. _____

13. $\dfrac{\frac{1}{4} + \frac{5}{9}}{\frac{7}{18} - \frac{2}{27}}$ 14. $\dfrac{\frac{3}{7} + \frac{1}{3}}{\frac{8}{15} - \frac{4}{21}}$ 13. _____

    14. _____

EXERCISE 3-12 SET B

NAME _____   DATE _____

SIMPLIFY EACH COMPLETELY: ANSWERS

1. $\dfrac{\frac{3}{5}}{\frac{2}{3}}$ 2. $\dfrac{\frac{3}{4}}{\frac{5}{7}}$

1. _____

2. _____

3. $\dfrac{\frac{5}{6}}{\frac{1}{3}}$ 4. $\dfrac{\frac{5}{8}}{\frac{3}{4}}$

3. _____

4. _____

5. $\dfrac{2\frac{5}{12}}{4\frac{1}{16}}$ 6. $\dfrac{3\frac{7}{20}}{5\frac{2}{15}}$

5. _____

6. _____

7. $\dfrac{6 + \frac{1}{3}}{5 - \frac{1}{3}}$ 8. $\dfrac{5 + \frac{3}{4}}{2 - \frac{3}{4}}$

7. _____

8. _____

9. $\dfrac{\frac{5}{6} + \frac{1}{6}}{\frac{2}{3} + \frac{3}{4}}$ 10. $\dfrac{\frac{3}{5} + \frac{2}{5}}{\frac{4}{5} + \frac{2}{3}}$

9. _____

10. _____

11. $\dfrac{1\frac{3}{10} + 2\frac{1}{2}}{18}$ 12. $\dfrac{4\frac{1}{8} + 2\frac{3}{4}}{18}$

11. _____

12. _____

13. $\dfrac{\frac{3}{5} + \frac{5}{6}}{\frac{8}{15} - \frac{5}{12}}$ 14. $\dfrac{\frac{4}{9} + \frac{1}{4}}{\frac{9}{16} - \frac{5}{12}}$

13. _____

14. _____

3-13 Comparing Fractions

In a certain survey 3/8 of the people questioned preferred the color red, while 4/11 like the color blue. According to the survey, which color had the greater preference?

To solve this problem we must determine if 3/8 > 4/11 or if 4/11 > 3/8. Those two fractions are very close in value so you can not simply get the answer by inspection. What you must do is change each fraction to fractions with the same denominator, by either reducing them to lower terms or raising them to higher terms. Once the fractions have the same denominator, you can compare them.

In the problem above, 3/8 and 4/11 can't be reduced, so we must raise them to higher terms. In this case the LCD is 88.

$$\frac{3}{8} \xrightarrow{\times 11} \frac{33}{88}$$

$$\frac{4}{11} \xrightarrow{\times 8} \frac{32}{88}$$

Changing each fraction to the LCD we notice that 33/88 > 32/88.

$$\text{Thus } \frac{3}{8} > \frac{4}{11}.$$

> **To compare fractions** convert the fractions into fractions with the same denominator before deciding if they are equal or not equal.

Example 1: Compare the fractions 4/5 and 5/6.

$$\frac{4}{5} \xrightarrow{\times 6} \frac{24}{30}$$

Since the fractions are not reducible, we must raise them to higher terms. The LCD is 30 in this case.

$$\frac{5}{6} \xrightarrow{\times 5} \frac{25}{30}$$

Change each fraction to a denominator of 30.

25/30 > 24/30

Determine which fraction is larger by looking at the like fractions.

$$\text{Thus } \frac{5}{6} > \frac{4}{5}.$$

Example 2: Compare the fractions 14/16 and 21/24.

Both fractions are reducible.

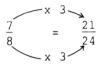   But 7/8 = 7/8.

Thus 14/16 = 21/24.

Example 3: Arrange the following in order from largest to smallest:

$$\frac{7}{8}, \frac{3}{4}, \frac{11}{12}, \frac{5}{6}$$

The fractions are not reducible, so we raise them to higher terms. The LCD is 24 in this case.

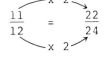

Change each fraction to fractions with denominators of 24.

Determine the order by comparing the like fractions.

22/24 > 21/24 > 20/24 > 18/24.

Thus 11/12 > 7/8 > 5/6 > 3/4.

PROBLEM 1: Compare these fractions:

$\frac{10}{15}$ and $\frac{14}{21}$

Answer: They are equal.

$\frac{10}{15}$ reduces to $\frac{2}{3}$
(÷ by 5)

$\frac{14}{21}$ reduces to $\frac{2}{3}$
(÷ by 7)

PROBLEM 2: Arrange these fractions in order from largest to smallest:

$\frac{5}{9}, \frac{2}{3}, \frac{11}{18}$

Answer: 2/3, 11/18, 5/9

The LCD is 18.

$\frac{5}{9} = \frac{10}{18}, \frac{2}{3} = \frac{12}{18}, \frac{11}{18} = \frac{11}{18}$

EXERCISE 3-13 SET A

NAME _____   DATE _____

REPLACE THE ? WITH =, >, OR <:   ANSWERS

1. $\frac{1}{2}$? $\frac{3}{6}$ 2. $\frac{2}{3}$? $\frac{4}{6}$ 1. _____

2. _____

3. $\frac{5}{6}$? $\frac{3}{6}$ 4. $\frac{4}{7}$? $\frac{3}{7}$ 3. _____

4. _____

5. $\frac{5}{6}$? $\frac{6}{7}$ 6. $\frac{7}{8}$? $\frac{8}{9}$ 5. _____

6. _____

7. $\frac{5}{4}$? $\frac{4}{3}$ 8. $\frac{6}{5}$? $\frac{7}{6}$ 7. _____

8. _____

ARRANGE IN ORDER FROM LARGEST TO SMALLEST:

9. $\frac{1}{2}$, $\frac{3}{8}$, $\frac{9}{16}$ 9. _____

10. $1\frac{3}{5}$, $1\frac{1}{2}$, $1\frac{2}{3}$ 10. _____

11. $\frac{5}{4}$, $\frac{3}{2}$, $\frac{4}{5}$, $\frac{2}{3}$ 11. _____

12. 2, $\frac{9}{4}$, $\frac{15}{7}$, $\frac{11}{6}$ 12. _____

13. $\frac{11}{12}$, $\frac{13}{15}$, $\frac{11}{18}$ 13. _____

EXERCISE 3-13   SET B

NAME _____   DATE _____

REPLACE THE ? WITH =, >, OR <:   ANSWERS

1. $\frac{1}{2}$? $\frac{4}{8}$ 2. $\frac{2}{5}$? $\frac{4}{10}$ 1. _____

                                                        2. _____

3. $\frac{5}{8}$? $\frac{3}{8}$ 4. $\frac{5}{6}$? $\frac{4}{6}$ 3. _____

                                                        4. _____

5. $\frac{6}{7}$? $\frac{7}{8}$ 6. $\frac{4}{5}$? $\frac{6}{7}$ 5. _____

                                                        6. _____

7. $\frac{4}{3}$? $\frac{5}{4}$ 8. $\frac{6}{5}$? $\frac{5}{4}$ 7. _____

                                                        8. _____

ARRANGE IN ORDER FROM LARGEST TO SMALLEST:

9. $\frac{5}{8}$, $\frac{7}{16}$, $\frac{1}{2}$ 9. _____

10. $1\frac{2}{5}$, $1\frac{2}{3}$, $1\frac{3}{4}$ 10. _____

11. $\frac{4}{3}$, $\frac{4}{5}$, $\frac{3}{2}$, $\frac{3}{4}$ 11. _____

12. 2, $\frac{7}{3}$, $\frac{15}{7}$, $\frac{11}{6}$ 12. _____

13. $\frac{5}{12}$, $\frac{7}{15}$, $\frac{7}{18}$ 13. _____

3-14 Applications Involving Fractions

In this section you will encounter word problems that involve operations with fractions and mixed numbers. Again, you should read each problem carefully, analyze the problem to decide what operations should be used, and solve for the answer.

Example 1: At a convention of 252 delegates, 168 of them favored the new charter. What fraction of the delegates were in favor of the new charter?

Analyze: You need to find what part 168 is of the total of 252 delegates.

Solve: $\dfrac{168}{252} = \dfrac{2 \times \cancel{2} \times \cancel{2} \times \cancel{3} \times \cancel{7}}{3 \times \cancel{2} \times \cancel{2} \times \cancel{3} \times \cancel{7}} = \dfrac{2}{3}$

"ALL THESE FIFTHS ARE GETTING ME DIZZY."

Example 2: How many encyclopedias that are 1 1/4 inches thick can you place on a shelf that is 40 inches long?

Analyze: What do you do with the 1 1/4 and the 40? If you are stumped, try using simpler numbers. Use a whole number such as 2 for the 1 1/4. With a 40 inch shelf and 2 inch books, you would get an answer of 20 by dividing the 2 into the 40. So you would divide the 1 1/4 into the 40 to solve the original problem. (Note: 1 1/4 divided into 40 is the same as 40 ÷ 1 1/4.)

Solve: $40 \div 1\,1/4 = 40 \div \dfrac{5}{4} = \cancel{40}^{8} \times \dfrac{4}{\cancel{5}_{1}} = 32$

Example 3: You buy 1/2 lb of fudge, 3/4 lb of taffy, and 1/8 lb of caramels. What is the total amount purchased?

Analyze: To find the total here, you must add the amounts.

Solve: $\dfrac{1}{2} + \dfrac{3}{4} + \dfrac{1}{8} = \dfrac{4}{8} + \dfrac{6}{8} + \dfrac{1}{8} = \dfrac{11}{8} = 1\dfrac{3}{8}$ lb

Example 4: How much larger is a 5/8 inch socket than a 17/32 inch socket?

Analyze: You need to find the difference between the two measurements, so you must subtract the two.

Solve: $\frac{5}{8} - \frac{17}{32} = \frac{20}{32} - \frac{17}{32} = \frac{3}{32}$

Example 5: John ate 1/4 of a pie. Mike ate 1/2 of what was left. How much of the pie remained uneaten?

Analyze: John ate 1/4, so 3/4 was left. Mike ate 1/2 of what was left. "Of" means "times," so you have to multiply 1/2 times 3/4 to get the amount Mike ate. Then, you must subtract that answer from the 3/4 to get the amount that was uneaten. (Hint: It may help to draw a picture.)

Solve: $1 - \frac{1}{4} = \frac{3}{4}$ (amount left after John ate)

$\frac{1}{2} \times \frac{3}{4} = \frac{3}{8}$ (amount Mike ate)

$\frac{3}{4} - \frac{3}{8} = \frac{6}{8} - \frac{3}{8} = \frac{3}{8}$ (amount that was uneaten)

PROBLEM 1: Robert Matern ate 83 hamburgers in 2 1/2 hours during a contest in May of 1973. How many hamburgers did he eat per hour?

Answer: $33\frac{1}{5}$

(Note: "Per" means to divide.)

burgers per hour
↓ ↓ ↓
83 ÷ 2 1/2

$= 83 \div \frac{5}{2} = 83 \times \frac{2}{5}$

$= \frac{166}{5} = 33\frac{1}{5}$

PROBLEM 2: You purchased 80 shares of stock at 40 7/8 dollars per share. A year later you sold all the shares for 51 3/8 dollars per share. How much profit did you make on the sale?

Answer: $840
paid:
$80 \times 40\frac{7}{8} = 80 \times \frac{327}{8}$
= $3270

received:
$80 \times 51\frac{3}{8} = 80 \times \frac{411}{8}$
= $4110

profit:
$4110 − $3270 = $840

EXERCISE 3-14   SET A

NAME _____   DATE _____

                                                                  ANSWERS

1. Which makes the smallest hole: a 5/16 inch, 1/4 inch,
   or 9/32 inch bit?   1. _____

2. Which is the largest: a 1/2 inch, 3/8 inch, or 7/16
   inch staple?   2. _____

3. All 26,000 tickets were sold for the big game. What
   fraction of the fans were "no-shows," if only 20,800
   actually attended?   3. _____

4. If 18 people are coming to dinner and you plan to allow
   3/4 lb of meat per person, how many pounds of meat do
   you need to buy?   4. _____

5. How many 3 1/2 ft pieces of rope can be cut from a
   49 ft coil?   5. _____

6. A pole is 32 feet long. If 3/8 of the pole is underground,
   how many feet are above ground?   6. _____

7. Jim weighed 134 3/4 pounds at age 14 and 236 1/2 pounds
   at age 18. How much weight did he gain in the 4 years?   7. _____

8. At a recent fill-up, your car took 18 7/10 gallons of gas.
   If your tank holds 20 1/2 gallons, how much gas was in the
   tank?   8. _____

9. The three sections of a book shelf are 18 3/8 inches, 28 3/4
   inches, and 30 1/2 inches. How long is the entire shelf? 9. _____

10. What is the cost of 2 3/4 pounds of onions at 28¢ per lb? 10. _____

11. On a trip of 1830 miles you filled up your gas tank four
    times with 14 1/2, 15 1/5, 14 1/10, and 16 1/5 gallons of
    gas. How many miles did your car travel per gallon of gas? 11. _____

12. A package of hot dogs weighs 16 ounces and contains 12
    hot dogs. What is the weight of each hot dog?   12. _____

13. Each lap of an indoor running track is 1/11 of a mile.
    How many laps do you need to run to cover a 1/2 mile?   13. _____

14. A share of INCEL stock sells for 49 3/8 dollars. How
    much would 160 shares of that stock cost?   14. _____

15. Fran weighed 145 lb and lost 3/4 lb a day for six
    consecutive days. What was her weight after the six days? 15. _____

16. A crate containing 24 boxes of candy, each weighing 1 3/4
    lb, has a total weight of 44 5/8 lb. What is the weight of
    the empty crate that holds the boxes of candy?   16. _____

EXERCISE 3-14   SET B

NAME _____   DATE _____

                                                         ANSWERS

1. If you sleep 1/3 of the day and work for 5/12 of the day,
    what fraction of the day is left for other activities?   1. _____

2. If 2800 of the 3500 voters favored Proposition D, what
    fraction of the voters favored the proposition?   2. _____

3. If you ride your bike at 12 1/2 miles per hour, how far
    would you ride in 2 3/4 hours?   3. _____

4. Lee Hong ate 89 1/2 sausages in 6 minutes on May 3, 1972.
    How many sausages did she eat per minute?   4. _____

5. How many 1/4 pound hamburger patties can be made from
    38 1/2 pounds of hamburger?   5. _____

6. 3/4 of a pie is divided evenly between six people. How
    much of the pie does each person receive?   6. _____

7. After you trim 5/8 lb of fat off a 4 3/4 lb pork roast,
    how much meat is left?   7. _____

8. During a tennis match, Wendy's weight went from 112 1/2
    to 109 1/4 pounds. How much weight did she lose?   8. _____

9. Bob was 62 3/4 inches tall on his 13th birthday. He grew
    3 1/2 inches the next year and 2 7/8 inches the year after
    that. How tall was he on his 15th birthday?   9. _____

10. What is the cost of 7 2/3 lb of apples at 51¢ a pound?   10. _____

11. A recipe for a wedding punch calls for 12 quarts of cham-
    pagne. If the champagne comes in bottles that are 4/5 of
    a quart, how many bottles of champagne would be needed?   11. _____

12. How many pieces of material that are 6 1/4 yards long can
    be cut from a bolt of material that is 43 3/4 yards long? 12. _____

13. You buy 32 shares of ACME stock at 29 3/8 dollars per
    share, and 760 shares of COMTEC stock at 33 1/8 dollars
    per share. What is the total of the purchases?   13. _____

14. During a 8 hour work day, you take three coffee breaks,
    1/4 hour each, and a lunch break, 3/4 hour long. How
    many hours are you actually working in a 5 day week?   14. _____

15. A full section of pipe is 18 1/4 feet long. If 22 full
    sections and one-half of another section are laid end
    to end, what is the total length of the pipe?   15. _____

Chapter 3 Summary

Concepts

You may refer to the sections listed below to review how to:

1. define proper and improper fractions. (3-1)

2. raise fractions to higher terms by multiplying the numerator and the denominator, by the same number. (3-2)

3. reduce fractions by finding divisors of both the numerator and the denominator or by using factoring. (3-2)

4. convert between improper fractions and mixed numbers. (3-3)

5. multiply fractions by reducing, cross canceling, and multiplying the resulting numerators and denominators. (3-4)

6. divide fractions by inverting the divisor and changing to multiplication. (3-5)

7. multiply and divide mixed numbers by first changing them into improper fractions. (3-6)

8. find fractional parts of a number by using the word "of" as an indication of multiplication. (3-7)

9. add or subtract fractions with the same denominators. (3-8)

10. find the least common denominator by factoring each denominator. (3-10)

11. add or subtract unlike fractions by converting each to a fraction with the LCD as its denominator. (3-9), (3-10)

12. add or subtract mixed numbers by either changing them to improper fractions or by operating separately with the fractional and whole parts. (3-11)

13. simplify complex fractions by dividing the answer of the fractions above the line by the answer of those below the line. (3-12)

14. compare the size of fractions by changing them into fractions with the same denominators. (3-13)

15. solve word problems involving operations with fractions. (3-14)

Terminology

This chapter's important terms and their corresponding page numbers are:

canceling: the process of dividing a numerator and denominator by the same number when multiplying fractions. (109)

complex fraction: a fraction with more than one fraction line. (149)

denominator: the part of the fraction written below the fraction line. (103)

fraction: a portion of a whole, indicated by the quotient of whole numbers. (103)

fraction line: the line in the middle of a fraction. (103)

improper fraction: a fraction with a value that is not less than one. (104)

least common denominator: the smallest whole number that is evenly divisible by the denominators of other fractions. (137)

like fractions: fractions that have the same denominators. (133)

mixed number: the sum of a whole number and a proper fraction. (113)

numerator: the part of a fraction written above the fraction line. (103)

proper fraction: a fraction with a value that is less than one. (104)

raise a fraction to higher terms: the process of multiplying the numerator and the denominator of a fraction by the same number. (107)

reciprocal: the fraction formed by interchanging the numerator and denominator of a given fraction. (121)

reduce a fraction to lower terms: the process of dividing the numerator and denominator of a fraction by a number that divides into both evenly. (108)

reduce to lowest terms: the process of reducing a fraction until no whole number except 1 divides evenly into both its numerator and denominator. (108)

Student Notes

Chapter 3 Practice Test A

ANSWERS

1. Find the number represented by the ?: $\frac{6}{7} = \frac{?}{63}$

    1. _____

2. Express $17\frac{5}{6}$ as an improper fraction.

    2. _____

3. $\frac{5}{6} \times \frac{2}{3}$

4. $\frac{5}{9} \div \frac{4}{9}$

    3. _____

    4. _____

5. $2\frac{3}{4} \times 1\frac{5}{6}$

6. $\frac{1}{8} + \frac{5}{8}$

    5. _____

    6. _____

7. $\frac{2}{3} + \frac{5}{12} - \frac{3}{4}$

8. $\frac{7}{15} + \frac{11}{18} + \frac{3}{10}$

    7. _____

    8. _____

9. $\frac{4}{15} - \frac{5}{27}$

10. $3\frac{2}{3} - 2\frac{5}{6}$

    9. _____

    10. _____

11. $4\frac{5}{6} + 1\frac{4}{5}$

12. $\dfrac{4 + \frac{1}{3}}{4 - \frac{3}{4}}$

    11. _____

    12. _____

13. Arrange in order from largest to smallest: $\frac{4}{3}, \frac{8}{7}, \frac{7}{6}$

    13. _____

14. A team won three-fourths of its games during the season. If the team played 28 games, how many games did it win?

    14. _____

15. Find the cost per ounce of 3 1/3 ounces of cheese that sells for 90¢.

    15. _____

163

Chapter 3 Practice Test B

ANSWERS

1. Find the number represented by the ?: $\frac{78}{123} = \frac{26}{?}$

1. _____

2. Express $\frac{55}{15}$ as a reduced mixed number.

2. _____

3. $\frac{5}{8} \times \frac{3}{7}$ 4. $\frac{3}{7} \div \frac{4}{7}$

3. _____

4. _____

5. $3\frac{5}{6} \times 2\frac{5}{9}$ 6. $\frac{3}{10} + \frac{1}{10}$

5. _____

6. _____

7. $\frac{3}{10} + \frac{1}{2} + \frac{2}{5}$ 8. $\frac{7}{18} + \frac{5}{14} + \frac{8}{21}$

7. _____

8. _____

9. $\frac{7}{12} - \frac{3}{20}$ 10. $2\frac{1}{2} - 1\frac{5}{8}$

9. _____

10. _____

11. $3\frac{1}{7} + 4\frac{2}{3}$ 12. $\dfrac{5 + \frac{1}{4}}{5 - \frac{2}{3}}$

11. _____

12. _____

13. Arrange in order from largest to smallest: $\frac{7}{9}, \frac{9}{11}, \frac{2}{3}$

13. _____

14. Seven-eighths of those registered at a resort were senior citizens. If a total of 616 people were registered, how many were senior citizens?

14. _____

15. Find the cost per ounce of 5 2/3 ounces of syrup that sells for 68¢.

15. _____

Chapter 3 Supplementary Exercises

Section 3-1
Classify the following as proper or improper fractions:

1. $\dfrac{7}{8}$
2. $\dfrac{8}{7}$
3. $\dfrac{8}{1}$
4. $\dfrac{12}{5}$
5. $\dfrac{9}{3}$
6. $\dfrac{1}{3}$

Section 3-2
Reduce the following completely:

1. $\dfrac{3}{6}$
2. $\dfrac{3}{15}$
3. $\dfrac{10}{16}$
4. $\dfrac{12}{18}$
5. $\dfrac{45}{75}$
6. $\dfrac{24}{30}$
7. $\dfrac{70}{84}$
8. $\dfrac{42}{96}$
9. $\dfrac{120}{168}$
10. $\dfrac{294}{336}$
11. $\dfrac{52}{65}$
12. $\dfrac{51}{85}$

Section 3-3
Change to mixed numbers:

1. $\dfrac{11}{5}$
2. $\dfrac{19}{2}$
3. $\dfrac{13}{6}$

Change to improper fractions:

4. 6
5. $1\dfrac{3}{4}$
6. $3\dfrac{2}{3}$

Section 3-4

1. $\dfrac{2}{3} \times \dfrac{5}{7}$
2. $\dfrac{1}{4} \times 7$
3. $\dfrac{3}{8} \times \dfrac{2}{5}$
4. $9 \times \dfrac{2}{3}$
5. $\dfrac{5}{30} \times \dfrac{7}{8}$
6. $\dfrac{2}{3} \times \dfrac{1}{7} \times \dfrac{5}{9}$
7. $\dfrac{14}{10} \times \dfrac{5}{21} \times \dfrac{15}{65}$
8. $\dfrac{10}{7} \times \dfrac{21}{30} \times \dfrac{6}{11}$
9. $\dfrac{60}{80} \times 16 \times \dfrac{144}{192}$

Section 3-5

1. $\dfrac{3}{5} \div \dfrac{1}{2}$
2. $\dfrac{10}{11} \div \dfrac{2}{3}$
3. $\dfrac{6}{8} \div \dfrac{2}{3}$
4. $\dfrac{5}{8} \div 3$
5. $15 \div \dfrac{5}{6}$
6. $\dfrac{6}{7} \div \dfrac{24}{21}$
7. $\dfrac{14}{21} \div \dfrac{16}{27}$
8. $\dfrac{28}{42} \div \dfrac{35}{30}$
9. $\dfrac{18}{20} \div \dfrac{54}{45}$

Section 3-6

1. $2\dfrac{1}{2} \times 5\dfrac{2}{3}$
2. $10 \times 4\dfrac{1}{5}$
3. $6\dfrac{3}{4} \times 2\dfrac{5}{32}$
4. $12\dfrac{5}{8} \div 13\dfrac{1}{4}$
5. $18\dfrac{7}{16} \div 6$
6. $30 \div 5\dfrac{5}{6}$

Section 3-7

1. $\frac{3}{5}$ of 170
2. $\frac{1}{6}$ of 2274
3. $\frac{3}{8}$ of 37
4. $\frac{5}{6}$ of $49\frac{1}{2}$
5. $\frac{1}{3}$ of $8\frac{1}{4}$
6. $\frac{3}{4}$ of $17\frac{2}{3}$

7. In a recent experiment, 5/6 of those tasting coffee preferred Brand X. If 1632 people participated in the experiment, how many preferred Brand X?

8. In a 1/3 off sale, what would be the sale price for a refrigerator that regularly sells for $915?

9. A drum holds 47 1/2 gallons of oil. If the drum is 3/4 full, how much oil does the drum contain?

10. 5/8 of the animals in a pet shop are fish and 1/2 of the fish are guppies. If the pet shop has a total of 1600 animals, how many are guppies?

Section 3-8

1. $\frac{5}{16} + \frac{2}{16}$
2. $\frac{1}{8} + \frac{3}{8}$
3. $\frac{7}{20} + \frac{7}{20}$
4. $\frac{5}{8} - \frac{2}{8}$
5. $\frac{35}{17} - \frac{8}{17}$
6. $\frac{11}{48} + \frac{31}{48} - \frac{9}{48}$

Section 3-9

1. $\frac{1}{5} + \frac{3}{10}$
2. $\frac{1}{2} + \frac{2}{3}$
3. $\frac{3}{4} + \frac{7}{8}$
4. $\frac{3}{4} - \frac{2}{3}$
5. $\frac{7}{5} - \frac{2}{3}$
6. $\frac{5}{8} - \frac{3}{16}$
7. $\frac{3}{4} - \frac{2}{5} - \frac{1}{10}$
8. $\frac{1}{2} + \frac{1}{3} + \frac{1}{4} + \frac{1}{5}$
9. $\frac{7}{24} + \frac{5}{6} - \frac{1}{12}$

Section 3-10

1. $\frac{7}{15} + \frac{5}{12}$
2. $\frac{5}{8} + \frac{3}{28}$
3. $\frac{7}{10} + \frac{11}{25}$
4. $\frac{23}{24} - \frac{13}{60}$
5. $\frac{37}{75} - \frac{11}{60}$
6. $\frac{1}{12} + \frac{7}{8} - \frac{11}{30}$

Section 3-11

1. $4\frac{3}{4} + 6\frac{1}{2}$
2. $38\frac{17}{32} + 29\frac{9}{16}$
3. $12 + 8\frac{7}{18}$
4. $5\frac{2}{3} - 2\frac{1}{4}$
5. $13 - 5\frac{7}{8}$
6. $119\frac{4}{15} - 95\frac{3}{10}$

Section 3-12

1. $\dfrac{\frac{2}{7}}{\frac{4}{7}}$

2. $\dfrac{4\frac{3}{5}}{6\frac{7}{15}}$

3. $\dfrac{\frac{3}{7} + \frac{2}{7}}{\frac{5}{6} - \frac{4}{15}}$

4. $\dfrac{3 + \frac{1}{3}}{2 - \frac{1}{3}}$

5. $\dfrac{\frac{3}{8} + 7\frac{1}{10}}{12}$

6. $\dfrac{\frac{2}{3} + \frac{3}{4}}{\frac{5}{6} - \frac{7}{18}}$

Section 3-13

Arrange in order from largest to smallest:

1. $\dfrac{11}{21}, \dfrac{4}{9}, \dfrac{5}{12}$

2. $\dfrac{7}{10}, \dfrac{11}{15}, \dfrac{13}{18}$

3. $\dfrac{15}{8}, \dfrac{11}{6}, \dfrac{13}{7}$

Section 3-14

1. What fraction of a quart is 24 ounces? (1 qt = 32 oz)

2. In a recent trip, Anne spent 2 1/2 hours driving to the airport, 5 hours flying, 3/4 of an hour waiting for her baggage, and 1 1/3 hours riding a taxi to her destination. How many hours did Anne's trip take?

3. Because of traffic problems, the trip to work on Monday took 3 1/2 hours. However, on Tuesday, the same trip took only 1 2/3 hours. How much more time was spent driving on Monday than on Tuesday?

4. The roasting time for a 16 to 20 pound turkey is 1/4 hour per pound. How long will it take to roast an 18 pound turkey?

5. Henry practices the violin 3/4 of an hour a day. If he practices 6 days a week, how many hours will he have practiced in four weeks?

6. A recipe for 4 people requires 2 1/2 pounds of hamburger. If you plan to have 10 people for dinner, how many pounds of hamburger should you use?

7. How many 3/4 quart cans of grape juice must be used to completely fill a 6 quart punch bowl?

8. A pole that is 50 feet long is driven vertically into the bottom of a lake to act as a support for a pier. If 2/3 of the pole is below the water level, how many feet are above the water?

9. The Leaning Tower of Pisa in Italy tilts another 1/4 inch from center each year. At this rate how much further will it be leaning in 25 years?

10. Frank, Mike, and 93,998 other golfers each own one share in the Poppy Hills Golf Course. If Mike sells his share to Frank, what fraction of the total shares does Frank own?

Chapter 4

Decimals

After finishing Chapter 4 you should be able to do the following:

1. read and write decimal numbers.
2. add, subtract, multiply, and divide decimal numbers.
3. round off decimal numbers.
4. convert fractions to decimals and vice versa.
5. compare the size of decimals.
6. solve problems that involve both fractions and decimals.
7. solve word problems involving decimals.

On the next page you will find a pretest for this chapter. The purpose of the pretest is to help you determine which sections in this chapter you need to study in detail and which sections you can review quickly. By taking and correcting the pretest according to the instructions on the next page you can better plan your pace through this chapter.

Chapter 4 Pretest

Take and correct this test using the answer section at the end of the book. Those problems that give you difficulty indicate which sections need extra attention. Section numbers are in parentheses before each problem.

ANSWERS

(4-1)   1. Write 27.016 in words.   1. _____

(4-2)   2. Round off 3,907.6197 to the nearest thousandth.   2. _____

(4-3)   3. 16.7 + 14 + 0.017 + 3.0065 + 674.6   3. _____

(4-4)   4. 843.4 - 78.687   (4-5) 5. 91.23 x 67.5   4. _____

5. _____

(4-6)   6. 34.19 ÷ 5.26   6. _____

(4-7)   7. Find 23 ÷ 7.351 rounded to the nearest tenth.   7. _____

(4-8)   8. 5.274 x 5,000,000   (4-9) 9. 436.4 ÷ 200,000   8. _____

9. _____

(4-9)   10. Express 0.62 as a reduced fraction.   10. _____

(4-9)   11. Express $3\frac{1}{16}$ as an exact decimal.   11. _____

(4-10) 12. Arrange in order from largest to smallest:
            0.05, 0.049, 0.005, 0.5   12. _____

(4-11) 13. $(1\frac{1}{3})^2 + 5.4 \times 2\frac{2}{3}$ 13. _____

(4-12) 14. You are working part time at a shoe store and earn $3.96 per hour plus time-and-a-half for overtime. If during one week you work 20 regular hours and 4 overtime hours, what do you earn for the week?   14. _____

(4-13) 15. What is the cost of a 9 feet by 12 feet piece of carpet, if the price including installation and tax is $20.75 per square yard?   15. _____

4-1 Reading and Writing Decimals

I am sure you have seen numbers such as .5, 15.32, 0.638, 127.009. These kinds of numbers are called **decimal numbers** and the dot used in them is called the **decimal point**. Decimal numbers give another way to express proper fractions and mixed numbers. Let me explain.

In our number system we have learned place values for whole numbers. In our system we notice that the place values get smaller as we go from left to right. Each place value is equal to the previous value divided by 10. So any digit to the right of the ones place must have a value that is less than one--a fraction.

NUMBERS THAT ARE GLOOMY AND MISERABLE ARE <u>DECIMAL NUMBERS</u>.

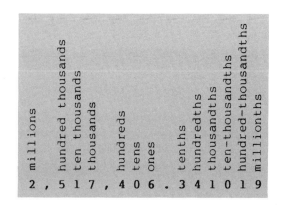

For example, in .576:

        the 5 means 5 tenths (5/10),

        the 7 means 7 hundredths (7/100),

        the 6 means 6 thousandths (6/1000).

Numbers to the right of the decimal point are read like whole numbers and are given a value according to the position of the last digit. Let me explain.

```
tenths
 hundredths
  thousandths
   ten-thousandths
.7             is read "seven tenths." (7/10)
.2 3           is read "twenty-three hundredths." (23/100)
.1 7 7         is read "one hundred seventy-seven thousandths." (177/1000)
.0 0 4 8       is read "forty-eight ten-thousandths." (48/10,000)
```

Notice that place values to the right of the decimal point end in "ths."

If we have numbers to the right and left of the decimal point, we really have a mixed number. For example,

5.3 is read "five and three tenths."

- five → whole number
- and → decimal point
- three tenths → fractional part

16.08 is read "sixteen and eight hundredths."
172.056 is read "one hundred seventy-two and fifty-six thousandths."

We can summarize **reading numbers** that contain a decimal point as follows:

> 1. Read the whole number part before the decimal point.
> 2. Read an "and" for the decimal point.
> 3. Read the number to the right of the decimal point along with the place value determined by the last digit of the number.

Note: Decimal numbers that have no whole number part, such as .6, .37, or .0063, are sometimes written with a zero in the ones place--0.6, 0.37, or 0.0063. The zero in the ones place does two things: it calls attention to the decimal point and it emphasizes that the number is less than one; it has zero ones.

Example 1: Represent 20,056.0073 in words.

twenty thousand, fifty-six and seventy-three ten-thousandths

Example 2: Write one thousand, five and seven hundredths using numbers.

1,005.07

Example 3: In the number 23.75 what does the:

7 represent?   7 tenths (7/10)
5 represent?   5 hundredths (5/100)
3 represent?   3 ones (3)
2 represent?   2 tens (20)

PROBLEM 1: Write 215.0705 in words.   Answer: two hundred fifteen and seven hundred five ten-thousandths

PROBLEM 2: Write using numbers: Answer: 2,055.12

two thousand fifty-five
and twelve hundredths

EXERCISE 4-1   SET A

NAME _____   DATE _____

WRITE IN WORDS: ANSWERS

1. 0.5   1. _____

2. 0.17   2. _____

3. 0.039   3. _____

4. 5.0007   4. _____

5. 16.35   5. _____

6. 426.9   6. _____

7. 6.1236   7. _____

WRITE USING NUMBERS:

8. seven tenths   8. _____

9. twelve hundredths   9. _____

10. nine and three thousandths   10. _____

11. forty-five and six ten-thousandths   11. _____

12. one hundred and sixteen hundredths   12. _____

13. three hundred fifty-six and two
    hundred seven thousandths   13. _____

14. five thousand, twenty-three and three
    thousand, five hundred seventeen ten-thousandths   14. _____

15. two million, eighty thousand and one
    thousand eighty-six millionths   15. _____

IN THE NUMBER 167.2354:

16. What does the 2 represent?   16. _____

17. What does the 5 represent?   17. _____

18. What does the 4 represent?   18. _____

19. What does the 6 represent?   19. _____

EXERCISE 4-1   SET B

NAME _____   DATE _____

WRITE IN WORDS: ANSWERS

1. 0.7   1. _____

2. 0.29   2. _____

3. 0.076   3. _____

4. 6.0007   4. _____

5. 17.82   5. _____

6. 429.7   6. _____

7. 5.2174   7. _____

WRITE USING NUMBERS:

8. five tenths   8. _____

9. sixteen hundredths   9. _____

10. four and nine thousandths   10. _____

11. sixty-one and four ten-thousandths   11. _____

12. one thousand and five hundredths   12. _____

13. nine hundred ninety-nine and six
    hundred nine thousandths   13. _____

14. eight thousand, eleven and one thousand,
    four hundred, eighteen ten-thousandths   14. _____

15. three million, four thousand and one
    hundred two millionths.   15. _____

IN THE NUMBER 5,128.3467:

16. What does the 4 represent?   16. _____

17. What does the 6 represent?   17. _____

18. What does the 1 represent?   18. _____

19. What does the 7 represent?   19. _____

4-2 Rounding Off Decimals

The more digits there are after the decimal point in a number, the more difficult it is to read and work with that number. In many situations, an approximate or rounded off number would suffice. Rounded off decimal numbers have a value that is close to the original number.

Decimal numbers are rounded off using a similar procedure to the one we used to round off whole numbers in Chapter 1.

> **Rounding Off Decimal Numbers**
>
> 1. Look at the digit to the right of the place that you are rounding off to.
>
> 2. If that digit is 5 or more, round off by increasing the digit in the place you are rounding to by one, and discarding all the digits to the right of that place.
>
> 3. If that digit is less than 5, round off by discarding all the digits to the right of the place you are rounding to.

Let me make that clear by explaining some examples.

Example 1: Round off 3.47 to the nearest tenth (to one **decimal place**).

$$3.47 \doteq 3.5$$

tenths place ⟶ ⟵ digit to the right is more than 5

So round off by increasing the tenths place by 1 (changing the 4 to a 5), and discarding the digits to its right.

Example 2: Round off 17.4236 to two decimal places.

$$17.4236 \doteq 17.42$$

second decimal place ⟶ ⟵ digit to the right is less than 5

So round off by discarding the digits to the right of the second decimal place.

Example 3:   Round off 0.5895 to the nearest thousandth.

$$0.5895 \doteq 0.590$$

thousandths place ⟶ ↑ ↳ digit to the right is 5

So round off by increasing the thousandths place by one and discarding the digits to the right of the thousandths place.

0.5895 ≐ 0.590   (Note: The zero is left at the end of the answer. Since the problem asks you to round off to the thousandths place, there should be a digit in the thousandths place.)

PROBLEMS: Round off to the nearest:

	whole number	tenth	hundredth	thousandth
1. 4.7658	_____	_____	_____	_____
2. 15.0313	_____	_____	_____	_____
3. 105.47532	_____	_____	_____	_____
4. 68.599512	_____	_____	_____	_____
5. 0.60051	_____	_____	_____	_____

Answers: 1. 5,   4.8,   4.77   4.766

2. 15,   15.0   15.03   15.031

3. 105,   105.5,   105.48   105.475

4. 69,   68.6,   68.60,   68.600

5. 1,   .6,   .60,   .601

PROBLEMS: Round off 5672.418 to the:

Answers: (Watch out! You are not rounding off the decimal values.)

6. nearest ten.   6. 5670

7. nearest hundred.   7. 5700

8. nearest thousand.   8. 6000

176

EXERCISE 4-2 SET A

NAME _____   DATE _____

ROUND OFF TO THE NEAREST TENTH: ANSWERS

1. 36.72   1. _____

2. 4.65   2. _____

3. 125.482   3. _____

4. 89.971   4. _____

5. 8.95   5. _____

6. 0.149   6. _____

ROUND OFF TO THE NEAREST HUNDREDTH:

7. 18.724   7. _____

8. 5.475   8. _____

9. 792.038218   9. _____

10. 0.8964   10. _____

11. 7.195   11. _____

12. 1.0249   12. _____

ROUND OFF 562.01846 TO THE NEAREST:

13. thousandth   13. _____

14. ten-thousandth   14. _____

15. ten   15. _____

16. hundred   16. _____

ROUND OFF 46.96352:

17. to 2 decimal places.   17. _____

18. to 3 decimal places.   18. _____

19. to 4 decimal places.   19. _____

EXERCISE 4-2   SET B

NAME _____   DATE _____

ROUND OFF TO THE NEAREST TENTH: ANSWERS

1. 43.84   1. _____

2. 116.75   2. _____

3. 3.271   3. _____

4. 9.973   4. _____

5. 37.95   5. _____

6. 0.349   6. _____

ROUND OFF TO THE NEAREST HUNDREDTH:

7. 95.623   7. _____

8. 0.365   8. _____

9. 706.019216   9. _____

10. 7.6973   10. _____

11. 24.295   11. _____

12. 100.0349   12. _____

ROUND OFF 382.01937 TO THE NEAREST:

13. thousandth   13. _____

14. ten-thousandth   14. _____

15. hundred   15. _____

16. ten   16. _____

ROUND OFF 52.69524:

17. to 2 decimal places.   17. _____

18. to 3 decimal places.   18. _____

19. to 4 decimal places.   19. _____

4-3 Adding Decimals

The method for adding decimal numbers is very similar to adding whole numbers. You want to make sure that you add the digits with the same place value.

> **Adding decimals:**
>
> 1. Place the numbers in a column so that their decimal points are lined up.
> 2. Add the digits that have the same place value.
> 3. Place the decimal point in the answer below the other decimal points.

For example, to add 5.3 + .076 + 12.21:

```
        ┌──── Line up the decimal points.
        ↓
     5.3
      .076      Add the digits in each column and
   + 12.21     place the decimal point in the answer
    17.586     below the other decimal points.
```

If any of the numbers you are trying to add are whole numbers without a decimal point, you can simply place a decimal point at the end of the whole number.

For example: 26   =   26.
               5   =   5.
              37,612 =   37,612.

Let us look at a few examples.

Example 1: 52 + 3.01 + 0.035 + 1.58 = ?

```
        ┌──── Line up the decimal points.
        ↓
    52.
     3.01       (Notice a decimal point was placed
      .035      at the end of 52 before adding.)
   + 1.58
    56.625     Add each column and place the decimal
               point in the answer below the other
               decimal points.
```

Example 2: 307.52 + 136 + .65 + 28 + 1.17 + 0.2 = ?

```
    307.52
    136.
      .65
     28.
      1.17
   +  0.2
    473.54
```

Try the following problems.

PROBLEM 1: 56 + 354.89 + 7.98 + 0.02 = ?

Answer: 418.89

```
     56.
    354.89
      7.98
+     0.02
    418.89
```

PROBLEM 2: 0.01 + 0.002 + 0.0003 + 0.00004 = ?

Answer: 0.01234

```
    0.01
    0.002
    0.0003
+   0.00004
    0.01234
```

PROBLEM 3: 465.9087 + 7.8001 + 1235.2 + 45 + 3.2 = ?

Answer: 1757.1088

```
    465.9087
      7.8001
   1235.2
     45.
+     3.2
   1757.1088
```

PROBLEM 4: What is wrong with the solution below?

2.345 + 90.8 + 1234.009 + 0.823 + .2 = ?

```
       2.345
      90.8
    1234.009
       0.823
+       .2
    1237.187
```

Answer:
The decimal points were not lined up in a vertical column.

PROBLEM: 5 What is wrong with the solution below?

307 + 47.6 = ?

```
     .307
+  47.6
   47.907
```

Answer:
The decimal point for 307 was placed at the beginning of the number instead of at the end.

EXERCISE 4-3   SET A

NAME _____   DATE _____

ADD THE FOLLOWING: ANSWERS

1. 287.04   2.   436.09   1. _____
      4.5   7.6
    + 29.   + 80.   2. _____

3.   0.05   4.   0.056   3. _____
     0.096   0.7
   + 0.1   + 0.48   4. _____

5.   81.26   6.   19.05   5. _____
     43.99   30.26
     16.57   12.47   6. _____
   + 4.06   + 6.98

7. 586.94 + 306.87   7. _____

8. 0.25 + 0.016 + 0.9   8. _____

9. 12.6 + 14 + 126.423   9. _____

10. 307.0007 + 16.25 + 526 + 18.04   10. _____

11. 809 + 3.65 + 19.0702 + 76.43   11. _____

12. 1287.52 + 4700.072 + 29.1103 + 6.5 + 81 12. _____

13. 123 + 86.057 + 1283.001 + 4.7 + 68.0119 13. _____

EXERCISE 4-3 SET B

NAME _____   DATE _____

ADD THE FOLLOWING: ANSWERS

1. 173.07   2. 180.03   1. _____
      3.5   2.6
  + 45.   +   11.   2. _____

3. 0.072   4. 0.012   3. _____
    0.01   0.07
  + 0.0036   + 0.0058   4. _____

5. 14.52   6. 24.27   5. _____
    17.21   18.06
    23.65   30.29   6. _____
  + 10.08   + 16.84

7. 425.63 + 174.52   7. _____

8. 0.57 + 0.4 + 0.085   8. _____

9. 43 + 17.4 + 485.207   9. _____

10. 301.0005 + 19.76 + 443 + 6.99   10. _____

11. 8.0123 + 46.78 + 94 + 0.091   11. _____

12. 1706.58 + 4.217 + 89 + 316.4 + 15.0103   12. _____

13. 62.019 + 4.302 + 8256 + 9000.5 + 7.3   13. _____

4-4 Subtracting Decimals

Subtracting or finding the difference between decimal numbers is similar to adding decimal numbers in that you want to operate on digits with the same place value. To do this you must line up the decimal points before you do the subtraction.

For example: 27.973 - 2.241 = ?

1. Line up the decimal points.

2. Subtract and place the decimal point in the answer below the other points.

```
  27.973
-  2.241
  25.732
```

In that example, both numbers had 3 decimal digits. Subtraction is easier when both numbers have the same number of decimal digits. In order to have that happen with other subtraction problems, we sometimes have to give an alternate representation for a decimal number.

Let me show you what I mean.

The number .7 can be written in many different ways:

```
    .7
=   .70
=   .700
=   .7000
=   .70000
    etc.
```

This is true since .70 = 70/100, which reduces to 7/10; and .700 = 700/1000, which also reduces to 7/10. Similarly all the other representations reduce to 7/10.

Those zeros that are attached after the decimal point do not change the value of the decimal number; they simply give other equivalent numbers. By attaching zeros we likewise get:

```
    52                  1.89
=   52.             =   1.890
=   52.0            =   1.8900
=   52.00           =   1.89000
=   52.000          =   1.890000
    etc.                etc.
```

Example 1: .5762 - .34 = ?

```
        ┌──── Line up the decimal points.
        ↓
     .5762
   - .3400      Change the .34 to .3400 to get the same
     .2362     number of digits after the decimal point.
```

Example 2: 56.4 - 3.27 = ?

```
    56.40      Change the 56.4 to 56.40 to get the same
  -  3.27      number of digits after the decimal point.
    53.13
```

Example 3: 189 - 13.567 = ?

```
   189.000     Change the 189 to 189.000 to get the same
  - 13.567     number of digits after the decimal point.
   175.433
```

This then is a summary of **subtracting decimal numbers.**

> 1. Get the same number of digits after the decimal point by attaching zeros after the decimal point.
> 2. Line up the decimal points of both numbes.
> 3. Subtract as usual.
> 4. Place the decimal point in the answer below the other decimal points.

PROBLEM 1:   23.62 - 19.048 = ?   Answer: 4.572

```
                        23.620
                      - 19.048
                         4.572
```

PROBLEM 2:   57.2457 - 20.7 = ?   Answer: 36.5457

```
                        57.2457
                      - 20.7000
                        36.5457
```

PROBLEM 3:   901 - 23.452 = ?   Answer: 877.548

```
                       901.000
                      - 23.452
                       877.548
```

EXERCISE 4-4   SET A

NAME _____   DATE _____

SUBTRACT THE FOLLOWING: ANSWERS

1.   8.647   2.   17.65   1. _____
   - 5.216   - 8.41
                                                 2. _____

3.   47.7   4.   84.3   3. _____
   - 26.25   - 72.16
                                                 4. _____

5.   316   6.   485   5. _____
   - 158.017   - 206.035
                                                 6. _____

7.   52.093   8.   60.082   7. _____
   - 18.4   - 41.7
                                                 8. _____

9.   785.1852   10.   643.1926   9. _____
   - 97   - 195
                                                10. _____

11. 750 - 32.5   11. _____

12. 420 - 14.7   12. _____

13. 89.2876 - 71.6   13. _____

14. 806.1 - 392.0753   14. _____

15. 1532 - 27.6   15. _____

16. 4275 - 186.7284   16. _____

EXERCISE 4-4 SET B

NAME _____   DATE _____

SUBTRACT THE FOLLOWING: ANSWERS

1. 5.789   2. 36.86   1. _____
   - 2.356   - 5.23
                                                     2. _____

3. 85.6   4. 66.4   3. _____
   - 37.24   - 28.17
                                                     4. _____

5. 175   6. 294   5. _____
   - 143.128   - 189.317
                                                     6. _____

7. 48.086   8. 49.067   7. _____
   - 15.1   - 18.2
                                                     8. _____

9. 386.1972   10. 820.0751   9. _____
   - 57   - 639
                                                    10. _____

11. 870 - 46.8   11. _____

12. 526 - 120.76   12. _____

13. 43.2765 - 24.4   13. _____

14. 976.5 - 288.0176   14. _____

15. 1716 - 3.14159   15. _____

16. 1287 - 32.4   16. _____

4-5 Multiplying Decimals

To understand how to multiply decimal numbers we must look at the fractions which the decimal numbers represent.

For example: .3 x .71 = ?

$$.3 = \frac{3}{10} \text{ and } .71 = \frac{71}{100}$$

$$\text{So } .3 \times .71 = \frac{3}{10} \times \frac{71}{100}$$

$$= \frac{213}{1000}$$

$$= .213$$

If every time you wanted to multiply decimal numbers you had to change them to fractions, it could become a very involved process. So as we have done previously, we will use that example to obtain a shorter way to multiply decimals.

Let us look at that example again.

.3 x .71 = .213

```
   .3   has 1 decimal digit.
  .71   has 2 decimal digits.
 .213   has 3 decimal digits.
```

The number of decimal digits in the product is the total of the number of decimal digits in the numbers being multiplied.

We can now do the above example as follows:

Write the numbers above each other, ignoring the decimal points, and multiply the numbers as if they were whole numbers.

```
    .71  ←—— has 2 decimal digits.
   x .3  ←—— has 1 decimal digit.
   .213  ←—— so the answer has a total of 3 decimal digits.
```

Example 1: 15.31 x .07 = ?

```
   15.31  ←—— has 2 decimal digits.
   x .07  ←—— has 2 decimal digits.
  1.0717  ←—— so the answer has a total of 4 decimal digits.
```

Example 2: 19.86 xx .089 = ?

```
    1 9.8 6   ←— has 2 decimal digits.
  x   .0 8 9  ←— has 3 decimal digits.
    1 7 8 7 4
  1 5 8 8 8
  1.7 6 7 5 4 ←— so the answer has a total of 5 decimal digits.
              (Note: To place the decimal in the answer, start
              from the right and move 5 places to the left.)
```

Example 3: .0002 x .007 = ?

```
     .0 0 0 2    ←— has 4 decimal digits.
   x   .0 0 7    ←— has 3 decimal digits.
  .0 0 0 0 0 1 4 ←— so the answer has a total of 7 decimal digits.
                 (Note: 5 zeros are used as place holders in
                 front of the 14 to give the 7 decimal digits.)
```

PROBLEM 1: 43.6 x 2.7 = ? Answer: 117.72

```
      4 3.6
    x   2.7
    3 0 5 2
    8 7 2
  1 1 7.7 2
```

PROBLEM 2: 6.075 x 3.14 = ? Answer: 19.0755

```
        6.0 7 5
      x   3.1 4
      2 4 3 0 0
      6 0 7 5
    1 8 2 2 5
    1 9.0 7 5 5 0
```

PROBLEM 3: 5.3 x .006 = ? Answer: 0.0318

```
          5.3
      x   .0 0 6
         .0 3 1 8
```

PROBLEM 4: What is wrong with the solution below?

```
       43
    x .05
     21.5
```

Answer:

The decimal point should be placed two digits from the *right* of the number.

EXERCISE 4-5   SET A

NAME _____   DATE _____

MULTIPLY THE FOLLOWING: ANSWERS

1.   14.37   2.   21.75   1. _____
     x   3   x   5
                                                                  2. _____

3.   8.08   4.   6.06   3. _____
     x 7.1   x 5.1
                                                                  4. _____

5.   694.4   6.   503.7   5. _____
     x   2.6   x   4.8
                                                                  6. _____

7.   2.0014   8.   4.0042   7. _____
     x   .025   x   .035
                                                                  8. _____

9.   16.34 x 4   9. _____

10. 26.5 x 0.0008   10. _____

11. 7.019 x .04   11. _____

12. 50.48 x 5.25   12. _____

13. 148.34 x 2.017   13. _____

14. 609.24 x .0125   14. _____

EXERCISE 4-5 SET B

NAME _____   DATE _____

MULTIPLY THE FOLLOWING: ANSWERS

1. 16.36   2. 17.25   1. _____
    x   4   x   3
                                           2. _____

3. 7.07   4. 9.09   3. _____
    x 8.1   x 6.1
                                           4. _____

5. 185.6   6. 294.5   5. _____
    x 3.7   x 4.9
                                           6. _____

7. 5.0028   8. 7.0016   7. _____
    x .015   x .045
                                           8. _____

9. 20.15 x 3   9. _____

10. 15.5 x 0.0006   10. _____

11. 9.096 x .07   11. _____

12. 43.28 x 7.15   12. _____

13. 34.26 x 3.016   13. _____

14. 507.23 x .0175   14. _____

4-6 Dividing Decimals

Dividing decimals is done using similar methods as used in dividing whole numbers. The only difference is in placing the decimal point in the answer. *If the divisor is a whole number, the decimal point is simply placed in the answer above its position in the dividend and the division is done ignoring the decimal point.* The reasoning here is that the divisor is dividing into the whole number part and the fractional part with the decimal point used to separate both parts in the answer.

Example 1: 52.8 ÷ 4 = ?

```
              13.2
whole number divisor → 4 ) 52.8
              4
              ―
              12
              12
              ――
               8
               8
```

Place the decimal point above the decimal point in the dividend. That separates the whole and fractional parts of the answer.

If the divisor is not a whole number, you can make it a whole number by moving its decimal point to the right. You must, however, also move the decimal point the same number of places to the right in the dividend.

Example 2: 2.38 ÷ .7 = ?

```
           3.4
   .7 ) 2.3̣8
        2 1
        ―――
          2 8
          2 8
```

1. Move the decimal point to change the the divisor into a whole number.
2. Move the point the same in the dividend.
3. Place the decimal point in the answer above the moved decimal point.
4. Do the division ignoring the decimal points.

What you are really doing when you move the decimal point one place to the right is multiplying the divisor and the dividend by 10.

$$2.38 \div .7 = \frac{2.38}{.7} \underset{\times 10}{\overset{\times 10}{=}} \frac{23.8}{7} = 23.8 \div 7$$

You have transformed the original problem into an equivalent problem, using the same methods as raising fractions to higher terms. Similar reasoning can be used to justify movement of the decimal point in other division problems.

Example 3: 1.30647 ÷ 4.07 = ?

```
              .3 2 1
   4.0 7 ) 1.3 0̣ 6 4 7
           1 2 2 1
           ―――――
               8 5 4
               8 1 4
               ―――
                 4 0 7
                 4 0 7
```

1. Move the decimal point to change the divisor into a whole number.
2. Move the point the same in the dividend.
3. Place the decimal point in the answer above the moved decimal point.
4. Do the division ignoring the decimal points.

Example 4: 21 ÷ .0028 = ?

```
              7 5 0 0.
    .0 0 2 8 ⌐2 1.0 0 0 0
              1 9 6
              1 4 0
              1 4 0
                  0
                  0
                  0
                  0
```

1. Move the decimal point to change the divisor into a whole number.
2. Move the point the same in the dividend. (Notice four zeros were attached.)
3. Place the decimal point in the answer above the moved decimal point.
4. Do the division ignoring the decimal points.

Example 5: Find the exact answer for .23 ÷ 18.4.

(Finding the exact answer means to continue dividing until there is no remainder. The answer is not to be rounded off.)

```
                .0 1 2 5
    1 8.4 ⌐.2 3 0 0 0
            1 8 4
              4 6 0
              3 6 8
                9 2 0
                9 2 0
```

Zeros were attached one at a time until there was no remainder in the division.

To find the exact answer in a division problem, it may be necessary to attach zeros after the decimal point as we did that last example.

PROBLEM 1: 510.17 ÷ 6.002 = ?   Answer: 85

```
                              8 5.
            6.0 0 2 ⌐5 1 0.1 7 0
                    4 8 0 1 6
                      3 0 0 1 0
                      3 0 0 1 0
```

PROBLEM 2: Find the exact answer for   Answer: 3.125

30 ÷ 9.6

```
                          3.1 2 5
              9.6 ⌐3 0.0 0 0 0
                   2 8 8
                     1 2 0
                       9 6
                       2 4 0
                       1 9 2
                         4 8 0
                         4 8 0
```

192

EXERCISE 4-6   SET A

NAME _____   DATE _____

FIND THE EXACT ANSWER IN EACH DIVISION: ANSWERS

1. 4 ⟌ 50.4   2. 6 ⟌ 82.2   1. _____

                                                       2. _____

3. 7.2 ⟌ 58.608   4. 3.7 ⟌ 34.447   3. _____

                                                       4. _____

5. 0.65 ⟌ 48.1   6. 0.54 ⟌ 51.3   5. _____

                                                       6. _____

7. 84 ⟌ 1.554   8. 66 ⟌ 1.155   7. _____

                                                       8. _____

9. .01636 ÷ 4.09   10. .02807 ÷ 8.02   9. _____

                                                       10. _____

11. 21.08 ÷ .0034   12. 21.84 ÷ .0026   11. _____

                                                       12. _____

13. 243.36 ÷ 6.24   14. 488.36 ÷ 8.42   13. _____

                                                       14. _____

15. 1.1 ÷ 3.2   16. 1.4 ÷ 6.4   15. _____

                                                       16. _____

EXERCISE 4-6 SET B

NAME _____   DATE _____

FIND THE EXACT ANSWER IN EACH DIVISION   ANSWERS

1. $3\overline{)50.7}$ 2. $5\overline{)68.5}$ 1. _____

                                                                   2. _____

3. $6.4\overline{)46.272}$ 4. $8.3\overline{)76.028}$ 3. _____

                                                                   4. _____

5. $0.74\overline{)18.5}$ 6. $0.62\overline{)27.9}$ 5. _____

                                                                   6. _____

7. $58\overline{)3.741}$ 8. $72\overline{)1.044}$ 7. _____

                                                                   8. _____

9. .06018 ÷ 7.08   10. .05134 ÷ 6.04   9. _____

                                                                  10. _____

11. 13.16 ÷ .0028   12. 35.28 ÷ .0036   11. _____

                                                                  12. _____

13. 206.64 ÷ 3.28   14. 355.68 ÷ 4.56   13. _____

                                                                  14. _____

15. 1.3 ÷ 3.2   16. 2.8 ÷ 6.4   15. _____

                                                                  16. _____

4-7 Dividing Decimals— Rounding Off Answers

Frequently when dividing decimal numbers, the answer seems to go on and on. Consider the problem:

```
        1.7 7 7 7 ...  = 1.7̄
    9 | 1 6.0 0 0 0
        9                (The dash over
        ‾‾               the 7 indicates
        7 0              that it repeats
        6 3              forever in the
        ‾‾‾              decimal.)
        7 0
        6 3
        ‾‾‾
        7 0
        6 3
```

You could attach zeros forever and never get an exact answer. Because of this and because long decimal answers are difficult to work with, the answers to division problems are frequently rounded off to a number that is close to the exact answer.

YOU HAVE TO BE SHARP TO ROUND OFF NUMBERS.

So 16 ÷ 9 = 1.7777 . . . , but rounded off to the nearest

tenth it equals 1.8,
hundredth it equals 1.78,
thousandth it equals 1.778.

Depending on the degree of accuracy asked for, the answers will vary. Yet each answer is close to the actual answer of 1.7̄.

Example 1: Rounded off to the nearest hundredth, 4.29 ÷ .045 = ?

In order to round off the answer to the nearest hundredth, you must carry out the division one digit past the hundredths place. Zeros were attached so that the division could be done.

```
              9 5.3 3 3    ≐   95.33   (rounded off to the
    .0 4 5 | 4.2 9 0 0 0 0                hundredths place)
             4 0 5
             ‾‾‾‾‾
               2 4 0
               2 2 5
               ‾‾‾‾‾
                 1 5 0
                 1 3 5
                 ‾‾‾‾‾
                   1 5 0
                   1 3 5
                   ‾‾‾‾‾
                     1 5 0
                     1 3 5
                     ‾‾‾‾‾
```

Example 2: Round off the answer to three decimal places.

$$7.8 \div 5.06 = ?$$

Perform the division one digit past the third decimal place.

```
              1.5 4 1 5  = 1.542  (rounded to the third
   5.06⟌7.80,0 0 0 0             decimal place)
         5 0 6
         2 7 4 0
         2 5 3 0
           2 1 0 0
           2 0 2 4
               7 6 0
               5 0 6
               2 5 4 0
               2 5 3 0
```

Example 3: Find the quotient correct to the nearest tenth.

$$16.302 \div 7.6 = ?$$

Carry out the division one digit past the tenths place.

```
          2.1 4  ≑ 2.1  (correct to the
   7.6⟌16.3,0 2          nearest tenth)
       1 5 2
         1 1 0
           7 6
           3 4 2
           3 0 4
```

In order to round off the answer correctly when dividing, remember to carry out the division process one place past the degree of accuracy asked for.

PROBLEM 1: Find $1.54 \div 4.27$ correct to two decimal places.   Answer: 0.36

```
                .3 6 0
   4.27⟌1.54,0 0 0
         1 2 8 1
           2 5 9 0
           2 5 6 2
               2 8 0
                   0
```

PROBLEM 2: Rounded to the nearest thounsandth,   Answer: .058

$$.175 \div 3.023 = ?$$

```
                .0 5 7 8
   3.023⟌.175,0 0 0 0
          1 5 1 1 5
            2 3 8 5 0
            2 1 1 6 1
              2 6 8 9 0
              2 4 1 8 4
```

EXERCISE 4-7   SET A

NAME _____   DATE _____

FIND THE ANSWERS ROUNDED TO THE NEAREST TENTH: ANSWERS

1. 8.5 ÷ 7   2. 8.4 ÷ 9   1. _____

                                              2. _____

3. 3.006 ÷ .36   4. 3.096 ÷ .48   3. _____

                                              4. _____

5. 608 ÷ 2.4   6. 854 ÷ 4.2   5. _____

                                              6. _____

FIND THE ANSWERS ROUNDED TO 2 DECIMAL PLACES:

7. 8.43 ÷ 6   8. 7.24 ÷ 5   7. _____

                                              8. _____

9. 9.1 ÷ .289   10. 6.3 ÷ .361   9. _____

                                              10. _____

FIND THE ANSWERS ACCURATE TO THE THOUSANDTHS PLACE:

11. 11 ÷ 66   12. 49 ÷ 81   11. _____

                                              12. _____

13. 28.89 ÷ 7.205   14. 31.6 ÷ 6.308   13. _____

                                              14. _____

EXERCISE 4-7 SET B

NAME _____ DATE _____

FIND THE ANSWERS ROUNDED TO THE NEAREST TENTH: ANSWERS

1. 7.6 ÷ 9   2. 9.6 ÷ 5   1. _____

                                                                    2. _____

3. 2.511 ÷ .54   4. 3.706 ÷ .68   3. _____

                                                                    4. _____

5. 798 ÷ 1.8   6. 728 ÷ 2.1   5. _____

                                                                    6. _____

FIND THE ANSWERS ROUNDED TO 2 DECIMAL PLACES:

7. 2.56 ÷ 7   8. 6.79 ÷ 9   7. _____

                                                                    8. _____

9. 5.2 ÷ .169   10. 8.6 ÷ .121   9. _____

                                                                    10. _____

FIND THE ANSWERS ACCURATE TO THE THOUSANDTHS PLACE:

11. 21 ÷ 36   12. 35 ÷ 75   11. _____

                                                                    12. _____

13. 24.7 ÷ 8.206   14. 26.49 ÷ 4.408   13. _____

                                                                    14. _____

4-8 Multiplying and Dividing by Numbers that End in Zeros

If you multiply and divide numbers that end in zeros using the methods explained so far in this chapter, you may find yourself spending a lot of time working with those zeros but still obtaining an answer with the decimal point misplaced. Look at these examples using those methods.

Example 1: 5.37 x 20,000 = ?

```
              5.3 7
    x 2 0 0 0 0
              0 0 0
            0 0 0
          0 0 0
        0 0 0
      1 0 7 4
    1 0 7 4 0 0.0 0
```

Thus 5.37 x 20,000 = 107,400.

Example 2: 83.4 ÷ 100,000 = ?

```
                    .0 0 0 8 3 4
    1 0 0 0 0 0 | 8 3.4 0 0 0 0 0
                  8 0 0 0 0 0
                    3 4 0 0 0 0
                    3 0 0 0 0 0
                      4 0 0 0 0
                      4 0 0 0 0
```

Most of the effort in those examples was involved in taking care of zeros. There are shorter and more efficient ways to handle those zeros. Let me show them to you.

Multiplying by Numbers that End in Zeros

Consider Example 1 again. You can move the multiplier to the right as we did in Section 1-8.

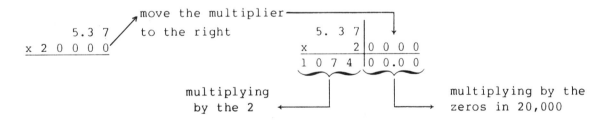

Example 3: 0.764 x 46,000 = ?

```
        0.7 6 4 |
    x     4 6 | 0 0 0
        4 5 8 4 | 0 0 0
      3 0 5 6   |
      3 5 1 4 4.0 0 0
```

Notice that the multiplier was moved to the right before the multiplication was done.

Thus 0.764 x 46,000 = 35,144

PROBLEM 1: $53.5 \times 10^5 = ?$ Answer: 5,350,000

$$10^5 = 10 \times 10 \times 10 \times 10 \times 10 = 100,000$$

```
     5 3.5
  x  1 0 0 0 0 0
  5 3 5 0 0 0 0.0
```

Dividing by Numbers that End in Zeros

When trying to divide numbers, the more digits in the divisor, the more involved the division process can be. When the divisor contains a lot of zeros as in Example 2, those zeros seem to make the problem even more difficult than it really is. There is a short cut to handle a divisor that ends in zeros. You can again move decimal points to get an equivalent problem.

Example 4: $83.4 \div 100,000 = ?$

$$83.4 \div 100,000 = \frac{83.4}{100,000}$$ Change the division into a fraction.

$$= \frac{.00083.4}{1.00,000}$$ Move the decimal to the left, eliminating zeros in the denominator (divisor). Move the decimal the same number of places to the left in the numerator (dividend).

$$= \frac{.000834}{1}$$

$$= .000834$$

Example 5: $2860 \div 20,000 = ?$

$$2860 \div 20,000 = \frac{.2860}{2.0000}$$ Move decimal points four places to the left to eliminate the zeros in the denominator (divisor).

$$= \frac{.286}{2}$$ Do the division with the new divisor of 2.

$$= .143$$

From these examples you can see that *if the divisor ends in zeros, you can eliminate those zeros by moving the decimal point to the left in both the divisor and the dividend before you complete the division process.*

PROBLEM 2: $139.2 \div 87,000 = ?$ Answer: .0016

$$= \frac{.139.2}{87.000} = \frac{.0016}{87 \overline{).1392}}$$

```
     87
    ---
     522
     522
```

PROBLEM 3: $57.6 \div 10^4 = ?$ Answer: .00576

$$10^4 = 10 \times 10 \times 10 \times 10 = 10,000$$

$$57.6 \div 10^4 = \frac{.0057.6}{1.0000} = .00576$$

EXERCISE 4-8   SET A

NAME _____   DATE _____

MULTIPLY THE FOLLOWING: ANSWERS

1. 4.36 x 1,000   2. 5.23 x 10,000   1. _____

                                                                 2. _____

3. 47.617 x 20,000   4. 53.109 x 4,000   3. _____

                                                                 4. _____

5. 0.093 x 42,000   6. 0.072 x 63,000   5. _____

                                                                 6. _____

7. 3.127 x 10^5 8. 4.173 x 10^6 7. _____

                                                                 8. _____

FIND THE EXACT ANSWERS:

9. 54.3 ÷ 100   10. 75.6 ÷ 100   9. _____

                                                                 10. _____

11. 700 ÷ 2,000   12. 1200 ÷ 3000   11. _____

                                                                 12. _____

13. 1957.3 ÷ 370,000   14. 1939.6 ÷ 260,000   13. _____

                                                                 14. _____

15. 376.5 ÷ 10^5 16. 408.7 ÷ 10^6 15. _____

                                                                 16. _____

EXERCISE 4-8   SET B

NAME _____   DATE _____

MULTIPLY THE FOLLOWING   ANSWERS

1. 8.17 x 100,000   2. 6.04 x 1000   1. _____

                                                  2. _____

3. 92.575 x 3000   4. 84.019 x 200,000   3. _____

                                                  4. _____

5. 0.072 x 54,000   6. 0.051 x 76,000   5. _____

                                                  6. _____

7. 7.263 x 10^6 8. 6.108 x 10^5 7. _____

                                                  8. _____

FIND THE EXACT ANSWERS:

9. 82.3 ÷ 100   10. 64.5 ÷ 100   9. _____

                                                  10. _____

11. 2400 ÷ 6000   12. 2100 ÷ 7000   11. _____

                                                  12. _____

13. 1982.4 ÷ 420,000   14. 1263.6 ÷ 360,000   13. _____

                                                  14. _____

15. 510.3 ÷ 10^6 16. 827.4 ÷ 10^5 15. _____

                                                  16. _____

4-9 Changing Between Fractions and Decimals

To **change a fraction into a decimal** you must remember that the fraction line indicates that you divide the numerator by the denominator.

Example 1: Express 3/4 as an exact decimal.

$$\frac{3}{4}_{\text{—divisor}} = 4\overline{\smash{)}\begin{array}{r}.75\\3.00\\\underline{2\ 8}\\20\\\underline{20}\end{array}}$$

Thus 3/4 is the decimal .75.

Example 2: Express 8/12 as a decimal rounded to two decimal places.

$$\frac{8}{12}_{\text{—divisor}} = 12\overline{\smash{)}\begin{array}{r}.666\ldots = 0.\overline{6} \doteq 0.67\\8.000\\\underline{7\ 2}\\80\\\underline{72}\\80\\\underline{72}\end{array}}$$

Thus $8/12 \doteq 0.67$.

Example 3: Express 2 3/5 as a decimal number.

$$2\frac{3}{5} = \frac{13}{5} = 5\overline{\smash{)}\begin{array}{r}2.6\\13.0\\\underline{10}\\3\ 0\\\underline{3\ 0}\end{array}}$$

Thus $2\frac{3}{5} = 2.6$.

PROBLEM 1: Express 7/8 as an exact decimal.   Answer: 0.875

$$8\overline{\smash{)}\begin{array}{r}.875\\7.000\\\underline{6\ 4}\\60\\\underline{56}\\40\\\underline{40}\end{array}}$$

PROBLEM 2: Express $4\frac{3}{17}$ as a decimal accurate to the tenths place.   Answer: 4.2

$$4\frac{3}{17} = \frac{71}{17} = 17\overline{\smash{)}\begin{array}{r}4.17\\71.00\\\underline{68}\\3\ 0\\\underline{1\ 7}\\1\ 30\\\underline{1\ 19}\end{array}}$$

Some commonly used relationships between fractions and decimals are listed in the chart below.

Fraction—Decimal Equivalents

$\frac{1}{2} = .5 \qquad \frac{1}{4} = .25 \qquad \frac{3}{4} = .75 \qquad \frac{1}{3} = .\overline{3} \doteq .33 \qquad \frac{2}{3} = .\overline{6} \doteq .67$

Changing Decimals to Fractions

To **change a decimal number into a fraction** you must remember the place values of digits to the right of the decimal point. As you go to the right from the decimal point, they are tenths, hundredths, thousandths, ten-thousandths, etc. Using those place values and the way you read decimal numbers in our number system, you can easily convert decimals to fractions.

For example: .9 reads "nine tenths" = $\frac{9}{10}$

.35 reads "thirty-five hundredths" = $\frac{35}{100} = \frac{7}{20}$ (reduced)

.052 reads "fifty-two thousandths" = $\frac{52}{1000} = \frac{13}{250}$ (reduced)

If there is a whole number before the decimal point, you really have a mixed number.

For example: 5.3 reads "five and three tenths" = $5\frac{3}{10}$

18.07 reads "eighteen and seven hundredths" = $18\frac{7}{100}$

PROBLEMS: Change the following to reduced fractions or mixed numbers:

3. .25

4. .172

5. 3.7

6. 45.02

7. 100.003

Answers:

3. $.25 = \frac{25}{100} = \frac{1}{4}$

4. $.172 = \frac{172}{1000} = \frac{43}{250}$

5. $3.7 = 3\frac{7}{10}$

6. $45.02 = 45\frac{2}{100} = 45\frac{1}{50}$

7. $100.003 = 100\frac{3}{1000}$

EXERCISE 4-9   SET A

NAME _____   DATE _____

CHANGE TO EXACT DECIMALS   ANSWERS

1. $\dfrac{3}{5}$ 2. $\dfrac{7}{14}$ 1. _____

                                                 2. _____

3. $\dfrac{5}{8}$ 4. $\dfrac{1}{16}$ 3. _____

                                                 4. _____

5. $\dfrac{41}{32}$ 6. $\dfrac{37}{32}$ 5. _____

                                                 6. _____

CHANGE TO DECIMALS ROUNDED TO 3 DECIMAL PLACES:

7. $\dfrac{6}{11}$ 8. $\dfrac{5}{6}$ 7. _____

                                                 8. _____

9. $1\dfrac{6}{7}$ 10. $2\dfrac{4}{9}$ 9. _____

                                                 10. _____

11. $23\dfrac{2}{3}$ 12. $37\dfrac{7}{9}$ 11. _____

                                                 12. _____

CHANGE TO REDUCED FRACTIONS OR MIXED NUMBERS:

13. 0.6   14. 0.4   13. _____

                                                 14. _____

15. 0.62   16. 0.74   15. _____

                                                 16. _____

17. 0.0425   18. 0.0475   17. _____

                                                 18. _____

19. 4.001   20. 3.009   19. _____

                                                 20. _____

EXERCISE 4-9   SET B

NAME _____   DATE _____

CHANGE TO EXACT DECIMALS   ANSWERS

1. $\dfrac{4}{5}$ 2. $\dfrac{9}{12}$ 1. _____

                                                            2. _____

3. $\dfrac{3}{8}$ 4. $\dfrac{5}{16}$ 3. _____

                                                            4. _____

5. $\dfrac{39}{32}$ 6. $\dfrac{35}{32}$ 5. _____

                                                            6. _____

CHANGE TO DECIMALS ROUNDED TO 3 DECIMAL PLACES:

7. $\dfrac{5}{9}$ 8. $\dfrac{4}{7}$ 7. _____

                                                            8. _____

9. $2\dfrac{3}{13}$ 10. $3\dfrac{5}{11}$ 9. _____

                                                           10. _____

11. $44\dfrac{1}{6}$ 12. $32\dfrac{1}{7}$ 11. _____

                                                           12. _____

CHANGE TO REDUCED FRACTIONS OR MIXED NUMBERS:

13. 0.2   14. 0.6   13. _____

                                                           14. _____

15. 0.86   16. 0.98   15. _____

                                                           16. _____

17. 0.0375   18. 0.0275   17. _____

                                                           18. _____

19. 1.007   20. 2.003   19. _____

                                                           20. _____

4-10 Comparing Decimals

In Section 3-13 you learned to compare the size of fractions by changing them to fractions with common denominators. Decimal numbers that have the same number of decimal places have the same place value. They actually represent fractions with a common denominator.

For example, consider the decimals

.34, .07, .18, .91 :

$$.34 = \frac{34}{100}$$

$$.07 = \frac{7}{100}$$

$$.18 = \frac{18}{100}$$

$$.91 = \frac{91}{100}$$

They have the same number of decimal places and represent fractions with common denominators.

```
507
50.7
5.07
.507
.0507
 ?
```

IF DECIDING WHICH IS LARGEST GIVES YOU TROUBLE, YOU MAY HAVE MISSED THE POINT.

The method, then, for **comparing the size of decimal numbers** is to represent each with the same number of decimal places. Those with a larger number after the decimal point will have a larger fractional part.

Example 1: Which is larger: .370 or .307?

They both have the same number of decimal places, and since 370 > 307, .370 is larger than .307.

Example 2: Arrange in order from largest to smallest:

0.1, 0.0195, 0.019, 0.109, 0.19

Since the greatest number of decimal places is four places in the number 0.0195, we want to get four decimal places in each number by attaching zeros after the decimal points. Then, since 1900 > 1090 > 1000 > 195 > 190, the order from largest (on top) to smallest is:

```
   0.19   = .1900
   0.109  = .1090
   0.1    = .1000
   0.0195 = .0195
   0.019  = .0190
```

Example 3: Arrange in order from largest to smallest:

1.23, 1.203, 1, 1.3, 1.2

Since the greatest number of decimal places is three places in the number 1.203, we want to get three decimal places in each number by attaching zeros after the decimal points. Then, since 1.300 > 1.230 > 1.203 > 1.000, the order from largest (on top) to smallest is:

```
1.3   = 1.300
1.23  = 1.230
1.203 = 1.203
1.2   = 1.200
1     = 1.000
```

PROBLEM 1: Which is larger: .078 or .08?

Answer: .08

.078 = .078
.08 = .080

Since 80 is greater than 78, then .08 is larger.

PROBLEM 2: Arrange in order from largest to smallest: 3.4, 4.3, 4.03, 4.003, 3.0004

Answer: (largest on top)

```
4.3    = 4.3000
4.03   = 4.0300
4.003  = 4.0030
3.4    = 3.4000
3.0004 = 3.0004
```

PROBLEM 3: Which is largest:

$\frac{5}{8}$, 0.8, $\frac{5}{6}$, or 0.75?

Answer: $\frac{5}{6}$

Change each to a decimal.

$\frac{5}{6} = .8\overline{3}$

.8 = .800

.75 = .750

$\frac{5}{8} = .625$

$.8\overline{3}$ is the largest decimal number.

208

EXERCISE 4-10 SET A

NAME _____ DATE _____

WHICH IS LARGER: ANSWERS

1. 0.5 or 0.6?   1. _____

2. 0.57 or 0.75?   2. _____

3. 2.076 or 1.067?   3. _____

4. 5.092 or 5.009?   4. _____

5. 0.6 or 0.06?   5. _____

6. 0.162 or 0.1062?   6. _____

7. 3.4002 or 3.402?   7. _____

8. 76.125 or 76.1025?   8. _____

9. $\frac{2}{3}$ or .66?   9. _____

10. $3\frac{1}{4}$ or 3.125?   10. _____

ARRANGE IN ORDER FROM LARGEST TO SMALLEST:

11. 0.7, 0.76, 0.076, 0.706, 0.6   11. _____

12. 0.54, 0.4, 0.5, 0.504, 0.054   12. _____

13. 3.12, 2.13, 2, 3.102, 2.3   13. _____

14. $\frac{3}{4}$, $\frac{3}{5}$, .7, .65, .076   14. _____

15. $1\frac{1}{3}$, 1.8, $1\frac{1}{2}$, 1.258, 1.625   15. _____

EXERCISE 4-10   SET B

NAME _____   DATE _____

WHICH IS LARGER: ANSWERS

1. 0.9 or 0.8?   1. _____

2. 0.89 or 0.98?   2. _____

3. 3.1205 or 3.0125?   3. _____

4. 12.076 or 12.706?   4. _____

5. 0.07 or 0.7?   5. _____

6. 0.129 or 0.1029?   6. _____

7. 3.5007 or 3.507?   7. _____

8. 84.306 or 84?   8. _____

9. $\frac{3}{8}$ or .4?   9. _____

10. $4\frac{1}{3}$ or 4.033?   10. _____

ARRANGE IN ORDER FROM LARGEST TO SMALLEST:

11. 0.37, 0.037, 0.07, 0.3, 0.307   11. _____

12. 0.9, 0.901, 0.091, 0.9001, 0.19   12. _____

13. 4.3, 4.03, 3.4, 3.04, 3.4003   13. _____

14. $\frac{4}{5}$, .755, .82, $\frac{5}{8}$, .5   14. _____

15. 2.66, $2\frac{2}{3}$, 2.7, $2\frac{7}{8}$, 2.78   15. _____

4-11 Operating with Both Fractions and Decimals

The last skill to develop in this unit involves problems that combine both fractional and decimal numbers. When doing problems of this type, you must first remember to use the proper **order of operations** as covered in Section 2-8.

> 1. Do any operations inside parentheses.
> 2. Do the powers and square roots.
> 3. Do the multiplications and divisions from left to right.
> 4. Do the additions and subtractions from left to right.

Secondly, you must decide whether to do the problem using only fractions, only decimals, or a combination of fractions and decimals. Consider the problem: 5 3/4 x 18.8; it can be done in three different ways.

1. Using only fractions:
$$5\frac{3}{4} \times 18.8 = 5\frac{3}{4} \times 18\frac{8}{10}$$
$$= 5\frac{3}{4} \times 18\frac{4}{5}$$
$$= \frac{23}{\cancel{4}_2} \times \frac{\cancel{94}^{47}}{5}$$
$$= \frac{1081}{10}$$
$$= 108\frac{1}{10}$$

2. Using only decimals:
$$5\frac{3}{4} \times 18.8 = 5.75 \times 18.8$$

```
      5.7 5
   x  1 8.8
      4 6 0 0
    4 6 0 0
    5 7 5
    1 0 8.1 0 0   = 108.1
```

3. Using both fractions and decimals:
$$5\frac{3}{4} \times 18.8 = \frac{23}{4} \times \frac{18.8}{1}$$
$$= \frac{23}{\cancel{4}_1} \times \frac{\cancel{18.8}^{4.7}}{1}$$
$$= \frac{108.1}{1}$$
$$= 108.1$$

As you can see, we obtained the same answer no matter which way we did the problem. In deciding which method to use, the following suggestions may be helpful.

Operating with both fractions and decimals:

1. If the fractions can be converted into exact decimals, work the problem using decimal numbers.
2. If the fractions can not be converted into exact decimals, work the problem using fractions.

Example 1: $(1.5)^2 + 4\frac{3}{5} \times 6 = ?$

Note: 3/5 is the exact decimal 0.6:

$$5\overline{)3.0} \quad .6$$
$$\underline{3\ 0}$$

$$(1.5)^2 + 4\frac{3}{5} \times 6 \;=\; (1.5)^2 + 4.6 \times 6$$
$$=\; 2.25 \;+\; 4.6 \times 6$$
$$=\; 2.25 \;+\; 27.6$$
$$29.85$$

Example 2: $3 - 6\frac{2}{3} \div 2.4 = ?$

Note: $\frac{2}{3} = .\overline{6}$ is not an exact decimal, work with fractions.

$$3 - 6\frac{2}{3} \div 2.4 \;=\; 3 - \frac{20}{3} \div 2\frac{4}{10}$$
$$=\; 3 - \frac{20}{3} \div 2\frac{2}{5}$$
$$=\; 3 - \frac{20}{3} \div \frac{12}{5}$$
$$=\; 3 - \frac{\cancel{20}^5}{3} \times \frac{5}{\cancel{12}_3}$$
$$=\; 3 - \frac{25}{9}$$
$$=\; \frac{27}{9} - \frac{25}{9} = \frac{2}{9}$$

Even with those suggestions some trial and error might be necessary to determine the easiest way to do a particular problem.

PROBLEM 1: $32.2 \times (3\frac{1}{2} - 2\frac{3}{7}) = ?$ Answer: 34.5

$$32.2 \times (\frac{7}{2} - \frac{17}{7}) =$$
$$32.2 \times (\frac{49}{14} - \frac{34}{14}) =$$
$$\cancel{32.2}^{2.3} \times \frac{15}{\cancel{14}_1} = 34.5$$

212

EXERCISE 4-11   SET A

NAME _____   DATE _____

PERFORM THE INDICATED OPERATIONS: ANSWERS

1. $14.4 \div 4\frac{1}{2}$ 2. $19.2 \div 1\frac{3}{5}$ 1. _____

                                                                    2. _____

3. $3\frac{3}{4} \times 5.6$ 4. $1\frac{5}{8} \times 7.2$ 3. _____

                                                                    4. _____

5. $(1\frac{1}{2})^2 + 2\frac{2}{3} \times 1.2$ 6. $(2\frac{1}{3})^2 + 3\frac{1}{3} \times 3.1$ 5. _____

                                                                    6. _____

7. $0.25 \times (2 + 4 \div 12)$ 8. $0.75 \times (2 + 3 \div 15)$ 7. _____

                                                                    8. _____

9. $\frac{1}{2} \times 3^3 - 5.3 \times .031$ 10. $\frac{1}{4} \times 5^2 - 7.2 \times .15$ 9. _____

                                                                   10. _____

11. $80 \times \frac{1}{4} \div 5 + 8.25$ 12. $80 \div \frac{1}{4} \times 5 + 8.25$ 11. _____

                                                                   12. _____

13. $\frac{1}{3} \times \sqrt{16} + 10 \div 2.5$ 14. $\frac{1}{6} \times \sqrt{49} + 21 \div 3.5$ 13. _____

                                                                   14. _____

15. $\frac{3}{5} \times 10^3 \div (1.25 \times 10^2)$ 16. $\frac{3}{4} \times 10^4 \div (.8 \times 10^2)$ 15. _____

                                                                   16. _____

EXERCISE 4-11   SET B

NAME _____   DATE _____

PERFORM THE INDICATED OPERATIONS: ANSWERS

1. $22.4 \div 3\frac{1}{2}$ 2. $36 \div 2\frac{2}{5}$ 1. _____

                                                                    2. _____

3. $4\frac{3}{5} \times 6.5$ 4. $2\frac{1}{4} \times 5.2$ 3. _____

                                                                    4. _____

5. $(1\frac{1}{5})^2 + 1\frac{1}{6} \times 4.8$ 6. $(3\frac{1}{2})^2 + 2\frac{1}{2} \times 3.7$ 5. _____

                                                                    6. _____

7. $0.5 \times (3 + 2 \div 6)$ 8. $0.75 \times (4 + 4 \div 16)$ 7. _____

                                                                    8. _____

9. $\frac{3}{4} \times 8^2 - 4.2 \times .35$ 10. $\frac{1}{2} \times 7^2 - 8.5 \times .46$ 9. _____

                                                                   10. _____

11. $100 \times \frac{1}{5} \div 4 + 6.35$ 12. $100 \div \frac{1}{5} \times 4 + 6.35$ 11. _____

                                                                   12. _____

13. $\frac{2}{3} \times \sqrt{25} + 36 \div 4.5$ 14. $\frac{5}{6} \times \sqrt{4} + 39 \div 6.5$ 13. _____

                                                                   14. _____

15. $\frac{3}{4} \times 10^3 \div (.6 \times 10^2)$ 16. $\frac{1}{2} \times 10^2 \div (.008 \times 10^4)$ 15. _____

                                                                   16. _____

4-12 Applications Involving Decimals

"BALANCING MY CHECKBOOK IS FUN."

Word problems involving decimals can also be solved using the read, analyze, and solve procedure. When working with decimals, you may have to round off answers, especially when working with dollars and cents.

Example 1: Johnny Salo in 1929 ran 3665 miles from New York to Los Angeles in 79 days. How many miles did he average per day?

Analyze: Miles *per* day means miles *divided* by days.

Solve:
```
              46.392  ≐ 46.39 miles
           _____
        79 | 3665.000
            316
            ___
            505
            474
            ___
             31 0
             23 7
             ____
              7 30
              7 11
              ____
                190
                158
```

Example 2: At the start of the month you had a checking account balance of $25.36. During the month you made a deposit of $245.08 and wrote checks for $12.75, $25.00, $132.50, $18.64, and $5.76. What was the balance in your account at the end of the month?

Analyze: You must add the checks and subtract that amount from the total of the balance and deposits.

Solve:

	checks:	$ 12.75	initial balance:	$ 25.36
		25.00	deposit:	+ 245.08
		132.50		$270.44
		18.64	checks:	− 194.65
	+	5.76	final balance:	$ 75.79
		$194.65		

Thus the checking account balance was $75.79.

Example 3:   Hurts Car Rental charges $16.75 a day and 16.5¢ a mile. How much would you pay for one day if you drove 225 miles?

Analyze: Find the cost for the miles driven by multiplying 16.5¢ times 225 miles. Add that result to $16.75.

Solve: 16.5¢ = $.165 (changing the cents to dollars)

```
      mileage    $.165          total    $16.75
               x  225           cost:   + 37.13
                  825                    53.88
                  330
                  330
               $37.125 ≐ $37.13 (rounded off the the nearest
                                 cent-two decimal places)
```

Example 4:   You work for a department store and earn $4.56 per hour and time-and-a-half for overtime. If during one week you work 40 regular hours and 12 overtime hours, what are your earnings for the week?

Analyze:

1. Get regular earnings by multiplying 40 x $4.56.

2. Determine hourly overtime wage. Time-and-a-half means your hourly wage plus ½ of that wage. You can find that overtime wage by multiplying the regular wage by 1.5.

3. Get overtime earnings by multiplying 12 x $6.84.

4. Add the regular and overtime earnings together.

Solve:

1. 40 x $4.56 = $182.40

2. 1.5 x $4.56
   = $6.84 per hour

3. 12 x $6.84 = $82.08

4. $182.40 + $82.08
   = $264.48

Example 5:   At the beginning of the week when you filled up your gas tank your odometer read 52,756.8. At the end of the week you filled up again. Your odometer then read 53,172.1 and you put in 18.2 gallons of gas. How many miles was your car traveling for each gallon of gas, to the nearest tenth?

Analyze:

1. Determine the number of miles driven by subtracting the odometer readings.

2. Miles for each gallon implies that you divide the number of miles by the number of gallons.

Solve:

1.   53172.1
   - 52756.8
     415.3

2. 415.3 ÷ 18.2 ≐ 22.8

EXERCISE 4-12   SET A

NAME _____   DATE _____

ANSWERS

1. You started the week with $22.56 balance in your checking account. During the week you wrote checks for $47.35, $10.00, $239.50, and $22.98. You also made a deposit of $415.25. What was your balance at the end of the week?   1. _____

2. What was the balance in your checking account after beginning with $56.37, writing checks for $67.89, $45.00, $20.50, $176.67, and $8.75, and depositing $405.67?   2. _____

3. Which is the better buy: a $62.95 tire that lasts for 50,000 miles or the same type of tire costing $57.50 that lasts for 40,000 miles.   3. _____

4. Which is the better buy: 7 oz of toothpaste priced at $1.37, or 5.5 oz of the same brand priced at $1.04?   4. _____

5. The odometer reading in your car when you filled it up was 39,247.1. The next time you filled up your car it read 39,762.4 and you put in 18.3 gallons of gas. How many miles was your car traveling per gallon?   5. _____

6. Find the miles you travel per gallon of gasoline if the odometer readings between two consecutive fill-ups were 24,476.5 and 24,977.8 and the car used 20.9 gallons of gas.   6. _____

7. During a week you work 40 regular hours and 7 overtime hours. If your regular wage is $4.96 per hour and you get time-and-a-half for overtime, what is your income for the week?   7. _____

8. If you earn $4.20 per hour and time-and-a-half for overtime what would you earn in a week in which you work 25 regular hours and 5 overtime hours?   8. _____

9. If a long distance phone call costs $.85 for the first 3 minutes and $.17 for each additional minute, how much does a 30 minute call cost?   9. _____

10. On your utility bill, you are charged $.0425 per kilowatt-hour of electricity and $.197 for each thermal unit of natural gas. If you use 250 kilowatt-hours of electricity and 180 thermal units of natural gas, what is your bill?   10. _____

11. If cigarettes cost $.95 a pack and Roland smokes 2 packs a day, how much will he spend on cigarettes in one year (365 days)?   11. _____

12. If term life insurance costs $6.78 per $1000 of coverage each year, how much would a $25,000 policy cost a year?   12. _____

EXERCISE 4-12   SET B

NAME _____   DATE _____

                                                        ANSWERS

1. What was the balance in your checking account, after
   beginning with $65.45, writing checks for $30.00, $23.67,
   $122.75, $7.50, $8.99, and $17.36, and depositing $285.88?   1. _____

2. At a drug store you buy the following: 2 bottles of multiple
   vitamins at $3.95 a bottle, 3 toothbrushes at $1.07 each,
   and 4 boxes of cough drops costing $.49 each. If the sales
   tax on the purchase is $.91, what is the total cost?   2. _____

3. By comparing the cost per ounce, determine which is the better
   buy: a 12.5 oz can of soup priced at 63¢, or a 10.75 oz can
   of the same soup priced at 55¢.   3. _____

4. If gasoline costs 47.2¢ per liter and you use 160 liters
   each month, how much do you spend on gas in a year?   4. _____

5. Find the miles you travel per gallon of gasoline if the
   odometer readings between two consecutive fill-ups were
   19,876.4 and 20,259.6 and the car used 19.9 gallons of gas.   5. _____

6. During a year you spent the following on your car: gas and
   oil, $1490.26; minor repairs, $295.47; and tires, $275.75.
   If you drove 19,500 miles that year, what was the cost of
   driving your car for a mile?   6. _____

7. If you earn $5.38 per hour and time-and-a-half for overtime,
   what would you earn in a week in which you work 40 regular
   hours and 11 overtime hours?   7. _____

8. The rainfall measured in inches in the San Francisco Bay
   Area usually looks like this: Jan. (4.4), Feb. (3.0),
   Mar. (2.5), Apr. (1.6), May (0.4), Jun. (0.1), Jul. (0.0),
   Aug. (0.0), Sept. (0.2), Oct. (1.0), Nov. (2.3), Dec. (4.0).
   What is the average monthly rainfall in the San Francisco
   Bay area?   8. _____

9. If a long distance call costs $1.25 for the first 3 minutes
   and $.32 for each additional minute, how much does a 45
   minute call cost?   9. _____

10. If your family drinks a gallon of milk a day and milk costs
    $1.95 a gallon, how much do you spend on milk in a year?   10. _____

11. If the yearly cost for $1000 of term life insurance is
    $7.16, how much would a $35,000 policy cost a year?   11. _____

12. You buy 2 3/4 yards of material at $4.80 per yard and pay
    $.87 sales tax. How much change do you get from a $20 bill? 12. _____

4-13 Applications Involving Perimeters and Areas

"MAY I HAVE DISTANCE."

Perimeter

The word **perimeter** means the measure around an object. So to find the perimeter of a 2-dimensional figure, you must find the total distance around the figure. For example:

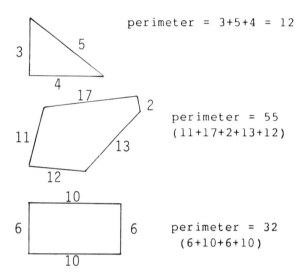

perimeter = 3+5+4 = 12

perimeter = 55
(11+17+2+13+12)

perimeter = 32
(6+10+6+10)

The figure in the last example is a rectangle. The opposite sides of a rectangle have the same measurement. Therefore, in problems that involve rectangles, only two of the measurements are given, the length and the width. For example, what is the perimeter of a 46" by 27" rectangle?

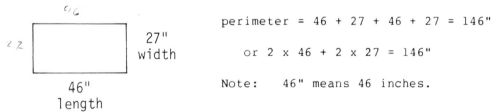

perimeter = 46 + 27 + 46 + 27 = 146"

or 2 x 46 + 2 x 27 = 146"

Note: 46" means 46 inches.

Area

The **area** of a figure is the measure of the amount of space in the interior of the figure. Areas are measured in square units, such as square inches or square feet. The area simply tells you how many of those square units can be placed inside the figure. If you count the number of squares in the rectangle below you will see that it has an area of 12 square units.

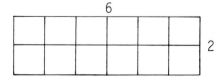

12 square units can be placed inside a 6 by 2 rectangle. Its area is 12 square units.

If every time you wanted to find the area of a rectangle, you had to draw and count squares, it could take a lot of time. There is an easy method for finding the **area of a rectangle**:

> **AREA = LENGTH x WIDTH**

Thus a 6 by 2 rectangle has an area of 6 x 2 = 12 square units. A 46" by 27" rectangle has an area of 46 x 27 = 1242 square inches.

Example 1: You are painting two walls that are 9' by 16' and two walls that are 9' by 20'. (9' means 9 feet.) If a gallon of paint covers 350 sq ft, how many gallons should you buy?

Analyze:

1. Find the total wall area by adding the areas of the four walls.

2. Find the number of gallons needed by dividing the total area by 350.

3. **Since you can't buy 1.85 gallons, you will need 2.**

Solve:

1. 9 x 16 = 144 sq ft
   9 x 16 = 144 sq ft
   9 x 20 = 180 sq ft
   9 x 20 = <u>180 sq ft</u>
             648 sq ft

2. 648 ÷ 350 = 1.85

3. 1.85 = 2 gallons

Example 2: Molding to make a picture frame costs $4.75 a foot. The glass to cover the picture costs $.84 a square foot. What would it cost for the molding and the glass to make an 18" by 18" frame?

Analyze:

1. Find the cost of the molding by multiplying the perimeter by $4.75.

2. Find the cost of the glass by multiplying the area by $.84.

3. Find the total cost by adding the cost of the molding and the glass.

Solve: (Note: 18" = 1½ ft)

1. perimeter: = 1½+1½+1½+1½
             = 6'
   cost: 6 x $4.75 = $28.50

2. area: = 1½ x 1½ = 1.5 x 1.5
       = 2.25 sq ft
   cost: 2.25 x $.84 = $1.89

3. $28.50
   <u>+ 1.89</u>
   $30.39

PROBLEM 1: How much will it cost to fence a 56' by 70' yard and plant the inside with sod (grass), if the fencing costs $3.95 a foot and the sod costs $.49 a square foot?

Answer: $2916.20
fence: 56 + 70 + 56 + 70 = 252'
       252 x $3.95 = $995.40
sod: 56 x 70 = 3290 sq ft
     3920 x $.49 = $1920.80
total: $995.40 + $1920.80
      = $2916.20

EXERCISE 4-13 SET A

NAME _____   DATE _____

ROUND DOLLAR AMOUNTS TO THE NEAREST CENT: ANSWERS

1. [rectangle 9' by 4']   Find the perimeter and area.   1. _____

2. [square 15" by 15"]   Find the perimeter and area.   2. _____

3. How much would it cost to carpet a 9' by 12' room if the
    complete cost for the carpet is $23.65 a square yard?   3. _____

4. How much would it cost to carpet a 12' and 15' room if the
    complete cost for the carpet is $19.99 a square yard?   4. _____

5. You are fixing up your rectangular shaped back yard which
    is 30 ft. by 33 ft. You are going to build a fence competely
    around the area and cover the ground with sod. If the
    fencing costs $4.15 a foot and the sod costs $.39 a square
    foot, what will the project cost you?   5. _____

6. Molding to make a fancy picture fra[me costs $]4.65 a foot.
    The glass to cover the picture cos[ts ...]
    How much would it cost for the mol[ding and glass for]
    a 42" by 36" frame?

7. If floor tile costs $.49 a square [foot, how much would it]
    cost to tile a 9' by 11' kitchen[?]

8. If mirror tile costs $.99 a squar[e foot, how much would it]
    cost to tile a strip of wall tha[t ...]

9. How much would it cost to make [the foundation for]
    a house, if the cost of constru[ction is $... a square]
    foot?

10. How much would it cost to lay a [... square foot]
    patio, if it costs $2.35 per sq[uare foot?]

11. An acre is 43,560 square feet. [If an acre of property]
    costs $32,000, what is the cos[t ...]

EXERCISE 4-13   SET B

NAME _____   DATE _____

ROUND DOLLAR AMOUNTS TO THE NEAREST CENT: ANSWERS

1. [rectangle 12" × 7"]   Find the perimeter and area.   1. _____

2. [square 18' × 18']   Find the perimeter and area.   2. _____

3. How much would it cost to put linoleum in a 9' by 15'
   kitchen, if linoleum costs $11.99 a square yard?   3. _____

4. How much would it cost to cover a 24' by 19' area with
   Astro-turf, if Astro-turf costs $1.56 a square foot?   4. _____

5. You are enclosing a 46' by 52' play area with a redwood
   fence costing $5.12 a foot. If you also cover the area
   with gravel that costs $.17 a square foot, what is the
   total cost of the project?   5. _____

6. The framing material to make a 60" by 36" window frame
   costs $2.25 a foot. If the glass costs $.95 a square
   foot, how much will the window cost?   6. _____

7. If ceramic tile costs $1.37 a square foot, how much will
   it cost to tile a 6½' by 10¼' entry hall?   7. _____

8. If carpet tiles cost $1.29 a square foot, how much will
   it cost to carpet two 8' by 10' bedrooms?   8. _____

   ow much would it cost to build a 24' and 36' second
       v addition to a home, if the cost of construction
         58 a square foot?   9. _____

            uld it cost to replace a 6' by 8' piece of
               iding glass door if the glass costs $1.17
                  nd the labor charge is $17.25?   10. _____

                  square feet, how many acres are in a
                     is 576 feet by 363 feet? If the
                        is $1398, how much will that
                                                              11. _____

Chapter 4 Summary

<u>Concepts</u>

You may refer to the sections listed below to review how to:

1. read and write decimal numbers using the place values to the right of the decimal point. (4-1)

2. round off decimal numbers using the "five or more" rule. (4-2)

3. add and subtract decimal numbers by first lining up their decimal points. (4-3), (4-4)

4. multiply decimal numbers by positioning the decimal point according to the total number of decimal digits in the numbers being multiplied. (4-5)

5. divide decimal numbers by moving the decimal points to make the divisor a whole number. (4-6), (4-7)

6. use short-cuts for multiplying and dividing numbers that end in zeros. (4-8)

7. convert fractions to decimals by dividing the numerator by the denominator. (4-9)

8. convert decimals to fractions by making use of the way decimal numbers are read in our number system. (4-9)

9. compare the size of decimal numbers by writing each with the same number of decimal places. (4-10)

10. solve problems involving operations on both fractions and decimals by working the problem using only fractions, only decimals, or a combination of the two. (4-11)

11. solve word problems involving decimals. (4-12), (4,13)

12. find perimeters by adding the lengths of sides. (4-13)

13. find areas of rectangles by multiplying length times width. (4-13)

Terminology

This chapter's important terms and their corresponding page numbers are:

area: the measure of the number of square units that covers the interior of a 2-dimensional closed figure. (219)

decimal digit: a digit to the right of the decimal point. (187)

decimal number: a number containing a decimal point. (171)

decimal place: the position of a digit to the right of the decimal point. (175)

decimal point: a dot (.) between two digits separating the whole number part and the fractional part of a number. (171)

perimeter: the measure of the outer boundary of a 2-dimensional closed figure. (219)

Student Notes

Chapter 4 Practice Test A

ANSWERS

1. Write the following using numbers:
   four hundred seven and seven thousandths.   1. _____

2. Round off 746.325 to the nearest hundredth.   2. _____

3. 45.6 + 0.0017 + 75 + 197.7 + 3.12   3. _____

4. 65.3 - 27.296   5.   45.07 x 17.96   4. _____

                                                      5. _____

6. 341.22 ÷ 4.84   6. _____

7. Find 59 ÷ 5.293 rounded to the nearest hundredth. 7. _____

8. 4.862 x 500,000   9.   573.3 ÷ 3,000,000   8. _____

                                                      9. _____

10. Express 0.026 as a reduced fraction.   10. _____

11. Express $2\frac{5}{16}$ as an exact decimal.   11. _____

12. Arrange in order from largest to smallest:
    1.27, 1.0127, 1.1027, 1.2   12. _____

13. $5\frac{5}{6}$ x 4.2 + $(3\frac{2}{3})^2$ 13. _____

14. You started the week with a $58.23 balance in your
    checking accoung. During the week you wrote checks for
    $10.25, $115.16, $74.00, $3.95, and $8.99, and deposited
    $215.66. What was your balance at the end of the week?   14. _____

15. What would it cost to fence in a 20' by 12' play area and
    cover it with gravel, if the wire fencing costs $3.85
    a foot and the gravel costs $.47 a square foot?   15. _____

Chapter 4 Practice Test B

ANSWERS

1. Write 84.56 in words.   1. _____

2. Round off 56.374 to the nearest tenth.   2. _____

3. 1.25 + 756 + .0018 + 34.76 + 152.5   3. _____

4. 57.5 − 19.458   5. 87.6 × 14.5   4. _____

                                                        5. _____

6. 15.264 ÷ 5.76   6. _____

7. Find 45 ÷ 6.237 rounded to the nearest hundredth.   7. _____

8. 9.2165 × 4,000,000   9. 476.5 ÷ 500,000   8. _____

                                                        9. _____

10. Express 0.38 as a reduced fraction.   10. _____

11. Express $5\frac{3}{8}$ as an exact decimal.   11. _____

12. Arrange in order from largest to smallest:
    0.306, 0.3, 0.036, 0.0306   12. _____

13. $(2\frac{1}{3})^3 + 4\frac{1}{6} \times 9.6$ 13. _____

14. How much would a 24 minute long-distance phone call cost,
    if you pay $1.27 for the first 3 minutes and $.26 for each
    additional minute?   14. _____

15. What is the cost to fence a 32' by 17' area and plant
    the inside with sod, if the fencing costs $6.98 a foot
    and the sod costs $.69 a square foot?   15. _____

Chapter 4 Supplementary Exercises

Section 4-1

Write in words:

1. 0.7
2. 5.4
3. 0.03
4. 16.01
5. 0.12
6. 3.035
7. 0.006
8. 122.07
9. 0.123
10. 2.0031

Write using numbers:

11. six tenths
12. seven hundredths
13. five and two thounsandths
14. sixty-four and sixteen thousandths
15. nine and eighty ten thousandths

Section 4-2

Round off to the nearest hundredth:

1. 5.732
2. 5.735
3. 5.738
4. 16.405
5. 16.403
6. 16.407
7. 141.9531
8. 0.7675
9. 452.1234
10. 67.095
11. 89.997
12. 89.993

Section 4-3

1. 4.3 + 7.6
2. 0.21 + 8.7
3. 6.38 + 75
4. 89 + 1.003
5. 7.65 + 13.74
6. 8.07 + 3.025
7. 1.0134 + 12 + 0.276
8. 17.6 + 18 + 19.007
9. 3.2 + 0.0012 + 23 + 570
10. 327 + 9.876 + 5.6 + 4.32
11. 47 + 8.06 + 2.763 + 4.5
12. 12.32 + 4.57 + 10.9 + 865

Section 4-4

1. 9.79 − 4.35
2. 6.878 − 5.193
3. 45.75 − 17.6
4. 124.127 − 79.07
5. 56.318 − 27.5
6. 6.007 − 2.03
7. 720.24 − 46.518
8. 6.5 − 3.824
9. 16.72 − 12.0231
10. 826 − 176.285
11. 42 − 21.76
12. 1763 − 1.763
13. 150.008 − 37.0992
14. 17 − 0.0329
15. 500.5 − 50.095

Section 4-5

1. 9.5 x 6
2. 8.75 x 3
3. 4.5 x 6.8
4. 42.13 x 5.06
5. 7.18 x 3.2
6. 20.24 x 0.765
7. 173.5 x 57.22
8. 5327.8 x 1.015
9. 47.8 x .0487
10. 423.001 x 6.07
11. 56.4 x 83.005
12. 64.73 x 42.75
13. 1624 x 0.0005
14. 99 x 0.065
15. 0.0132 x 7000

Section 4-6

Find the exact quotients:

1. 105 ÷ 6
2. 7 ÷ 8
3. 32.93 ÷ 3.7
4. 52 ÷ 1.3
5. 75.6 ÷ .0012
6. 107.6 ÷ 2.152
7. 45.7 ÷ 0.032
8. 6.9408 ÷ 5.76
9. 1.1016 ÷ .017

Section 4-7

Round off answers to the nearest tenth:

1. 15.7 ÷ 9
2. 23 ÷ 91
3. 6.293 ÷ 4.37
4. 459 ÷ 3.7
5. 12 ÷ 1.093
6. 43.5 ÷ 2.39
7. 47.9 ÷ 0.19
8. 4139.6 ÷ 24.5
9. 98.123 ÷ 23
10. 21.91 ÷ 2.017
11. 6.4352 ÷ 8.6
12. 178 ÷ 36.91

Section 4-8

Find the exact answers:

1. 5.7 x 100
2. 37.86 x 30,000
3. 13.8 x 10^5
4. 0.098 x 73,000
5. 463.187 x 21,000
6. 0.707 x 4,000,000
7. 5.7 ÷ 100
8. 29746.2 ÷ 3000
9. 948.5 ÷ 35,000
10. 161 ÷ 23,000
11. 28.9 ÷ 10^3
12. 76586 ÷ 200,000

Section 4-9

Change to exact decimals:

1. 3/4
2. 3/10
3. 7/8
4. 14/5
5. 5/32
6. 27/16

Change to reduced fractions:

7. 0.25
8. 0.8
9. 0.015
10. 0.228
11. 0.0002
12. 0.0075

Section 4-10

Arrange in order from largest to smallest:

1. 0.5, 0.505, 0.55, 0.5055
2. 1.7, 1.701, 1.07, 1.7001
3. 3.3, 3.33, 3.13, 3 1/3
4. 3/4, 19/25, .756, .0765

Section 4-11

1. $12.7 \times 3\frac{1}{4}$
2. $14\frac{1}{2} \div 3.7$
3. $7^2 \times 2\frac{1}{4} + 8.7$
4. $4.6 \times (2\frac{1}{3} - \frac{5}{6})$
5. $15\frac{1}{2} \times 0.89 - 3$
6. $(1\frac{5}{6})^2 + 4 \times 8.6$
7. $8\frac{2}{3} + 4.25 \div 3.4$
8. $47.75 + 8\frac{3}{4} \times 7\frac{1}{5}$
9. $156 - 15\frac{7}{8} - 3.125$

Section 4-12

1. What is the balance in your checking account if you start with a balance of $147.62 and write checks for $5.76, $13.85, and $124.99?

2. What is the cost per ounce of cheese that sells for $1.90 for 7.6 oz?

3. If Arts Rent-a-Car charges $14.99 a day and $.17 a mile, what would you pay if you drive 317 miles in two days?

4. If soft drinks cost $1.97 a six-pack and you drink 18 cans a month, how much would you spend on soft drinks a year?

5. At the start of a trip your car odometer reads 65,176.8 and at the end of the trip it reads 66,098.3. If you use 46 gallons of gas during the trip, how many miles does your car travel per gallon?

6. Working for Benny's Burgers you make $3.90 per hour plus time-and-a-half for overtime. If during one week you work 40 regular hours and 5 overtime hours, what is your total earnings for the week?

7. The Alcat Company's budget shows that $.34 out of each dollar is spent on wages. How much is spent on wages in a $317,000 budget?

8. The price for the same product at four different stores is $1.17, $1.58, $1.46, and $1.07. What is the average price charged for that item?

9. Which is the better buy: 4 ounces of instant coffee selling for $3.39 or 6 ounces of the same coffee selling for $4.99?

10. Which is the better buy: a 6 ounce container of mushrooms selling for $.79 or loose mushrooms of the same quality selling for $2.19 a pound?

Section 4-13

1. If carpet costs $24.99 a square yard, how much would it cost to carpet a room that is 12' by 12'?

2. What would the molding cost to make an 8" by 10" picture frame, if the molding costs $1.26 per foot?

3. Find the area and perimeter of a lot that is 150' by 225'.

4. If the lot in problem 3 costs $2.99 a square foot, how much would the entire lot cost?

5. What is the total price for a 3.2 acre parcel, if it sells for $1.75 a square foot? (1 acre = 43,560 sq ft)

6. How much would it cost to replace a 4' by 5' window, if the glass costs $1.29 a square foot and the labor charge is $22.75?

7. If a gallon of paint covers 450 square feet of surface area, how many gallons of paint should you buy to paint two 12' by 37' walls and two 12' by 56' walls?

Unit II Exam

Chapters 3 and 4 ANSWERS

1. Express $\dfrac{196}{42}$ as a reduced mixed number. 1. _____

2. Write 152.037 in words. 2. _____

3. Express $\dfrac{45}{77}$ as a decimal rounded to the nearest hundredth. 3. _____

4. $\dfrac{5}{8} \times \dfrac{14}{45}$ 5. $6\dfrac{3}{4} \div 4\dfrac{1}{2}$ 4. _____

   5. _____

6. $\dfrac{3}{8} - \dfrac{1}{10} + \dfrac{5}{6}$ 7. $27\dfrac{2}{3} - 19\dfrac{4}{5}$ 6. _____

   7. _____

8. $\dfrac{6 - \dfrac{3}{4}}{5\dfrac{1}{3} + \dfrac{1}{2}}$ 9. $8.75 + 84 + 107.3 - 3.674$ 8. _____

   9. _____

10. $17.36 \times 15{,}000$ 11. $1.3314 \div 0.015$ 10. _____

    11. _____

12. Three-fifths of the students in a math class did all the assignments during the semester. If 35 students are in the class, how many did not do all the assignments? 12. _____

13. A board, $11\dfrac{1}{2}$ feet long, is cut into 3 equal pieces. How long is each piece? 13. _____

14. The balance owed on a loan is $6356.16. After 12 payments of $176.56, how much is still owed on the loan? 14. _____

15. A pad of scratch paper contains 100 sheets of 4.25 by 5.5 inch paper. If you use both sides of each sheet, how much writing area does the scratch pad provide? 15. _____

Unit III

Ratio, Percent, Measurement, and Graphs

In the previous two units you have learned the essentials of basic arithmetic. You have learned how to do computations that involve whole numbers, fractions, and decimals. You have also applied that knowledge to the solution of various kinds of real life problems.

In this unit you will discover there are many other everyday problems which can be solved with further knowledge of arithmetic.

In this unit you will learn to work with ratios, proportions, and percents. You will learn how to read measuring devices and convert between U.S. and metric units. You will learn how to read and analyze statistical graphs. After finishing this unit you should be able to do the following:

1. set up ratios and solve porportions
2. solve percentage problems.
3. apply percents and proportions to real life problems.
4. work with the U.S. and metric systems of measurement.
5. analyze measuring devices and statistical graphs.

Note: Unless otherwise indicated in this unit, when the answers have more than three decimal digits, they will be rounded off to the nearest hundredth.

Unit III

Brain Buster

Consider a very large piece of paper that is 0.01 inches thick. If you continue to fold that piece of paper in half, how high will the stack be after 25 such folds? Express the answer to the nearest tenth of a mile.

(Yes! The height of the stack would be much more than a mile.)

Chapter 5

Ratio, Proportion, and Percent

After completing this chapter you should be able to do the following:

1. set up and reduce ratios.
2. solve proportions.
3. express percents as fractions or decimals.
4. convert decimals and fractions to percents.
5. solve percentage problems.
6. solve word problems involving percents and proportions.

On the next page you will find a pretest for this chapter. The purpose of the pretest is to help you determine which sections in this chapter you need to study in detail and which sections you can review quickly. By taking and correcting the pretest according to the instructions on the next page you can better plan your pace through this chapter.

Note: Unless otherwise indicated in this chapter, when the answers have more than three decimal digits, they will be rounded off to the nearest hundredth.

Chapter 5 Pretest

Take and correct this test using the answer section at the end of the book. Those problems that give you difficulty indicate which sections need extra attention. Section numbers are in parentheses before each problem.

ANSWERS

(5-1) 1. Express a ratio of 63 to 105 as a reduced fraction.

1. _____

(5-2) 2. Solve for N: $\dfrac{N}{30} = \dfrac{28}{35}$

(5-2) 3. Solve for A: $\dfrac{39}{9} = \dfrac{A}{45}$

2. _____

3. _____

(5-3) 4. Solve for P: $\dfrac{16}{7} = \dfrac{9}{P}$

(5-3) 5. Solve for Y: $\dfrac{.83}{Y} = \dfrac{2.52}{26.6}$

4. _____

5. _____

(5-5) 6. Express 4.5% as a decimal and as a reduced fraction.

6. _____

(5-6) 7. Express .114 as a reduced fraction and as a percent.

7. _____

(5-7) 8. Find 14.9% of 300.

8. _____

(5-7) 9. How much sales tax would be added to the cost of a $79.99 watch, if the sales tax rate is 6.5%? Find the tax accurate to the nearest cent.

9. _____

(5-8) 10. 4 is what percent of 25?

10. _____

(5-8) 11. 9 is 12% of what number?

11. _____

(5-8) 12. What is $32\tfrac{2}{3}$% of $520.80?

12. _____

(5-4) 13. If 3 1/4 pounds of oranges cost $1.18, how much should 5 pounds cost, to the nearest cent?

13. _____

(5-9) 14. If the price of a ring drops from $176 to $97, what is the percent decrease in price, to the nearest tenth?

14. _____

(5-10) 15. What monthly payments are needed to pay off a $2400 loan with an annual interest rate of 11.7% for 42 months. Find the payments rounded to the nearest cent.

15. _____

5-1 Ratios

If there were 18 males and 12 females in a class, you could compare the number of men to women by saying there is a **ratio** of 18 men to 12 women. You could represent that comparison in three different ways:

$$18 \text{ to } 12$$

$$18 : 12$$

$$\frac{18}{12}$$

The ratio of 18 to 12 is another way to represent the fraction 18/12. The three representations above are equal.

$$18 \text{ to } 12 = 18 : 12 = \frac{18}{12}$$

Depending on the situation, any of those three forms can be used. *A ratio, then, is simple a comparison of numbers that can be expressed as a fraction.*

I DO MY HOMEWORK LISTENING TO MY RATIO.

The first operation to perform on ratios is reducing them to lowest terms. The above ratio, 18 : 12, can be reduced to lower terms just as we reduced fractions.

$$18 : 12 = \frac{18}{12} \begin{smallmatrix} \div 6 \\ \\ \div 6 \end{smallmatrix} \frac{3}{2}$$

So $18 : 12 = \frac{3}{2} = 3 : 2$

By reducing the ratio you can get a better understanding of the original ratio. The ratio of 18 males to 12 females reduces to a ratio of 3 males to 2 females. This means that for every 3 males in the class there are 2 females.

Example 1: Express the ratio of the value of 4 dimes to 3 quarters.

You must first change each to the same unit (cents).

4 dimes = 40¢ and 3 quarters = 75¢

Thus the ratio of the respective values is:

$$40 : 75 = \frac{40}{75} = \frac{8}{15}$$

235

Example 2: A basketball team wins 16 games and loses 14 games. Find the reduced ratio of:

wins to losses. $16 : 14 = \dfrac{16}{14} = \dfrac{8}{7}$

losses to wins. $14 : 16 = \dfrac{14}{16} = \dfrac{7}{8}$

wins to total games played. $16 : 30 = \dfrac{16}{30} = \dfrac{8}{15}$

Notice the order of the numbers is critical.

Example 3: Find the reduced ratio of $5\frac{2}{3}$ to 2.5.

$$5\frac{2}{3} : 2.5 = \frac{5\frac{2}{3}}{2.5} = \frac{5\frac{2}{3}}{2\frac{1}{2}}$$

$$= \frac{\frac{17}{3}}{\frac{5}{2}}$$

$$= \frac{17}{3} \div \frac{5}{2}$$

$$= \frac{17}{3} \times \frac{2}{5}$$

$$= \frac{34}{15}$$

If you keep in mind that a ratio is simply a comparison of quantities expressible as a fraction, you should be able to do the following problems.

PROBLEM 1: Five people have to split $65 between them.

    a. What is the ratio of money to person?

    b. What does the ratio tell you?

Answers:

a. 13:1 (reduced from 65:5)

b. Each person should get $13.

PROBLEM 2: What is the reduced ratio $3\frac{1}{4}$ inches to 3 feet?

Answer: 13:144

(Note: 3 ft = 36 in.)

$$3\tfrac{1}{4} : 36 = \frac{3\tfrac{1}{4}}{36} = 3\tfrac{1}{4} \div 36$$

$$= \frac{13}{4} \div \frac{36}{1}$$

$$= \frac{13}{4} \times \frac{1}{36} = \frac{13}{144}$$

EXERCISE 5-1   SET A

NAME _____   DATE _____

A TEAM WINS 8 GAMES OUT OF 12 GAMES PLAYED.   ANSWERS
WHAT IS THE REDUCED RATIO OF:

1. wins to games played?   1. _____

2. wins to losses?   2. _____

3. losses to games played?   3. _____

A JAR CONTAINS 12 WHITE, 10 RED, AND 18 BLUE BALLS.
WHAT IS THE REDUCED RATIO OF:

4. white balls to blue balls?   4. _____

5. red balls to the total number of balls?   5. _____

6. blue balls to balls that are not blue?   6. _____

A BANK CONTAINS 7 DIMES, 20 NICKELS, AND 9 QUARTERS.
WHAT IS THE REDUCED RATIO OF THE VALUE OF:

7. nickels to dimes?   7. _____

8. dimes to nickels?   8. _____

9. dimes to quarters?   9. _____

10. quarters to total money in the bank?   10. _____

WHAT IS THE REDUCED RATIO OF THE FOLLOWING DISTANCES:

11. 18 inches to 45 inches?   11. $2/5$

12. 5 inches to 5 feet?   12. $1/1$

13. 8 feet to 10 yards?   13. $1/5$

14. $11\frac{3}{8}$ inches to $8\frac{1}{2}$ inches?   14. _____

15. $3\frac{3}{4}$ inches to $10\frac{1}{2}$ inches?   15. _____

16. 9 feet to $3\frac{3}{8}$ inches?   16. _____

17. $20\frac{1}{4}$ inches to 3 yards?   17. _____

EXERCISE 5-1   SET B

NAME _____   DATE _____

IF A CLASS OF 40 CONTAINS 18 FEMALES,   ANSWERS
WHAT IS THE REDUCED RATIO OF:

1. females to the total in the class?   1. _____

2. females to males?   2. _____

3. males to the total in the class?   3. _____

A JAR CONTAINS 8 WHITE, 16 RED, AND 6 BLUE BALLS.
WHAT IS THE REDUCED RATIO OF:

4. white balls to red balls?   4. _____

5. red balls to the total number of balls?   5. _____

6. blue balls to balls that are not blue?   6. _____

A WALLET CONTAINS 12 $1 BILLS, 4 $5 BILLS, AND 3 $20 BILLS.
WHAT IS THE REDUCED RATIO OF THE VALUE OF:

7. $1 bills to $5 bills?   7. _____

8. $5 bills to $1 bills?   8. _____

9. $5 bills to $20 bills?   9. _____

10. $20 bills to the total money in the wallet?   10. _____

WHAT IS THE REDUCED RATIO OF THE FOLLOWING DISTANCES:

11. 14 inches to 35 inches?   11. _____

12. 3 inches to 3 feet?   12. _____

13. 9 feet to 12 yards?   13. _____

14. $12\frac{1}{2}$ inches to $9\frac{3}{8}$ inches?   14. _____

15. $5\frac{1}{4}$ inches to $9\frac{7}{8}$ inches?   15. _____

16. 2 feet to $2\frac{5}{8}$ inches?   16. _____

17. $13\frac{1}{2}$ inches to 3 yards?   17. _____

5-2 Proportions

A **proportion** is a statement that one ratio is equal to another ratio.

For example, a ratio of 4 : 8 is equal to a ratio of 3 : 6.

$$4 : 8 = \frac{4}{8} = \frac{1}{2} \quad \text{and} \quad 3 : 6 = \frac{3}{6} = \frac{1}{2}$$

$$4 : 8 = 3 : 6$$

$$\frac{4}{8} = \frac{3}{6}$$

Those ratios form a proportion since they are equal to each other. In a proportion you will notice that if you **cross multiply** the terms of a proportion as shown below, those **cross-products** are equal.

$\frac{4}{8} = \frac{3}{6}$ 4 x 6 = 8 x 3 (both equal 24)

$\frac{3}{2} = \frac{18}{12}$ 3 x 12 = 2 x 18 (both equal 36)

This, then, is the **fundamental principle** for working with proportions:

> If you cross multiply in a proportion, both answers are equal.

Example 1: Do $\frac{4}{7}$ and $\frac{36}{63}$ form a proportion?

Cross multiply the ratios. $\frac{4}{7} = \frac{36}{63}$ 4 x 63 = 252
                                                                  7 x 36 = 252

The cross-products are equal, so the ratios form a proportion.

That fundamental principle of proportions enables you to solve problems in which one number of the proportion is not known. For example, if N represents the number that is unknown in the proportion below, we can find its value.

$$\frac{N}{12} = \frac{3}{4}$$

4 x N = 12 x 3   Cross multiply the proportion.

4 x N = 36   You know that N = 9 since 4 x 9 = 36.
                       However, other problems may not be so
                       easy. You should learn this method
                       for finding the value for N:

$\frac{4 \times N}{4} = \frac{36}{4}$ *Divide the terms on both sides of the equals sign by the number next to the unknown letter.*

1 x N = 9   In this problem we would divide both sides by 4

N = 9   That will leave the N on the left side
                       and the answer (9) on the right side.

239

Example 2: Solve for N.

$$\frac{2}{5} = \frac{N}{35}$$

5 x N = 2 x 35   Cross multiply.

5 x N = 70

$$\frac{5 \times N}{5} = \frac{70}{5}$$ Divide by the number next to the unknown letter; divide by 5.

N = 14

Example 3: Solve for N.

$$\frac{15}{N} = \frac{3}{13}$$

3 x N = 15 x 13   Cross multiply.

3 x N = 195

$$\frac{3 \times N}{3} = \frac{195}{3}$$ Divide by the number next to the unknown letter; divide by 3.

N = 65

PROBLEM 1: Solve for N.

$$\frac{6}{7} = \frac{102}{N}$$

Answer: N = 119

$$\frac{6}{7} = \frac{102}{N}$$

$$\frac{6 \times N}{6} = \frac{714}{6}$$

N = 119

PROBLEM 2: Solve for N.

$$\frac{4}{N} = \frac{6}{27}$$

Answer: N = 18

$$\frac{4}{N} = \frac{6}{27}$$

$$\frac{6 \times N}{6} = \frac{108}{6}$$

N = 18

EXERCISE 5-2 SET A

NAME _____   DATE _____

DO THE FOLLOWING FORM PROPORTIONS?   ANSWERS

1. $\frac{12}{21}$ and $\frac{8}{14}$ 2. $\frac{24}{20}$ and $\frac{12}{10}$ 1. _____

2. _____

3. $\frac{57}{38}$ and $\frac{114}{76}$ 4. $\frac{138}{184}$ and $\frac{69}{92}$ 3. _____

4. _____

5. $\frac{73}{53}$ and $\frac{53}{33}$ 6. $\frac{34}{62}$ and $\frac{26}{43}$ 5. _____

6. _____

SOLVE FOR N IN EACH PROPORTION:

7. $\frac{N}{4} = \frac{6}{8}$ 8. $\frac{N}{3} = \frac{12}{9}$ 7. _____

8. _____

9. $\frac{20}{N} = \frac{4}{7}$ 10. $\frac{21}{N} = \frac{3}{8}$ 9. _____

10. _____

11. $\frac{18}{15} = \frac{N}{100}$ 12. $\frac{27}{6} = \frac{N}{100}$ 11. _____

12. _____

13. $\frac{65}{52} = \frac{5}{N}$ 14. $\frac{49}{91} = \frac{7}{N}$ 13. _____

14. _____

15. $\frac{N}{18} = \frac{63}{27}$ 16. $\frac{N}{12} = \frac{35}{15}$ 15. _____

16. _____

17. $\frac{9}{14} = \frac{N}{266}$ 18. $\frac{8}{57} = \frac{N}{285}$ 17. _____

18. _____

EXERCISE 5-2   SET B

NAME _____   DATE _____

DO THE FOLLOWING FORM PROPORTIONS?   ANSWERS

1. $\frac{10}{35}$ and $\frac{14}{21}$ 2. $\frac{12}{33}$ and $\frac{8}{22}$ 1. _____

2. _____

3. $\frac{78}{234}$ and $\frac{39}{117}$ 4. $\frac{228}{266}$ and $\frac{114}{133}$ 3. _____

4. _____

5. $\frac{56}{85}$ and $\frac{58}{65}$ 6. $\frac{96}{78}$ and $\frac{78}{69}$ 5. _____

6. _____

SOLVE FOR N IN EACH PROPORTION:

7. $\frac{N}{3} = \frac{6}{9}$ 8. $\frac{N}{2} = \frac{12}{8}$ 7. _____

8. _____

9. $\frac{42}{N} = \frac{7}{8}$ 10. $\frac{27}{N} = \frac{3}{7}$ 9. _____

10. _____

11. $\frac{27}{30} = \frac{N}{100}$ 12. $\frac{9}{15} = \frac{N}{100}$ 11. _____

12. _____

13. $\frac{6}{114} = \frac{5}{N}$ 14. $\frac{9}{117} = \frac{7}{N}$ 13. _____

14. _____

15. $\frac{N}{21} = \frac{24}{9}$ 16. $\frac{N}{24} = \frac{15}{18}$ 15. _____

16. _____

17. $\frac{9}{14} = \frac{N}{238}$ 18. $\frac{3}{38} = \frac{N}{494}$ 17. _____

18. _____

5-3 More on Solving Proportions

In Section 5-2 we covered the basic steps for solving a proportion. By using those methods and your knowledge of fractions and decimals, you can now attempt problems that are a little more difficult.

Example 1: Do $1\frac{3}{4} : \frac{2}{5}$ and $2\frac{11}{12} : \frac{2}{3}$ form a proportion?

$$\frac{1\frac{3}{4}}{\frac{2}{5}} \; ? \; \frac{2\frac{11}{12}}{\frac{2}{3}}$$

$$1\frac{3}{4} \times \frac{2}{3} \; ? \; \frac{2}{5} \times 2\frac{11}{12} \qquad \text{Cross multiply.}$$

$$\frac{7}{\cancel{4}_{2}} \times \frac{\cancel{2}^{1}}{3} \; ? \; \frac{\cancel{2}^{1}}{\cancel{5}_{1}} \times \frac{\cancel{35}^{7}}{\cancel{12}_{6}}$$

$$\frac{7}{6} = \frac{7}{6}$$

Yes, they do form a proportion.

BEAUTY CONTEST JUDGES ARE UP ON THEIR PROPORTIONS.

Example 2: Solve for P.

$$\frac{2}{3} = \frac{P}{100}$$

$3 \times P = 200$ Cross multiply.

$$\frac{3 \times P}{3} = \frac{200}{3}$$

Divide by the number next to the unknown letter; divide by 3.

$P \stackrel{\raisebox{-1pt}{.}}{=} 66.67$

Example 3: Solve for N.

$$\frac{4.6}{3.75} = \frac{N}{7.875}$$

$3.75 \times N = 4.6 \times 7.875$ Cross multiply

$$\frac{3.75 \times N}{3.75} = \frac{36.225}{3.75}$$

Divide by 3.75.

$N = 9.66$

Example 4: Solve for B.

$$\frac{\frac{1}{2}}{B} = \frac{\frac{1}{4}}{100}$$

$\frac{1}{4} \times B = \frac{1}{2} \times 100$ Cross multiply.

$\frac{\frac{1}{4} \times B}{\frac{1}{4}} = \frac{50}{\frac{1}{4}}$ Divide by $\frac{1}{4}$.

$B = 50 \div \frac{1}{4}$

$B = 50 \times 4$

$B = 200$

PROBLEM 1: Do 16.4 : 5.6 and 18.4 : 7.6 form a proportion?

Answer: No.

The cross products are not equal.

$$\frac{16.4}{5.6} = \frac{18.4}{7.6}$$

$16.4 \times 7.6 = 124.64$

$5.6 \times 18.4 = 103.04$

PROBLEM 2: Solve for A.

$$\frac{A}{\frac{17}{3}} = \frac{21}{100}$$

Answer: A = 1.19

$100 \times A = \frac{17}{\cancel{3}} \times \cancel{21}^{7}$

$\frac{100 \times A}{100} = \frac{119}{100}$

$A = 1.19$

PROBLEM 3: Solve for B.

$$\frac{1.6}{B} = \frac{3.5}{6.39}$$

Answer: B ≐ 2.92

$3.5 \times B = 1.6 \times 6.39$

$\frac{3.5 \times B}{3.5} = \frac{10.224}{3.5}$

$B \doteq 2.92$

EXERCISE 5-3   SET A

NAME _____   DATE _____

DO THE FOLLOWING FORM PROPORTIONS?   ANSWERS

1. $2\frac{2}{3} : 1\frac{1}{5}$ and $1\frac{2}{3} : \frac{3}{4}$ 2. $1\frac{1}{4} : 3\frac{1}{3}$ and $1\frac{1}{2} : 4$ 1. _____

2. _____

3. 1.4 to 5.6 and .4 to 1.6   3. _____

4. 0.3 to 3.8 and 0.9 to 11.4   4. _____

SOLVE FOR THE UNKNOWN IN EACH PROPORTION:

5. $\dfrac{A}{16} = \dfrac{7}{100}$ 6. $\dfrac{A}{15} = \dfrac{9}{100}$ 5. _____

6. _____

7. $\dfrac{5}{8} = \dfrac{7}{N}$ 8. $\dfrac{4}{9} = \dfrac{7}{N}$ 7. _____

8. _____

9. $\dfrac{Y}{.35} = \dfrac{1.2}{.17}$ 10. $\dfrac{Y}{.12} = \dfrac{1.8}{.71}$ 9. _____

10. _____

11. $\dfrac{5.3}{B} = \dfrac{9}{100}$ 12. $\dfrac{4.6}{B} = \dfrac{7}{100}$ 11. _____

12. _____

13. $\dfrac{F}{\frac{1}{2}} = \dfrac{38}{\frac{1}{4}}$ 14. $\dfrac{F}{\frac{1}{4}} = \dfrac{36}{\frac{1}{2}}$ 13. _____

14. _____

15. $\dfrac{17}{2.5} = \dfrac{P}{100}$ 16. $\dfrac{28}{3.9} = \dfrac{P}{100}$ 15. _____

16. _____

17. $\dfrac{1\frac{3}{4}}{D} = \dfrac{\frac{7}{8}}{1\frac{1}{4}}$ 18. $\dfrac{2\frac{1}{2}}{D} = \dfrac{\frac{5}{6}}{1\frac{1}{6}}$ 17. _____

18. _____

EXERCISE 5-3 SET B

NAME _____   DATE _____

DO THE FOLLOWING FORM PROPORTIONS?   ANSWERS

1. $3\frac{1}{3} : 1\frac{1}{5}$ and $\frac{5}{9} : \frac{1}{5}$ 2. $1\frac{2}{5} : \frac{3}{4}$ and $\frac{2}{3} : \frac{5}{14}$ 1. _____

2. _____

3. 4 to 28.5 and .8 to 5.7   3. _____

4. 4.2 to 1.8 and 6.3 to 2.7   4. _____

SOLVE FOR THE UNKNOWN IN EACH PROPORTION:

5. $\dfrac{A}{18} = \dfrac{3}{100}$ 6. $\dfrac{A}{21} = \dfrac{11}{100}$ 5. _____

6. _____

7. $\dfrac{7}{8} = \dfrac{6}{N}$ 8. $\dfrac{6}{7} = \dfrac{5}{N}$ 7. _____

8. _____

9. $\dfrac{Y}{.56} = \dfrac{3.4}{.23}$ 10. $\dfrac{Y}{.49} = \dfrac{5.6}{.19}$ 9. _____

10. _____

11. $\dfrac{4.2}{B} = \dfrac{11}{100}$ 12. $\dfrac{5.3}{B} = \dfrac{13}{100}$ 11. _____

12. _____

13. $\dfrac{F}{\frac{1}{4}} = \dfrac{56}{\frac{1}{2}}$ 14. $\dfrac{F}{\frac{1}{2}} = \dfrac{72}{\frac{1}{4}}$ 13. _____

14. _____

15. $\dfrac{13}{5.7} = \dfrac{P}{100}$ 16. $\dfrac{19}{8.6} = \dfrac{P}{100}$ 15. _____

16. _____

17. $\dfrac{1\frac{1}{3}}{D} = \dfrac{9\frac{1}{3}}{1\frac{2}{5}}$ 18. $\dfrac{2\frac{1}{2}}{D} = \dfrac{1\frac{2}{7}}{\frac{1}{7}}$ 17. _____

18. _____

5-4 Applications Involving Proportions

In Sections 5-2 and 5-3 you learned how to solve proportions. In this section you will use that knowledge to solve other word problems. Before we do that, let us review the method for solving proportions.

Solve for N: $\dfrac{N}{6} = \dfrac{17}{2}$

$2 \times N = 6 \times 17$ Cross multiply.

$\dfrac{2 \times N}{2} = \dfrac{102}{2}$ Divide both sides by the number next to the unknown letter.

$N = 51$ Do the computation.

Example 1: During 15 minutes of television watching there is 2 minutes of commercials. If you watch T.V. for 65 minutes, how many minutes of commercials do you watch?

Analyze: The ratio of total time to commercial time is 15 : 2. In 65 minutes the ratio of total time to commercial time is 65 : ?. You can set up a proportion of total time to commercial time by writing 15 : 2 = 65 : ?.

Solve: $\dfrac{15}{2} = \dfrac{65}{N}$ **Note:** When setting up the proportion, both ratios must be in the same logical order:

$15 \times N = 2 \times 65$

$\dfrac{15 \times N}{15} = \dfrac{130}{15}$ $\dfrac{\text{total time}}{\text{comm. time}} = \dfrac{\text{total time}}{\text{comm. time}}$

$N = 8\dfrac{2}{3}$ minutes

Example 2: On a sunny day a 5 ft woman has a 2 ft shadow, while a church steeple has a 27 ft shadow. How tall is the steeple?

Analyze: The ratio of the woman's height to her shadow is 5 : 2. The ratio of the steeple's height to its shadow is ? : 27. You can set up a proportion of height to shadow as follows:; 5 : 2 = ? : 27.

Solve: $\dfrac{5}{2} = \dfrac{N}{27}$ $\dfrac{\text{height}}{\text{shadow}} = \dfrac{\text{height}}{\text{shadow}}$

$2 \times N = 5 \times 27$

$\dfrac{2 \times N}{2} = \dfrac{135}{2}$

$N = 67.5$ ft

Example 3: If $4\frac{3}{4}$ pounds of apples cost $1.29, how much will 10 pounds cost, rounded to the nearest cent?

Analyze: The ratios of pounds to cost are $4\frac{3}{4}$: $1.29 and 10 : ?. You can set up a proportion of pounds to cost by writing it as $4\frac{3}{4}$: 1.29 = 10: ?.

Solve:
$$\frac{4\frac{3}{4}}{1.29} = \frac{10}{N} \qquad \frac{\text{pounds}}{\text{cost}} = \frac{\text{pounds}}{\text{cost}}$$

$$4.75 \times N = 1.29 \times 10$$

$$\frac{4.75 \times N}{4.75} = \frac{12.9}{4.75}$$

$$N \doteq \$2.72 \text{ (rounded to the nearest cent)}$$

Example 4: A brass alloy contains only copper and zinc in the ratio of 4 parts copper to 3 parts zinc. If a total of 140 grams is made, how much copper is needed?

Analyze: Since the total amount of alloy (140 g) is known and the amount of copper is unknown, you need to set up a proportion with ratios of copper to total material. The ratio of copper to zinc is 4 to 3, so the ratio of copper to total material is 4 : 7. Thus the proportion is 4 : 7 = ? : 140.

Solve:
$$\frac{4}{7} = \frac{N}{140} \qquad \frac{\text{copper}}{\text{total}} = \frac{\text{copper}}{\text{total}}$$

$$7 \times N = 4 \times 140$$

$$\frac{7 \times N}{7} = \frac{560}{7}$$

$$N = 80 \text{ g}$$

PROBLEM 1: If every 1.5 inches on a map represents 7 miles of actual distance, how long is a section of freeway that is 6 inches long on the map?

Answer: 28 miles

$$\frac{1.5}{7} = \frac{6}{N} \qquad \frac{\text{inches}}{\text{miles}} = \frac{\text{inches}}{\text{miles}}$$

$$\frac{1.5 \times N}{1.5} = \frac{42}{1.5}$$

$$N = 28 \text{ mi}$$

EXERCISE 5-4   SET A

NAME _____   DATE _____

                                                                ANSWERS

1. If a 6' man has a 5' shadow, how tall is a pine tree
   that has a 37' shadow?   1. _____

2. If a 2 1/2' shrub has a 6' shadow, how tall is a pole
   that has an 89' shadow?   2. _____

3. On a scale drawing, a 27 mile stretch of freeway is
   represented by a 7" line. How long would a street be
   that is represented by a 3 1/4 inch line?   3. _____

4. If the scale on a map reads "1.5 inches = 2 miles," a
   creek represented by 7 inches would be how many miles long?   4. _____

5. If 6 lbs of fertilizer covers 1400 sq ft, how many
   pounds of fertilizer would you need to fertilize 2000
   sq ft?   5. _____

6. If 5 gallons of paint covers 1800 square feet of stucco,
   how many gallons of paint would you need to paint a 6120
   sq ft building?   6. _____

7. If the real estate tax is $688 for a $40,000 condominium,
   what is the value of a condominium that pays real estate
   tax of $835.   7. _____

8. If you drive 175 miles in 3 hours, how many miles can you
   expect to drive in 5.75 hours?   8. _____

9. If you run 6.2 miles in 40 minutes, how long would it take
   you to run 7.5 miles at the same pace?   9. _____

10. If a $2.00 bet on the winning horse in a race returns
    $2.80, how much money would a $5.00 bet have returned?   10. _____

11. If 12 1/2 ounces of cough syrup cost $2.95, at that rate
    how much should 16.76 ounces of cough syrup cost?   11. _____

12. If a dozen apples costs 89¢, how much should 50 apples
    cost?   12. _____

13. If a punch mix calls for 2 parts mix to 5 parts water, how
    much mix should be used to fill a 128 ounce punch bowl?   13. _____

14. Suppose for every 4 games a team wins, it loses 3 games.
    How many games will it win in a 56 game season?   14. _____

15. If you wanted to split 6300 acres in a ratio of 2 to 7,
    how many acres would be in each part?   15. _____

EXERCISE 5-4   SET B

NAME _____   DATE _____

                                                                  ANSWERS

1. If an upright yardstick casts a 2' shadow, how tall is a
    building that casts an 82' shadow?   1. _____

2. If a 5' woman has a 7' shadow, how tall is a flagpole
    with a 62' shadow?   2. _____

3. If a 5 1/4 inch line on a map represents a 9 mile road,
    a 3 1/2 inch line would represent how many miles?   3. _____

4. One inch on a world atlas represents 97 miles. The
    distance from San Francisco to Hawaii, 2076 miles,
    would be represented by how many inches?   4. _____

5. If an 8 oz bottle of a chemical is to be used on 1200
    sq ft of lawn, how many ounces should you use on 700
    sq ft of lawn?   5. _____

6. If a tape recorder register reads 316 after 40 minutes of
    recording, what should it read after an hour of recording?   6. _____

7. If the interest paid by a bank for one day on an $8650
    account is $2.16, how large is an account that earns
    $5.20 interest in one day?   7. _____

8. If it takes 16 qt jars to can 44 lbs of peaches, how
    many qt jars should it take to can 121 lbs of peaches?   8. _____

9. Your car uses 25 1/4 gallons of gasoline in 510 miles. How
    much should it use in 750 miles?   9. _____

10. If 4 7/8 pounds of meat cost $12.84, how much should
    3.26 pounds cost?   10. _____

11. A dozen donuts costs $2.98. At that rate, what should
    you pay for 5 donuts?   11. _____

12. If sound travels 1100 ft per second, how far will it
    travel in 35 minutes?   12. _____

13. If 5290 acres of land is to be split in the ratio of
    3 to 7, how many acres will be in each part?   13. _____

14. For every 5 people that complete a class, 3 do not finish.
    Out of 40 enrolled in a class, how many finish the class?   14. _____

15. If a team's ratio of wins to losses is 7 to 4, how many
    wins did it have in a 121 game season?   15. _____

5-5 Percents

A common standard of measurement is the **percent (%)**. "Percent" means "out of a hundred." An 85% test score means that out of 100 points you would get 85 points. If the sales tax in your state is 7%, it means that for every 100¢ (dollar), there is 7¢ sales tax.

> Remember: "Percent" means "out of a hundred."

So 25% means 25 out of 100.

$$25\% = \frac{25}{100} = .25$$

137% means 137 out of 100.

$$137\% = \frac{137}{100} = 1.37$$

6.5% means 6.5 out of 100.

$$6.5\% = \frac{6.5}{100} = .065$$

PERCENT: THE ODOR OF A PURSE.

From those examples, you can see that a percent can be easily expressed as a fraction or a decimal.

Converting Percents to Fractions

To convert a percent to a fraction, drop the % sign, put the number over 100, and reduce if possible.

Converting Percents to Decimals

To convert a percent to a decimal, drop the % sign and move the decimal point two places to the left.

Example 1: Express 30% as a fraction and as a decimal.

$$30\% = \frac{30}{100} = \frac{3}{10} \text{ (a reduced fraction)}$$

$$30\% = .30 \text{ (a decimal)}$$

Example 2: Express 125% as a mixed number and as a decimal.

$$125\% = \frac{125}{100} = \frac{5}{4} = 1\frac{1}{4} \text{ (a reduced mixed number)}$$

$$125\% = 1.25 \text{ (a decimal)}$$

Example 3: Express 9.4% as a decimal.

$$9.4\% = .094$$

Example 4: Express $3\frac{1}{3}\%$ as a fraction.

$$3\frac{1}{3}\% = \frac{3\frac{1}{3}}{100} = 3\frac{1}{3} \div 100$$

$$= \frac{10}{3} \div \frac{100}{1}$$

$$= \frac{\cancel{10}^{1}}{3} \times \frac{1}{\cancel{100}_{10}} = \frac{1}{30}$$

You may sometimes find it easier to express a percent as a fraction by converting the percent to a decimal, then converting the decimal to a fraction.

Example 5: Express 18 1/2% as a fraction.

Since 1/2 is easily expressed as the exact decimal .5,

$$18\ 1/2\% = 18.5\% = .185 \quad \text{(a decimal)}$$

$$.185 = \frac{185}{1000} = \frac{37}{200} \quad \text{(a reduced fraction)}$$

PROBLEM 1: Express 5.3% as a decimal and as a fraction.

Answers: .053 and 53/1000

decimal: $5.3\% = .053$

fraction: $0.053 = \frac{53}{1000}$

PROBLEM 2: Express $166\frac{2}{3}\%$ as a mixed number.

Answer: $1\frac{2}{3}$

$$= 166\frac{2}{3}\% = \frac{166\frac{2}{3}}{100}$$

$$= 166\frac{2}{3} \div 100 = \frac{500}{3} \div \frac{100}{1}$$

$$= \frac{\cancel{500}^{5}}{3} \times \frac{1}{\cancel{100}_{1}} = \frac{5}{3} = 1\frac{2}{3}$$

EXERCISE 5-5   SET A

NAME _____   DATE _____

CONVERT THE FOLLOWING TO REDUCED FRACTIONS OR MIXED NUMBERS: ANSWERS

1. 17%   2. 23%   1. _____

                                  2. _____

3. 5%   4. 6%   3. _____

                                  4. _____

5. 8.4%   6. 6.2%   5. _____

                                  6. _____

7. $9\frac{2}{3}$%   8. $7\frac{5}{6}$%   7. _____

                                  8. _____

9. 136%   10. 215%   9. _____

                                  10. _____

11. $43\frac{3}{4}$% 12. $27\frac{1}{4}$% 11. _____

                                  12. _____

CONVERT THE FOLLOWING DECIMALS:

13. 45%   14. 58%   13. _____

                                  14. _____

15. 5%   16. 8%   15. _____

                                  16. _____

17. 236%   18. 189%   17. _____

                                  18. _____

19. 26.5%   20. 20.4%   19. _____

                                  20. _____

21. $8\frac{1}{4}$%   22. $9\frac{3}{4}$%   21. _____

                                  22. _____

EXERCISE 5-5   SET B

NAME _____   DATE _____

CONVERT THE FOLLOWING TO REDUCED FRACTIONS OR MIXED NUMBERS: ANSWERS

1. 19%   2. 31%   1. _____
                                    2. _____

3. 8%   4. 2%   3. _____
                                    4. _____

5. 7.6%   6. 4.8%   5. _____
                                    6. _____

7. $6\frac{1}{3}$%   8. $9\frac{1}{6}$%   7. _____
                                    8. _____

9. 212%   10. 152%   9. _____
                                    10. _____

11. $57\frac{1}{4}$% 12. $87\frac{3}{4}$% 11. _____
                                    12. _____

CONVERT THE FOLLOWING TO DECIMALS:

13. 36%   14. 45%   13. _____
                                    14. _____

15. 4%   16. 3%   15. _____
                                    16. _____

17. 157%   18. 139%   17. _____
                                    18. _____

19. 15.8%   20. 16.3%   19. _____
                                    20. _____

21. $4\frac{3}{4}$%   22. $6\frac{1}{4}$%   21. _____
                                    22. _____

5-6 Converting Decimals and Fractions to Percents

In the previous section you learned how to change percents to either decimals or fractions. In this section you will learn to do the reverse process, that is, change decimals or fractions to percents.

Remember that "percent" means "out of a hundred." So reading how many hundredths are in a decimal number tells the percent.

$$.07 = \frac{7}{100} \text{ so } .07 = 7\%.$$

$$.9 = .90 = \frac{90}{100} \text{ so } .9 = 90\%.$$

$$1.23 = 1\frac{23}{100} = \frac{123}{100} \text{ so } 1.23 = 123\%.$$

Those examples point out a very easy way to convert a decimal to a percent.

> **Converting Decimals to Percents**
>
> To convert a decimal to a percent, move the decimal point two places to the right and attach a % sign.

For example:

$$.34 = 34\% \qquad .01 = 1\%$$
$$.005 = .5\% \qquad .0625 = 6.25\%$$
$$2.75 = 275\% \qquad 1.146 = 114.6\%$$

In Section 4-9 you learned to change a fraction to a decimal by dividing the denominator into the numerator of the fraction. Since the above shows how to change a decimal to a percent, putting the two steps together will convert a fraction to a percent.

> **Converting Fractions to Percents**
>
> To convert a fraction to a percent, divide the denominator of the fraction into the numerator to get a decimal number, then convert the decimal to a percent.

For example:

$$\frac{3}{4} = 4\overline{)3.00}^{\;.75} = 75\% \qquad \frac{2}{5} = 5\overline{)2.0}^{\;.4} = 40\%$$

$$\frac{37}{500} = 500\overline{)37.000}^{\;.074} = 7.4\% \qquad \frac{8}{12} \doteq 12\overline{)8.00000}^{\;.66666} \doteq 66.67\%$$

$$\frac{5}{7} \doteq 7\overline{)5.00000}^{\;.71428} \doteq 71.43\% \qquad \frac{13}{11} \doteq 1\overline{)13.00000}^{\;1.18181} \doteq 118.18\%$$

Since the decimal form of some of those examples had many decimal digits, they were rounded off to the nearest hundredth of a percent. You may likewise have to round off the answers to some of the problems and exercises in this section.

PROBLEM 1: On a test you got 63 out of 75 possible points. What percent did you get correct?

Answer: 84%

63 out of 75 = $\frac{63}{75}$

$$\begin{array}{r} .84 \\ 75 \overline{\smash{)}63.00} \\ \underline{60\ 0} \\ 3\ 00 \\ \underline{3\ 00} \end{array}$$

.84 = 84%

PROBLEM 2: Express a ratio of 5 to 6
as a fraction,
as a decimal,
and as a percent

Answers:

fraction: 5 to 6 = $\frac{5}{6}$

decimal: $6 \overline{\smash{)}5.00000}$ = .83333

percent: .83333 ≈ 83.33%

PROBLEM 3: Express a 0.063 as a fraction and as a percent.

Answers:

fraction: $\frac{63}{1000}$

percent: 0.063 = 6.3%

PROBLEM 4: Express 4/5 as a decimal and as a percent.

Answers:

decimal: $5 \overline{\smash{)}4.0}$ = .8
$\underline{4.0}$

percent: .8 = 80%

PROBLEM 5: Express $2\frac{3}{8}$ as a percent.

Answer: 237.5%

$2\frac{3}{8} = \frac{19}{8} = 8 \overline{\smash{)}19.000}$ = 2.375

2.375 = 237.5%

EXERCISE 5-6   SET A

NAME _____   DATE _____

DETERMINE THE MISSING FORMS BELOW:

	MIXED NUMBER OR FRACTION	DECIMAL	PERCENT
1.	$\frac{37}{50}$	_____	_____
2.	$\frac{3}{25}$	_____	_____
3.	_____	.02	_____
4.	_____	.08	_____
5.	_____	_____	5.7%
6.	_____	_____	1.9%
7.	$\frac{1}{5}$	_____	_____
8.	$\frac{7}{10}$	_____	_____
9.	_____	.4675	_____
10.	_____	.4425	_____
11.	$2\frac{1}{6}$	_____	_____
12.	$1\frac{5}{6}$	_____	_____
13.	_____	2.375	_____
14.	_____	1.025	_____
15.	_____	_____	$6\frac{5}{8}$%

EXERCISE 5-6   SET B

NAME _____   DATE _____

DETERMINE THE MISSING FORMS BELOW:

	MIXED NUMBER OR FRACTION	DECIMAL	PERCENT
1.	$\frac{13}{50}$	_____	_____
2.	$\frac{2}{25}$	_____	_____
3.	_____	.04	_____
4.	_____	.06	_____
5.	_____	_____	6.1%
6.	_____	_____	4.3%
7.	$\frac{3}{5}$	_____	_____
8.	$\frac{2}{5}$	_____	_____
9.	_____	.2375	_____
10.	_____	.2425	_____
11.	$1\frac{2}{3}$	_____	_____
12.	$2\frac{1}{3}$	_____	_____
13.	_____	1.125	_____
14.	_____	2.075	_____
15.	_____	_____	$5\frac{3}{8}$%

5-7 Percent of a Number

Percents are often used to find a part of a number or quantity. For example, you may encounter statements such as "60% of those surveyed", "35% discount", or "5.5% sales tax." In statements such as those, you are finding a part of a total amount. Again the word "of" indicates multiplication, just as it did in finding a fraction of a number in Section 3-7.

For example:

    60% of 5690   means   60% x 5690

    35% of $236   means   35% x $236

    5.5% of $179.99   means   5.5% x $179.99

Remember to change the percent into either a fraction or a decimal before you use it in multiplication.

Example 1: Find 25% of 76.

Method 1 (as a decimal)

25% = .25

25% of 76 = .25 x 76 = 19

Method 2 (as a fraction)

$25\% = \frac{25}{100} = \frac{1}{4}$

25% of 76 = $\frac{1}{4}$ x 76 = 19

Example 2: Find 6.5% of $275.44.

6.5% = .065

6.5% of $275.44 = .065 x $275.44

                      = $17.9036 $\doteq$ $17.90

                      (rounded to the hundredths)

PROBLEM 1: Find 60% of 3420.   Answer: 2052

                                            60% of 3420 =

                                            .6 x 3420 = 2052

PROBLEM 2: Find 43 1/4% of $54.72.   Answer: $23.67

                                            43 1/4% = 43.25% = .4325

                                            43 1/4% of $54.72 =

                                            .4325 x $54.72 =

                                            $ 23.6664 $\doteq$ $23.67

Matters such as taxes, commissions, discounts, down payments, bonuses, raises, deductions, etc., are often calculated as a percent *of* a certain number. The word **"of"** in this situation again means **"times"**; it indicates that you must multiply the percent times the number.

Example 3: If your state has a 5% sales tax, what would your tax be, to the nearest cent, on a $178.99 purchase?

Analyze: Solve:
The sales tax is 5%   5% x $178.99 =
  of the price.   .05 x $178.99 = $8.9495
That is, 5% x $178.99,
  and 5% = 0.05.   ≐ $8.95

Example 4: If you now earn $295.50 per week, a 9.2% raise would increase your earnings by how much?

Analyze: Solve:
The raise is 9.2% of   9.2% x $295.50 =
  your present salary.   .092 x $295.50 = $27.186
That is, 9.2% x $295.50,
  and 9.2% = .092.   ≐ $27.19

Example 5: A local department store is giving a 30% discount on all merchandise. How much of a discount would you receive on a VCR that regularly sells for $489.99?

Analyze: Solve:
The discount is 30% of   30% of $489.99 =
  the regular price.   0.3 x $489.99 = $146.997
That is, 30% x $489.99,
  and 30% = 0.3.   ≐ $147.00

PROBLEM 3: A realtor's commission is 3% of the sale price of a house. If a house sells for $98,000, what is the realtor's commission?

Answer: $2940
3% of $98,000 =
.03 x $98,000 =
    $2940

PROBLEM 4: How much would you need for the down payment on a $9,984 car, if the dealer requires 15% down?

Answer: $1497.60
15% of $9984 =
0.15 x $9984 =
    $1497.60

PROBLEM 5: 5.2% of the students withdrew from school during a semester. If the beginning enrollment was 4250, how many students withdrew?

Answer: 221 students
5.2% of 4250 =
.052 x 4250 = 221

EXERCISE 5-7   SET A

NAME _____   DATE _____

ROUND OFF DOLLAR AMOUNTS TO THE NEAREST CENT.   ANSWERS

1. 30% of 50   2. 60% of 65   1. _____

                                                             2. _____

3. 5% of 18.7   4. 6% of 17.7   3. _____

                                                             4. _____

5. 0.8% of $1476.59   6. 0.3% of $5,700,259   5. _____

                                                             6. _____

7. 16.4% of 600   8. 19.3% of 700   7. _____

                                                             8. _____

9. 78.32% of $76.99   10. 67.58% of $992.86   9. _____

                                                            10. _____

11. 5 3/4% of 82   12. 4 1/4% of 91   11. _____

                                                            12. _____

13. If the sales tax rate is 7.5%, what is the sales tax on
     a bike priced at $189.95?   13. _____

14. How much would a 33% discount amount to on a $159.99 suit? 14. _____

15. How much would you make from selling $980 worth of books,
     if you receive 8 1/4% commission on the sale?   15. _____

16. How much would you make on the sale of a $125,000 house,
     if you receive 3 1/4% commission on the sale?   16. _____

17. If a car dealer requires a 15% down payment, how much would
     you have to put down when purchasing a $6957 car?   17. _____

18. The Newark School District plans to increase budget amounts
     by 5.8% next year. If $85,200 is now spent on the library,
     how much will be budgeted for next year?   18. _____

EXERCISE 5-7   SET B

NAME _____   DATE _____

ROUND OFF DOLLAR AMOUNTS TO THE NEAREST CENT.   ANSWERS

1.   20% of 55   2.   40% of 65   1. _____

                                                                          2. _____

3.   4% of 27.6   4.   3% of 18.4   3. _____

                                                                          4. _____

5.   0.12% of $49,600,740   6.   6.9% of $376.36   5. _____

                                                                          6. _____

7.   14.6% of 500   8.   16.7% of 700   7. _____

                                                                          8. _____

9.   86.23% of $612.75   10. 74.32% of $82.59   9. _____

                                                                          10. _____

11. 3 1/4% of 58   12. 6 3/4% of 84   11. _____

                                                                          12. _____

13. What is the sales tax on a $345.62 rug, if the sales tax
     rate is 4.5%?   13. _____

14. What would a 40% discount amount to on the purchase of a
     $89.75 radio?   14. _____

15. How much would you make from selling $850 worth of tickets,
     if you receive an 11 1/4% commission on the sales?   15. _____

16. If you receive 2 1/4% commission on sales you make, how
     much would you make on sales of $650,790?   16. _____

17. How much is the 25% down payment on a truck that sells
     for $5789?   17. _____

18. ZETCO stocks experienced a 12.6% decrease during a week.
     If the stock sold for $35.00 a share at the start of the
     week, how much did it sell for at the end of the week?   18. _____

5-8 Percentage Problems

Consider this problem again: "On a test you got 63 out of 75 possible points. What percent did you get correct?" Since "percent" means "out of a hundred," we can consider this problem as a proportion: 63 out of 75 is what number out of 100? That is,

$$\frac{63}{75} = \frac{P}{100}$$ (Note: P is used to represent the percent or part out of 100.)

We can get the answer for the percent (P) by solving that proportion.

$$\frac{63}{75} = \frac{P}{100}$$

$$\frac{75 \times P}{75} = \frac{6300}{75}$$

$$P = 84$$

The amount you got correct was 63, the test was based on 75 points, and we discovered that the percent was 84. That relationship can be expressed by the **percent proportion:**

$$\frac{A}{B} = \frac{P}{100}$$
A is the amount
B is the base
P is the percent

The percent proportion can be used to solve different types of percentage problems. If you can identify the amount (A), the base (B), and the percent (P), you can utilize the percent proportion.

Here is how you can identify the **amount (A),** the **base (B),** and the **percent (P):**

1. The percent (P) is written with the word "percent" or the % sign.
2. The base (B) follows the word "of."
3. The amount (A) is the remaining number.

Example 1: 15 is what percent of 50?

1. Identify the A, B, P:

   15 is what percent of 50?
   ↓   ↓   ↓
   A   P   B
             (unknown) (follows "of")

2. Set up the percent proportion:

   $$\frac{A}{B} = \frac{P}{100}$$

   $$\frac{15}{50} = \frac{P}{100}$$ (Leave the unknown with its letter.)

3. Solve the proportion:

   $$\frac{50 \times P}{50} = \frac{1500}{50}$$

   $$P = 30$$

Note: When using the percent proportion, you do *not* move the decimal point to express the percent. The answer to Example 1 is simply 30%.

Example 2: 16 is 22% of what number?

1. Identify the A, B, P:

   16 is 22% of what number?
   ↓   ↓   ↓
   A   P   B
               (unknown)

2. Set up the percent proportion:

   $$\frac{A}{B} = \frac{P}{100}$$

   $$\frac{16}{B} = \frac{22}{100}$$

3. Solve the proportion:

   $$\frac{22 \times B}{22} = \frac{1600}{22}$$

   $$B \doteq 72.73$$

Example 3: 9.35% of 259.9 is how much?

1. Identify the A, B, P:

   9.35% of 259.9 is how much?
   ↓   ↓   ↓
   P   B   A
                    (unknown)

2. Set up the percent proportion:

   $$\frac{A}{B} = \frac{P}{100}$$

   $$\frac{A}{259.9} = \frac{9.35}{100}$$

3. Solve the proportion:

   $$\frac{100 \times A}{100} = \frac{2430.065}{100}$$

   $$A \doteq 24.30$$

PROBLEM 1: $41\frac{1}{4}$ is what percent of 30?   Answer: P = 137.5

$$\frac{41.25}{30} = \frac{P}{100}$$

$$\frac{30 \times P}{30} = \frac{4125}{30}$$

$$P = 137.5$$

PROBLEM 2: 19.8 is $7\frac{1}{3}$% of what number?   Answer: 270

$$\frac{19.8}{B} = \frac{7\frac{1}{3}}{100}$$

$$\frac{7\frac{1}{3} \times B}{7\frac{1}{3}} = \frac{1980}{7\frac{1}{3}}$$

$$B = 1980 \div \frac{22}{3}$$

$$B = 1980 \times \frac{3}{22}$$

$$B = 270$$

EXERCISE 5-8   SET A

NAME _____   DATE _____

ANSWERS

1. 9 is what percent of 15? 1. _____

2. 28 is what percent of 35? 2. _____

3. 20 is 60% of what number? 3. _____

4. 18 is 40% of what number? 4. _____

5. What is 6% of 50? 5. _____

6. What is 5% of 70? 6. _____

7. 3.38 is what percent of 13? 7. _____

8. 4.2 is what percent of 20? 8. _____

9. 18 is 8% of what number? 9. _____

10. 15 is 8% of what number? 10. _____

11. 32% of 148 is how much? 11. _____

12. 43% of 291 is how much? 12. _____

13. 63 is what percent of 17.5? 13. _____

14. 15 is what percent of 2.5? 14. _____

15. 24 is $5\frac{2}{3}$% of what number? 15. _____

16. 18 is $4\frac{2}{3}$% of what number? 16. _____

17. What is 17.6% of 45.6? 17. _____

18. What is 15.3% of 20.7? 18. _____

19. $\frac{3}{4}$ is what percent of $\frac{5}{8}$? 19. _____

EXERCISE 5-8   SET B

NAME _____   DATE _____

                                                          ANSWERS

1. 12 is what percent of 40?   1. _____

2. 10 is what percent of 25?   2. _____

3. 60 is 75% of what number?   3. _____

4. 18 is 40% of what number?   4. _____

5. What is 4% of 80?   5. _____

6. What is 7% of 60?   6. _____

7. 19.8 is what percent of 90?   7. _____

8. 13.2 is what percent of 55?   8. _____

9. 17 is 3% of what number?   9. _____

10. 24 is 4% of what number?   10. _____

11. 82% of 176 is how much?   11. _____

12. 53% of 247 is how much?   12. _____

13. 27.5 is what percent of 11?   13. _____

14. 76.8 is what percent of 12?   14. _____

15. 42 is $7\frac{2}{3}$% of what number?   15. _____

16. 36 is $8\frac{2}{3}$% of what number?   16. _____

17. What is 14.8% of 21.8?   17. _____

18. What is 12.7% of 43.6?   18. _____

19. $\frac{3}{8}$ is what percent of $\frac{3}{4}$?   19. _____

5-9 Applications Involving Percents

In Section 5-7 you saw how percents can be used to determine sales tax, down payments, raises, commissions, discounts, and the like. In this section you will apply those principles along with your other arithmetic skills to solve other problems that involve the use of percents.

Example 1: 68% of those polled were in favor of the Park Initiative. If 6,275 people were polled, how many were in favor of the initiative? How many were not in favor of the initiative?

Analyze:
1. The number in favor is 68% of the 6,275 polled.

2. Subtract those in favor from the total polled to get the number not in favor.

Solve:
1. 68% of 6275 =
   .68 x 6275 = 4267

2.   6275
    - 4267
      2008

Example 2: According to Schedule X of the 1985 Federal Tax Booklet, if your taxable income is over $35,490 but not over $43,190, your tax is $7813.50 plus 38% of the amount over $35,490. If your taxable income for 1985 was $42,000, how much tax did you owe the government?

Analyze:
1. Determine the amount earned over $35,490.

2. Find 38% of that amount.

3. Find the total tax by adding the result of step 2 to $7813.50.

Solve:
1. $42,000 - $35,490 = $6510

2. 38% of $6510 =
   .38 x $6510 = $2473.80

3. $7813.50
    + 2473.80
      $10,287.30

Example 3: A stereo has a regular price of $598.00. If you receive a 25% discount, how much would the stereo cost?

Analyze:
1. The discount is 25% of the regular price.

2. Get the cost by subtracting the discount from the regular price.

Solve:
1. 25% of $598.00 =
   .25 x $598.00 = $149.50

2. $598.00
    - 149.50
      $448.50

PROBLEM 1: What is the final cost of a $28.75 item, after 6.5% sales tax has been added?

Answer: $30.62
tax: .065 x $28.75 = $1.87
cost: $28.75 + $1.87 = $30.62

Using the Percent Proportion

In other types of percent problems you may have to use the percent proportion that was discussed in Section 5-8:

$$\frac{A}{B} = \frac{P}{100}$$

where A = the amount
B = the base
P = the percent

You will find it easier to use the percent proportion if you first translate the percent word problem into a condensed phrase of the form ___ is ___% of ___. By doing that, you can readily identify the A, B, and P.

Example 4: During the last 3 years the Stuffers won 75% of their basketball games. If they won 69 games, how many games did they play?

Analyze:
1. Restate the problem in a condensed form:

   <u>69</u> is <u>75%</u> of <u>?</u>

2. P = 75, B = unknown, A = 69

Solve: $\frac{A}{B} = \frac{P}{100}$

$\frac{69}{B} = \frac{75}{100}$

$\frac{75 \times B}{75} = \frac{6900}{75}$

B = 92

Example 5: A microwave oven was reduced from $450 to $390. To the nearest tenth, what percent was the oven reduced?
(Note: In a percent decrease or increase problem, the base is always the *original price,* the "base price.")

Analyze:
1. The reduction is 460 (450-390).

2. Restate the problem in a condensed form:

   <u>60</u> is <u>?</u>% of <u>450</u>

3. P = unknown, B = 450, A = 60

Solve: $\frac{A}{B} = \frac{P}{100}$

$\frac{60}{450} = \frac{P}{100}$

$\frac{450 \times P}{450} = \frac{6000}{450}$

P $\doteq$ 13.3

Example 6: From 1970 to 1980 the price of gold increased from $35 to $534 per ounce. What was the percent increase in price, to the nearest whole percent?

Analyze:
1. The increase is $499 (534 - 35)

2. Restate the problem in a condensed form:
   <u>499</u> is <u>?</u>% of <u>35</u>

3. P = unknown, B = 35, A = 499

Solve: $\frac{499}{35} = \frac{P}{100}$

$\frac{35 \times P}{35} = \frac{49900}{35}$

P $\doteq$ 1426

EXERCISE 5-9   SET A

NAME _____   DATE _____

ROUND DOLLAR AMOUNTS TO THE NEAREST CENT.   ANSWERS

1. What is the final price of a $679.79 typewriter, if
   6 1/2% sales tax is charged?   1. _____

2. How much would a $179.99 ring cost after a 40% discount?   2. _____

3. How much would a $763.45 item cost after a 12% discount?   3. _____

4. What is the final price of a $1397 computer system, if 7%
   sales tax is charged?   4. _____

5. Using the tax facts from Example 2 on page 267, what would
   the tax be for a 1985 taxable income of $37,275?   5. _____

6. After making a 25% down payment on a $3450 boat, how much
   do you still owe me on the boat?   6. _____

7. 32% of the 46,550 registered voters in a precinct voted on
   election day. How many registered voters did vote?   7. _____

8. If you earn $27,890 a year but a change in jobs causes a
   6.2% cut in pay, how much would you then earn per year?   8. _____

9. If you earn $652 a month and get a 12.5% raise, what will
   your new monthly salary be?   9. _____

10. What is your percent score on a test if you get 68 points
    out of 80 possible?   10. _____

11. What is the net pay on $700 after deductions of 6.13% for
    FICA, 11.5% for federal income tax, and 2% for state tax?   11. _____

12. If a team wins 84 out of 120 games, what percent of their
    games did they win?   12. _____

13. If a $200 dress sells for $140, what is the percent
    decrease in price?   13. _____

14. If your hourly wage increases from $5.60 to $6.44, what
    is the percent increase in your hourly wage?   14. _____

15. In a school election 52% voted for Honest Hank. If 325
    voted for Hank, how many voted in the election?   15. _____

16. During a survey 526 people were watching the "Night Show."
    If that is 20% of those surveyed, how many were surveyed?   16. _____

17. After you drive an $8000 car out of the lot, its value
    drops to $7500. What is the percent decrease in its value?   17. _____

EXERCISE 5-9   SET B

NAME _____   DATE _____

ROUND DOLLAR AMOUNTS TO THE NEAREST CENT.   ANSWERS

1. What is the final price of a $978.88 sofa, if 6 1/2% sales
   tax is charged?   1. _____

2. How much would a $678.99 T.V. cost after a 30% discount?   2. _____

3. How much would a $89.75 radio cost after a 25% discount?   3. _____

4. What is the final price of a $2367.59 motor cycle, if 7 1/2%
   sales tax is charged?   4. _____

5. Using the tax facts from Example 2 on page 267, what would
   the tax be for a 1985 taxable income of $36,492?   5. _____

6. After making a 15% down payment on a $9450 van, how much
   do you still owe on the van?   6. _____

7. 47% of the 16,500 registered voters in a precinct voted on
   election day. How many registered voters did vote?   7. _____

8. If you earn $1700 a month but a change in jobs causes a 4.3%
   cut in pay, how much would you earn per year after the cut? 8. _____

9. If you earn $12,790 a year and get a 13.7% raise, what will
   your new salary be?   9. _____

10. If you picked 6 out of 15 winners at the horse races, what
    percent did you guess correctly?   10. _____

11. What is the net pay on $589 after deductions of 6.13% for
    FICA, 7.8% for federal income tax, and 1.5% for state tax? 11. _____

12. If a quarterback completes 27 out of 40 passes, what is his
    percent completion?   12. _____

13. If a $350 painting sells for $287, what is the percent
    decrease in price?   13. _____

14. If your hourly wage increases from $5.00 to $5.48, what
    is the percent increase in your hourly wage?   14. _____

15. 60% of the students in a class get grades above a "C." If
    27 people get above a "C," how many were in the class?   15. _____

16. Suppose 1.5% of the light bulbs shipped from a factory are
    defective. If there are 18 defective light bulbs in a
    shipment, how many were shipped?   16. _____

17. If the price of a share of stock drops from $175 to $154
    in one day, what is the percent decrease in price?   17. _____

5-10 Applications Involving Interest

"THESE PROBLEMS ARE REALLY <u>INTERESTING</u>."

If you borrow money, you must pay back more than you borrowed. That extra amount that you pay back is called **interest.** The amount of interest you are charged is determined by taking a percent of the amount borrowed times the length of time of the loan. The amount borrowed is called the **principle** (P). The percent used is called the rate (R). The length of time for the loan is called the time (T).

Interest on a loan is determined using this formula:

> **INTEREST = PRINCIPLE x RATE x TIME**
>
> **I = P x R x T**

Interest calculated using that formula is called **simple interest,** since it is computed on the original principle during the time of the loan. We will not be considering other types of interest problems, such as compound interest, in this section.

Example 1: How much interest is charged on a $500 loan for one year using a 12% yearly interest rate?

Analyze:
1. Use the interest formula with:

   P = $500
   R = 12% = .12 per year
   T = 1 year

Solve:
I = P x R x T
 = $500 x .12 x 1
 = $60

In interest problems you must make sure that the rate and the time are expressed in the same units. If the rate (R) is a *yearly* rate, then the time (T) must be expressed in *years*. If the rate (R) is a *monthly* rate, then the time (T) must be expressed in *months*.

Example 2: What is the interest on $3200 at 1.5% per month for 2 years?

Analyze:
1. Use the interest formula with:

   P = $3200
   R = 1.5% = .015 per month
   T = 2 years = 24 months

Solve:

I = P x R x T
 = $3200 x .015 x 24
 = $1152

Once you know how to calculate the interest on a loan, you can determine how much you must pay each month to pay off the loan. Keep in mind that when you pay off a loan you must pay back the principle *and* the interest.

Example 3: Determine the monthly payments necessary to pay off a $6300 loan for 3 years at an annual rate of 13%.

Analyze:
1. Determine the interest using the interest formula.
2. Determine the total to be paid back by adding the principle and the interest.
3. Determine the monthly payments by dividing the total by the number of months. (3 yr = 36 mo)

Solve:
1. I = P x R x T
   = $6300 x .13 x 3
   = $2457
2. total = $6300 + $2457
   = $8757
3. payments = $8757 ÷ 36
   = $243.25

Example 4: You are purchasing a new car with a total price of $9500. The dealer gives you a 10% discount and a $1500 trade-in on your old car. To pay off the balance you take out a loan with a 12.6% yearly rate for 42 months. To the nearest cent, what should your monthly payments be to pay off the loan?

Analyze:
1. Determine the balance by subtracting the 10% discount and the $1500 trade-in from the total price of $9500.
2. Determine the interest. P = $7050, R = 12.6% = .126 per yr., T = 42 months = 3.5 yrs.
3. Get the total to be repaid by adding principle and interest.
4. Get monthly payments by dividing the total by the number of months.

Solve:
1. discount = .10 x 9500
   = $950
   balance = 9500 - 950
   - 1500 = $7050
2. I = P x R x T
   = $7050 x .126 x 3.5
   = $3109.05
3. total = 7050 + 3109.05
   = $10,159.05
4. payments = 10159.05 ÷ 42
   ≐ $241.88

PROBLEM 1: Determine to the nearest cent the monthly payment needed to pay off a loan of $700 at 9.5% per year in two years.

Answer: $34.71
I = 700 x .095 x 2
  = $133
total = 700 + 133
  = $833
payments = 833 ÷ 24
  ≐ $34.71

EXERCISE 5-10   SET A

NAME _____   DATE _____

FIND THE SIMPLE INTEREST TO THE NEAREST CENT FOR LOANS WITH
THE FOLLOWING PRINCIPLE (P), RATE (R), AND TIME (T):   ANSWERS

1. P = $500,   R = 15% per year,   T = 2 years   1. _____

2. P = $700,   R = 16% per year,   T = 3 years   2. _____

3. P = $2000,   R = 9% per year,   T = 3.5 years   3. _____

4. P = $5000,   R = 8% per year,   T = 2.5 years   4. _____

5. P = $7500,   R = 12.76% per year, T = 4 years   5. _____

6. P = $9600,   R = 14.75% per year, T = 5 years   6. _____

7. P = $4655,   R = 13.83% per year, T = 42 months   7. _____

8. P = $5994,   R = 14¼% per year,   T = 30 months   8. _____

9. P = $10,800, R = 1.5% per month, T = 5 years   9. _____

10. P = $12,700, R = 1.2% per month, T = 6 years   10. _____

FIND THE FOLLOWING ANSWERS ACCURATE TO THE NEAREST CENT:

11. You borrow $700 from "Insta-loan" at a yearly interest rate
    of 17.5% for 4 years. What will your monthly payment be
    to repay the loan?   11. _____

12. You borrow $8900 from the "Easy-Out" Loan Company at a
    yearly interest rate of 16.9% for 5 years. What will
    your monthly payment be to repay the loan?   12. _____

13. "John's T.V." gives you a 25% discount on a $987.48 T.V.
    and you pay 6¼% sales tax on the balance. If you take out
    a 2 year loan with an annual interest rate of 15.6%, what
    will your monthly payment be?   13. _____

14. What is the monthly payment needed to pay off a $7325.66
    loan for 30 months with a yearly interest rate of 16¼%?   14. _____

15. After trading in your old car, you still owe $6200 on a
    new car. You pay 7% sales tax on that balance and the dealer
    adds $355 for license fees and dealer preparation fees. If
    you take out a 5 year loan for the balance after those fees,
    with a yearly interest rate of 19.2% on the loan, what is
    your monthly payment?   15. _____

EXERCISE 5-10   SET B

NAME _____   DATE _____

FIND THE SIMPLE INTEREST TO THE NEAREST CENT FOR LOANS WITH
THE FOLLOWING PRINCIPLE (P), RATE (R), AND TIME (T):   ANSWERS

1. P = $800,   R = 16% per year,   T = 2 years   1. _____

2. P = $400,   R = 13% per year,   T = 3 years   2. _____

3. P = $4000,   R = 11% per year,   T = 4.5 years   3. _____

4. P = $600,   R = 12% per year,   T = 1.5 years   4. _____

5. P = $6700,   R = 13.26% per year, T = 4 years   5. _____

6. P = $7900,   R = 15.46% per year, T = 5 years   6. _____

7. P = $3467,   R = 19.2% per year,   T = 18 months   7. _____

8. P = $7899,   R = 17.6% per year,   T = 42 months   8. _____

9. P = $10,700, R = 1.3% per month,   T = 5 years   9. _____

10. P = $11,600, R = 1.4% per month,   T = 6 years   10. _____

FIND THE FOLLOWING ANSWERS ACCURATE TO THE NEAREST CENT:

11. You borrow $900 from the "Man at Atlantic Plan" for 3
    years at a yearly interest rate of 14.5%. What is the
    monthly payment needed to pay off the loan?   11. _____

12. You borrow $8900 from the "Speed-Cash" Loan Company for
    5 years at a yearly interest rate of 15.6%. What is the
    monthly payment needed to pay off the loan?   12. _____

13. "Atlantic Stereo" gives you 1/3 off on a $1653.99 stereo
    system and you pay 5½% sales tax on the balance. What is
    the monthly payment, if you take out a 2 year loan with
    an annual interest rate of 14.8%?   13. _____

14. What is the monthly payment needed to pay off a $6987.45
    loan for 42 months with a yearly interest rate of 14¼%?   14. _____

15. After trading in your old car, you still owe $5800 on a
    new car. You pay 6% sales tax on that balance and the dealer
    adds $285 for license and dealer preparation fees. If you
    take out a 4 year loan for the balance after those fees,
    with a yearly interest rate of 17.4% on the loan, what is
    your monthly payment?   15. _____

Chapter 5 Summary

Concepts

You may refer to the sections listed below to review how to:

1. set up and reduce ratios by expressing them as fractions. (5-1)

2. solve proportions by using cross multiplication. (5-2), (5-3)

3. solve word problems involving proportions. (5-4)

4. convert a percent to a fraction by dropping the percent sign, placing the number over 100, and reducing if possible. (5-5)

5. convert a percent to a decimal by dropping the percent sign and moving the decimal point two places to the left. (5-5)

6. convert a decimal to a percent by moving the decimal point two places to the right and attaching a percent sign. (5-6)

7. convert a fraction to a percent by first changing the fraction to a decimal. (5-6)

8. find a percent of a number by using the word "of" as an indication of multiplication. (5-7)

9. solve percentage problems by using the percent proportion:

    $$\frac{A}{B} = \frac{P}{100}$$ where A is the amount, B is the base, and P is the percent. (5-8)

10. solve word problems involving percents. (5-9)

11. calculate simple interest using the formula:

    INTEREST = PRINCIPLE x RATE x TIME (5-10)

Terminology

This chapter's important terms and their corresponding page numbers are:

cross multiplying: in a proportion, the process of multiplying the numerator of the first ratio by the denominator of the second ratio, and multiplying the denominator of the first ratio by the numerator of the second ratio. (239)

cross products: the answers from cross multiplying. (239)

interest: profit derived from money that has been loaned. (271)

percent: amount expressed as a part out of one hundred. (251)

percent proportion: the proportion, $\frac{A}{B} = \frac{P}{100}$, used in percentage problems. (263)

principle: amount of money being loaned. (271)

proportion: a statement that two ratios are equal. (239)

ratio: a comparison of two quantities that can be expressed as a fraction. (235)

simple interest: the interest calculated on the original principle during the time period of the loan. (271)

Student Notes

Chapter 5 Practice Test A

ANSWERS

1. Express a ratio of 84 to 315 as a reduced fraction.

2. Solve for N.

    $\dfrac{N}{21} = \dfrac{24}{9}$

3. Solve for B.

    $\dfrac{91}{B} = \dfrac{13}{100}$

4. Solve for Y.

    $\dfrac{5.5}{Y} = \dfrac{9.6}{18.7}$

5. Solve for P.

    $\dfrac{5}{8} = \dfrac{P}{100}$

6. Express 6.4% as a decimal and as a reduced fraction.

7. Express .132 as a reduced fraction and as a percent.

8. What is 16.8% of 420?

9. 72 is what percent of 120?

10. 19 is 7% of what number?

11. What is $47\tfrac{1}{3}$% of $150.60?

12. To the nearest cent, what is the final price of clothing totaling $64.25, after 6% sales tax is added?

13. On a sunny day a 5 ft boy casts a 6 ft shadow. At the same time a pole casts a shadow of 32 ft. How tall is the pole?

14. How much would you still owe on a $3620 bill after paying 15% of the bill?

15. From the "Easy-Out Finance Company" you borrow $2140 with a simple interest loan of 14.6% per year and plan to pay it back in 42 months. To the nearest cent, how much will your monthly payments be?

Chapter 5 Practice Test B

ANSWERS

1. Express a ratio of 130 to 273 as a reduced fraction.

   1. _____

2. Solve for N.

   $$\frac{N}{21} = \frac{15}{9}$$

3. Solve for B.

   $$\frac{77}{B} = \frac{14}{100}$$

   2. _____

   3. _____

4. Solve for Y.

   $$\frac{6.5}{Y} = \frac{7.6}{12.3}$$

5. Solve for P.

   $$\frac{7}{16} = \frac{P}{100}$$

   4. _____

   5. _____

6. Express 8.2% as a decimal and as a reduced fraction.

   6. _____

7. Express .146 as a reduced fraction and as a percent.

   7. _____

8. What is 17.3% of 120?

9. 63 is what percent of 175?

   8. _____

   9. _____

10. 17 is 9% of what number?

11. What is $54\frac{2}{3}$% of $91.20?

    10. _____

    11. _____

12. To the nearest cent, what is the final price of tools costing $84.16, after 7% sales tax is added?

    12. _____

13. If the ratio of chemical to water in a plant spray is 2 to 9, how much water should be added to 8.6 ounces of the chemical?

    13. _____

14. If you earn $525 a month and receive a 7.6% raise, how much will you then earn per month?

    14. _____

15. From the "Fast Finance Company" you borrow $3040 with a simple interest loan of 15.2% per year and plan to pay it back in 30 months. To the nearest cent, how much will your monthly payments be?

    15. _____

Chapter 5 Supplementary Exercises

Section 5-1

A soccer team won 20 games, lost 16, and tied 4. Find the reduced ratio of:

1. wins to losses
2. wins to ties
3. wins to total games
4. ties to total games
5. ties to losses
6. losses to total games
7. losses to wins
8. ties to wins

Section 5-2

Do the following form proportions?

1. $\dfrac{4}{6}$ and $\dfrac{24}{36}$
2. $\dfrac{16}{21}$ and $\dfrac{39}{52}$
3. $\dfrac{7}{12}$ and $\dfrac{35}{60}$
4. $\dfrac{10}{17}$ and $\dfrac{20}{27}$
5. $\dfrac{12}{18}$ and $\dfrac{18}{27}$
6. $\dfrac{56}{85}$ and $\dfrac{36}{65}$

Solve for N:

7. $\dfrac{N}{5} = \dfrac{21}{15}$
8. $\dfrac{14}{N} = \dfrac{7}{4}$
9. $\dfrac{9}{14} = \dfrac{N}{84}$
10. $\dfrac{25}{60} = \dfrac{N}{24}$
11. $\dfrac{N}{48} = \dfrac{35}{20}$
12. $\dfrac{28}{12} = \dfrac{N}{18}$
13. $\dfrac{24}{45} = \dfrac{56}{N}$
14. $\dfrac{78}{52} = \dfrac{6}{N}$
15. $\dfrac{N}{56} = \dfrac{7}{8}$
16. $\dfrac{12}{19} = \dfrac{N}{114}$
17. $\dfrac{132}{N} = \dfrac{11}{4}$
18. $\dfrac{N}{45} = \dfrac{133}{35}$
19. $\dfrac{N}{12} = \dfrac{84}{112}$
20. $\dfrac{21}{N} = \dfrac{99}{165}$
21. $\dfrac{25}{70} = \dfrac{N}{84}$

Section 5-3

Solve for N:

1. $\dfrac{N}{9} = \dfrac{7}{12}$
2. $\dfrac{16}{N} = \dfrac{17}{100}$
3. $\dfrac{6}{7} = \dfrac{N}{4}$
4. $\dfrac{14}{N} = \dfrac{4}{9}$
5. $\dfrac{21}{17} = \dfrac{6}{N}$
6. $\dfrac{4.7}{9.3} = \dfrac{N}{6.5}$
7. $\dfrac{19}{2.75} = \dfrac{13.4}{N}$
8. $\dfrac{N}{5.6} = \dfrac{.43}{2.15}$
9. $\dfrac{6.5}{N} = \dfrac{11}{3.7}$
10. $\dfrac{\frac{1}{2}}{6} = \dfrac{N}{\frac{1}{4}}$
11. $\dfrac{N}{6\frac{1}{4}} = \dfrac{8}{\frac{1}{2}}$
12. $\dfrac{5\frac{1}{3}}{N} = \dfrac{1\frac{1}{6}}{6\frac{2}{3}}$

Section 5-4

1. If 9 baseball tickets cost $47.97, how much would 37 tickets cost?
2. If you spend $445 on gasoline for five months, what would you expect to spend on gasoline for the year?
3. If a 3" line appears as a 7" line on an enlargement, how long would a $16\frac{1}{2}$" line appear on the same enlargement?
4. If the ratio of males to females in a class of 35 students is 3 to 4, how many females are in the class?

Section 5-5

Express as reduced fractions or mixed numbers:

Express as decimals:

1. 28%
2. 76%
3. 4%
4. 9%
5. 13.6%
6. 28.8%
7. 6.2%
8. 5.4%
9. 147%
10. 225%
11. $62\frac{1}{2}$%
12. $42\frac{1}{4}$%
13. 36%
14. 47%
15. 8%
16. 7%
17. 14.3%
18. 16.2%
19. 8.25%
20. 6.75%
21. $9\frac{1}{2}$%
22. $8\frac{1}{4}$%
23. 156%
24. 207%

Section 5-6

Express the following as percents:

1. $\frac{3}{5}$
2. $\frac{1}{4}$
3. $\frac{9}{10}$
4. $\frac{7}{8}$
5. $2\frac{4}{5}$
6. 0.07
7. 0.24
8. 0.155
9. 2.13
10. 3.175
11. 0.0625
12. 0.1775

Section 5-7

1. 25% of 77
2. 6% of 83.7
3. 19.2% of 505
4. 7.6% of 85.53
5. 0.75% of 2375
6. 8 1/4% of 60
7. 143% of 500
8. 126.7% of 447.6
9. 5 2/3% of 911
10. .0124% of $456.77
11. 34.6% of $509.55
12. 0.325% of $4000
13. If the sales tax rate is 4.25%, what is the tax on a $500.00 item?
14. A 20% discount on a $799.95 TV would give you how much off the original price?
15. How much would you make on $75,600 in sales, if you receive 1.275% commission on the sales?
16. 76% of the students in an elementary school preferred peanut butter and jelly sandwiches. If the school has 800 students, how many preferred peanut butter and jelly sandwiches?

Section 5-8

1. 12 is what percent of 25?
2. 42 is 40% of what number?
3. 83% of 176 is how much?
4. 64 is what percent of 70?
5. 26 is 18% of what number?
6. 6% of 120 is how much?
7. 16.5 is what percent of 376.8?
8. 94 is 60% of what number?
9. 46.8 is 5.6% of what number?
10. 6.25% of 225 is how much?
11. What is 19.8% of 72.5?
12. 403 is $19\frac{1}{4}$% of what number?
13. 20.5 is 137% of what number?
14. 806.3 is what percent of 523?
15. 0.05 is what percent of 0.005?
16. 0.003 is 6% of what number?

Section 5-9

1. What is the final price of a $375.99 TV after $7\frac{1}{4}$% sales tax is charged?

2. If you earn $275.65 a week and get a 5.8% raise, how much would you earn each month?

3. What would you still owe on an $8,595 car if you make a 25% down payment?

4. What is your take home pay if you earn $1740 a month and have deductions that total 19.6%

5. If a $399.99 vacuum cleaner is on sale for $299.99, what is the percent decrease in price?

According to Schedule Y of the 1985 Federal Tax Booklet, if your taxable income is over $16,650 but not over $21,020, your tax is $1,811.40 plus 18% of the amount over $16,650.

6. What is the tax on a 1985 taxable income of $18,950?
7. What is the tax on a 1985 taxable income of $20,000?
8. What is the tax on a 1985 taxable income of $17,777?

Section 5-10

1. Find the simple interest on $6000 at 14.7% per year for 4 years.

2. Find the simple interest on $4750 at 1.6% per month for 3 years.

3. Find the simple interest on $9763.85 at 15.7% per year for 30 months.

4. What is the monthly payment needed to pay off a $9276.27 loan in 54 months with a yearly interest rate of $16\frac{1}{4}$%?

5. You purchase a $627.99 stereo and $6\frac{1}{4}$% sales tax is added. What would your monthly payment be if you take out a 2 year loan at 13.9% per year?

6. You are buying a microwave oven costing $479.99. If 7% sales tax is added and you take out a 2 year loan with an annual interest rate of 14.9% for the total amount, what is your monthly payment?

Chapter 6

Measurement and Graphs

After finishing this chapter you should be able to do the following:

1. read measuring devices.
2. use the units of measurement and make conversions in the U.S. system.
3. use the units of measurement and make conversions in the metric system.
4. make conversions between the U.S. and the metric systems of measurement.
5. analyze statistical graphs.

On the next page you will find a pretest for this chapter. The purpose of the pretest is to help you determine which sections in this chapter you need to study in detail and which sections you can review quickly. By taking and correcting the pretest according to the instructions on the next page you can better plan your pace through this chapter.

Note: Unless otherwise indicated in this chapter, when the answers have more than three decimal digits, they will be rounded off to the nearest hundredth.

Chapter 6 Pretest

Take and correct this test using the answer section at the end of the book. Those problems that give you difficulty indicate which sections need extra attention. Section numbers are in parentheses before each problem.

ANSWERS

(6-1)   1. What amount is indicated by the arrow below?   1. _____

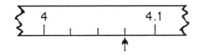

(6-2)   2. 2.7 mi = _?_ ft   (6-2)   3. 100 fl oz = _?_ cups   2. _____

3. _____

(6-2)   4. $1\frac{3}{4}$ tons = _?_ oz   (6-3)   5. 1 kl = _?_ liters   4. _____

5. _____

(6-3)   6. 1 g = _?_ cg   (6-4)   7. 9 m = _?_ cm   6. _____

7. _____

(6-4)   8. 1159 mm = _?_ m   (6-4)   9. 5.76 cg = _?_ mg   8. _____

9. _____

(6-5)   10. 50 in. = _?_ cm   (6-5)   11. 147.4 lb = _?_ kg   10. _____

11. _____

(6-5)   12. 3200 cl = _?_ gal   (6-5)   13. Which is longer: 12. _____
                                                                         120 yd or 100 m?

13. _____

(6-6)   14. What Centigrade temperature corresponds to 150° Fahrenheit?   14. _____

(6-7)   15. In the circle graph on page 315, what would the Smiths spend each month on mortgage payments, if they earn $25,500 per year?   15. _____

6-1 Reading Measuring Devices

We are always concerned with measurement. We frequently determine time, length, volume, weight, temperature, voltage, speed, etc. Each measurement includes a number followed by some kind of a unit, such as 52 seconds, 6.4 centimeters, 3½ quarts, 98.6 degrees, 9 volts, 55 miles per hour, etc. To measure quantities such as these, we use devices such as watches, measuring cups, scales, rulers, thermometers, etc. Before we study different units of measurement, let's examine some measuring devices and learn how to read them. I will use a thermometer to establish the principles needed to read various kinds of measuring devices.

To read a measuring device, you need to determine the value of the basic interval measured between adjacent marks of its scale.

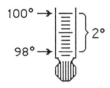

1. Find the amount measured between two labeled values of the scale. (Subtract the two labeled values.)

   100° − 98° = 2°

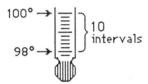

2. Find the number of equal intervals between the two labeled values. (Count the number of spaces.)

   There are 10 in this example.

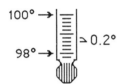

3. Calculate the basic interval of the measuring device. (Divide the result of step 1 by the result of step 2.)

   basic interval = 2° ÷ 10 = 0.2°

Once the basic interval of the scale has been established, you can now read the amounts measured by the thermometer.

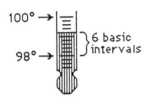

What temperature is indicated on the left?

   The temperature is 6 basic intervals above 98°.

   The temperature = 98° + 6 × 0.2°
                   = 98° + 1.2°
                   = 99.2°

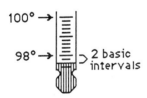

What temperature is indicated on the left?

   The temperature is 2 basic intervals below 98°.

   The temperature = 98° − 2 × 0.2°
                   = 98 − 0.4°
                   = 97.6°

Example 1: Find the measurements indicated by the arrows at A, B, and C on the ruler shown below.

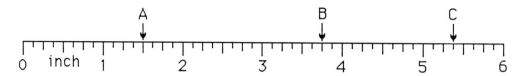

First, find the length measured by a basic interval on the ruler. Between 1 and 2 there are 8 equal intervals.

The basic interval = $1 \div 8 = \frac{1}{8}$ inch

A measures $1\frac{1}{2}$ in. It is 4 basic intervals above 1 inch.

$$1 + \frac{4}{8} = 1\frac{1}{2} \text{ in.}$$

B measures $3\frac{3}{4}$ in. It is 6 basic intervals above 3 inches.

$$3 + \frac{6}{8} = 3\frac{3}{4} \text{ in.}$$

C measures $5\frac{3}{8}$ in. It is 3 basic intervals above 5 inches.

$$5 + \frac{3}{8} = 5\frac{3}{8} \text{ in.}$$

Example 2: Find the measurements indicated by the arrows at A, B, and C on the voltmeter shown below.

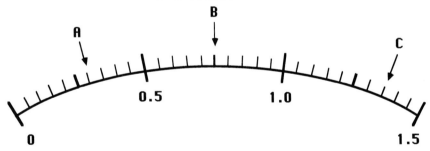

VOLTS

First, find the amount measured by a basic interval on the voltmeter. Between 0 and 0.5 volts, there are 10 equal intervals.
The basic interval = $0.5 \div 10 = 0.05$ volts

A measures 0.3 volts. It is 6 basic intervals above 0.

$$0 + 6 \times 0.05 = 0 + 0.3 = 0.3 \text{ volts}$$

B measures 0.75 volts. It is 5 basic intervals above 0.5

$$0.5 + 5 \times 0.05 = 0.5 + 0.25 = 0.75 \text{ volts}$$

C measures 1.35 volts. It is 7 basic intervals above 1.0.

$$1.0 + 7 \times 0.05 = 1.0 + 0.35 = 1.35 \text{ volts}$$

EXERCISE 6-1   SET A

NAME _____   DATE _____

Determine the length in centimeters (cm) indicated by the letters:

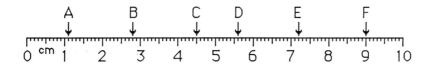

ANSWERS

1. A   2. B   1. _____

                                                                    2. _____

3. C   4. D   3. _____

                                                                    4. _____

5. E   6. F   5. _____

                                                                    6. _____

Determine the speed in miles per hour (mph) indicated by the letters:

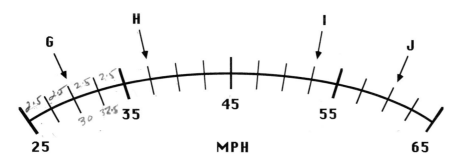

7. G   8. H   7. _____

                                                                    8. _____

9. I   10. J   9. _____

                                                                    10. _____

Determine the volume in milliliters (ml) indicated by the letters:

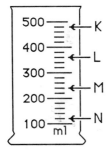   11. K   11. _____

                            12. L   12. _____

                            13. M   13. _____

                            14. N   14. _____

EXERCISE 6-1   SET B

NAME _____   DATE _____

Determine the length in inches (in.) indicated by the letters:

1. A   2. B   1. _____
                                        2. _____
3. C   4. D   3. _____
                                        4. _____
5. E   6. F   5. _____
                                        6. _____

Determine the weight in pounds (lb) indicated by the letters:

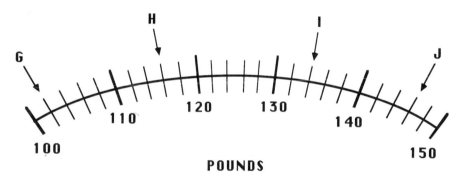

7. G   8. H   7. _____
                                        8. _____
9. I   10. J   9. _____
                                       10. _____

Determine the Centigrade temperature indicated by the letters:

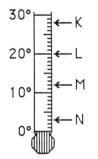

11. K 28°   11. _____
12. L 20°   12. _____
13. M 12°   13. _____
14. N 3°   14. _____

288

6-2 The U.S. Customary System of Measurement

In the United States we measure length with units such as inches, feet, yards, and miles; weight with units such as ounces, pounds, and tons; volume with units such as fluid ounces, pints, quarts, and gallons. These are the common units of the U.S. Customary System of Measurement. The relationship between these units are given in the chart below.

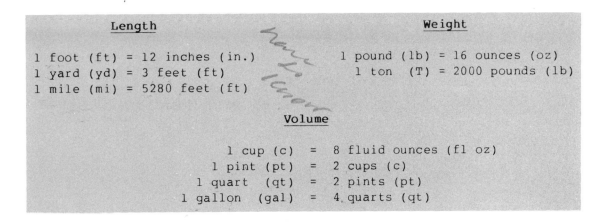

Length	Weight
1 foot (ft) = 12 inches (in.)	1 pound (lb) = 16 ounces (oz)
1 yard (yd) = 3 feet (ft)	1 ton (T) = 2000 pounds (lb)
1 mile (mi) = 5280 feet (ft)	

Volume	
1 cup (c)	= 8 fluid ounces (fl oz)
1 pint (pt)	= 2 cups (c)
1 quart (qt)	= 2 pints (pt)
1 gallon (gal)	= 4 quarts (qt)

You will find that it is often necessary to change a measurement in one unit into another unit. The method that I will use to explain how to do this is called the **unit-fraction** method.

The Unit-Fraction Method of Conversion

The unit-fraction method of converting units of measurement is based on the fact that multiplying a number by 1 does not change the value of the number. Consider this problem:

$$23 \text{ qt} = \underline{} \text{ gal}$$

The fraction $\frac{1 \text{ gal}}{4 \text{ qt}}$ has a value that is equal to 1, since 1 gal = 4 qt.

Such a fraction is called a unit-fraction and it can be used as follows:

$$23 \text{ qt} = 23 \text{ qt} \times 1$$
$$= 23 \text{ qt} \times \frac{1 \text{ gal}}{4 \text{ qt}}$$

We are now multiplying fractions, and can cancel units just as we did numbers.

$$23 \text{ qt} = 23 \, \cancel{\text{qt}} \times \frac{1 \text{ gal}}{4 \, \cancel{\text{qt}}} = \frac{23 \times 1 \text{ gal}}{4} = 5.75 \text{ gal}$$

Notice that since you wanted to eliminate "qt," it was put in the denominator of the unit fraction. Thus quarts (qt) canceled, leaving gallons (gal).

The **unit-fraction method** of changing units of measurement requires that you multiply by a fraction that has:

> 1. a value equal to 1. (The top measures the same as the bottom.)
> 2. a denominator using the given unit, so that it will cancel.
> 3. a numerator using the unit wanted in the answer.

Example 1:   40 lb = __?__ oz

$$40 \text{ lb} = 40 \text{ lb} \times 1$$

$$= 40 \text{ lb} \times \frac{16 \text{ oz}}{1 \text{ lb}}$$

$$= \frac{40 \times 16 \text{ oz}}{1}$$

$$= 640 \text{ oz}$$

Since we want to eliminate pounds (lb), that unit must be in the denominator of the unit-fraction, while its equivalent in ounces (oz) is placed in the numerator.

Example 2:   40 oz = __?__ lb

$$40 \text{ oz} = 40 \text{ oz} \times 1$$

$$= 40 \text{ oz} \times \frac{1 \text{ lb}}{16 \text{ oz}}$$

$$= \frac{40 \times 1 \text{ lb}}{16}$$

$$= 2.5 \text{ lb}$$

The unit-fraction has ounces (oz) on the bottom so that ounces (oz) cancel, leaving pounds (lb).

PROBLEM 1:   6.2 mi = __?__ ft

Answer: **32,736 ft**

$$6.2 \text{ mi} \times \frac{5280 \text{ ft}}{1 \text{ mi}} = \frac{6.2 \times 5280 \text{ ft}}{1}$$

$$= 32{,}736 \text{ ft}$$

PROBLEM 2: $56\frac{2}{5}$ ft = __?__ in.

Answer: **676.8 in.**

$$56\frac{2}{5} \text{ ft} \times \frac{12 \text{ in.}}{1 \text{ ft}} = 56\frac{2}{5} \times 12 \text{ in.}$$

$$= \frac{282}{5} \times 12 \text{ in.} = 676.8 \text{ in.}$$

PROBLEM 3:   30.46 pt = __?__ gal

Answer: $\doteq$ **3.81 gal**

a. change pints to quarts

$$30.46 \text{ pt} \times \frac{1 \text{ qt}}{2 \text{ pt}} = 15.23 \text{ qt}$$

b. change quarts to gallons

$$15.23 \text{ qt} \times \frac{1 \text{ gal}}{4 \text{ qt}} \doteq 3.81 \text{ gal}$$

EXERCISE 6-2   SET A

NAME _____   DATE _____

SOLVE FOR THE ? IN EACH PROBLEM: ANSWERS

1. 9 ft = _?_ in.   2. 15 lb = _?_ oz   1. _____

                                                            2. _____

3. 48 pt = _?_ qt   4. 72 fl oz = _?_ c   3. _____

                                                            4. _____

5. 12.6 yd = _?_ ft   6. 4.8 gal = _?_ qt   5. _____

                                                            6. _____

7. 9516.8 lb = _?_ T   8. 139.2 in. = _?_ ft   7. _____

                                                            8. _____

9. $4\frac{1}{6}$ mi = _?_ ft 10. $15\frac{1}{3}$ pt = _?_ c   9. _____

                                                            10. _____

11. $17\frac{1}{3}$ oz = _?_ lb 12. $18\frac{3}{8}$ in. = _?_ ft   11. _____

                                                            12. _____

13. 75 qt = _?_ c   14. 51,200 oz = _?_ T   13. _____

                                                            14. _____

15. $3\frac{1}{2}$ T = _?_ oz   16. $84\frac{3}{5}$ c = _?_ qt   15. _____

                                                            16. _____

17. 3.1 mi = _?_ yd   18. 4241.6 yd = _?_ mi   17. _____

                                                            18. _____

EXERCISE 6-2   SET B

NAME _____   DATE _____

SOLVE FOR THE ? IN EACH PROBLEM: ANSWERS

1. 5 cups = _?_ fl oz   2. 7 qt = _?_ pt   1. _____

                                                          2. _____

3. 81 in. = _?_ ft   4. 96 oz = _?_ lb   3. _____

                                                          4. _____

5. 4.75 tons = _?_ lb   6. 448.8 ft = _?_ in.   5. _____

                                                          6. _____

7. 50.7 ft = _?_ yd   8. 37.5 qt = _?_ gal   7. _____

                                                          8. _____

9. $18\frac{3}{8}$ lb = _?_ oz   10. $27\frac{2}{3}$ ft = _?_ in.   9. _____

                                                          10. _____

11. $9\frac{3}{4}$ ft = _?_ yd   12. $6\frac{3}{4}$ c = _?_ pt   11. _____

                                                          12. _____

13. 73,600 oz = _?_ T   14. 23 qt = _?_ c   13. _____

                                                          14. _____

15. $7\frac{5}{8}$ gal = _?_ fl oz   16. $131\frac{1}{5}$ fl oz = _?_ gal   15. _____

                                                          16. _____

17. 36.5 c = _?_ qt   18. 2.7 T = _?_ oz   17. _____

                                                          18. _____

6-3 The Metric System

The system of measurement used in most countries throughout the world, and used along with the U.S. Customary system in America, is the metric system. In the metric system there are basic units for length, weight, and volume.

"THAT GIRL IS SURE NICE. I'D LIKE TO METER."

Basic Units in the Metric System

1. For Length:

    meter (m)--a little longer than a yard

2. For Weight:

    gram (g)--about the weight of a paper clip

3. For Volume:

    liter (l)--a little more than a quart

Larger or smaller units in the system are obtained by multiplying or dividing those basic units by powers of 10. Prefixes are used to signify the change from one of the basic units.

Prefixes in the Metric System

* **kilo (k)** --means a thousand (1000)

  hecto (h)--means a hundred (100)

  deca (da)--means ten (10)

  deci (d) --means a tenth (1/10 or .1)

* **centi (c)**--means a hundredth (1/100 or .01)

* **milli (m)**--means a thousandth (1/1000 or .001)

Those with asterisks are the prefixes that are most commonly used, so we will concentrate on those in this chapter. *Putting a prefix together with a basic unit gives the other units in the metric system.* For example:

    kilogram means 1000 grams (1 kg = 1000 g)

    centimeter means $\frac{1}{100}$ meter (1 cm = .01 m)

    milliliter means $\frac{1}{1000}$ liter (1 ml = .001 liter)

When the prefix kilo is used, you have a unit that is larger than the basic unit. When centi or milli is used, you have a unit that is smaller than the basic unit. The first step in understanding the metric system is to become familiar with its units and to have a feel for the relative size of each unit. The chart below shows the metric units and their relative size.

	Larger Units			Basic Units	Smaller Units		
	kilo (1000)	hecto (100)	deca (10)	1	deci (.1)	centi (.01)	milli (.001)
length	km	hm	dam	meter	dm	cm	mm
weight	kg	hg	dag	gram	dg	cg	mg
volume	kl	hl	dal	liter	dl	cl	ml

Example 1: What is a mg?

mg means milligram. It is used to measure weight.

$$mg = \frac{1}{1000} g = .001 g$$

So a milligram is much lighter than a gram. It would take 1000 mg to make a gram (1000 mg = 1 g).

Example 2: What is a km?

km means kilometer. It is used to measure length or distance.

$$km = 1000 m$$

So a kilometer is much larger than a meter. It is 1000 meters.

Example 3: What is a cl?

cl means centiliter. It is used to measure volume.

$$cl = \frac{1}{100} liter = .01 liter$$

So a cl is less than a liter. It would take 100 cl to make a liter (100 cl = 1 liter).

PROBLEM 1: Arrange in order from largest to smallest: g, kg, mg, cg

Answer: kg, g, cg, mg

kg = 1000 g
cg = .01 g
mg = .001 g

PROBLEM 2: 1 kl = __?__ liters

Answer: 1000

kilo means 1000

EXERCISE 6-3   SET A

NAME _____   DATE _____

WRITE THE FOLLOWING IN WORDS: ANSWERS

1. km   2. kl   1. _____
                                                  2. _____

3. cg   4. cm   3. _____
                                                  4. _____

5. ml   6. mg   5. _____
                                                  6. _____

REPLACE THE ? WITH THE CORRECT NUMBER:

7. 1 km = _?_ m   8. 1 kg = _?_ g   7. _____
                                                  8. _____

9. 1 cl = _?_ liter   10. 1 cm = _?_ m   9. _____
                                                  10. _____

11. 1 mg = _?_ g   12. 1 ml = _?_ liter   11. _____
                                                  12. _____

13. 1 m = _?_ cm   14. 1 g = _?_ cg   13. _____
                                                  14. _____

15. 1 liter = _?_ ml   16. 1 m = _?_ mm   15. _____
                                                  16. _____

ARRANGE IN ORDER FROM LARGEST TO SMALLEST:

17. kg, cg, g   18. cm, m, mm   17. _____
                                                  18. _____

19. mm, km, cm   20. cl, ml, kl   19. _____
                                                  20. _____

EXERCISE 6-3   SET B

NAME _____   DATE _____

WRITE THE FOLLOWING IN WORDS: ANSWERS

   1. kg   2. km   1. _____

                                                       2. _____

   3. cl   4. cg   3. _____

                                                       4. _____

   5. mm   6. ml   5. _____

                                                       6. _____

REPLACE THE ? WITH THE CORRECT NUMBER:

   7. 1 kl = _?_ liters   8. 1 km = _?_ m   7. _____

                                                       8. _____

   9. 1 cg = _?_ g   10. 1 cl = _?_ liter   9. _____

                                                      10. _____

  11. 1 mm = _?_ m   12. 1 mg = _?_ g   11. _____

                                                      12. _____

  13. 1 liter = _?_ cl   14. 1 m = _?_ cm   13. _____

                                                      14. _____

  15. 1 g = _?_ mg   16. 1 liter = _?_ ml   15. _____

                                                      16. _____

ARRANGE IN ORDER FROM LARGEST TO SMALLEST:

  17. mg, g, cg   18. cm, m, km   17. _____

                                                      18. _____

  19. kl, ml, cl   20. mg, kg, cg   19. _____

                                                      20. _____

6-4 Conversions Within the Metric System

Now that you know the units of the metric system, you can learn to change a measurement using one unit into the same measurement using a different unit. I will first review how to do this using the unit-fraction method as covered in Section 6-2, and then using a short cut.

The Unit-Fraction Method

The unit-fraction method requires that you multiply a given measurement by a fraction that has:

1. a value equal to 1. (The top measures the same as the bottom.)
2. a denominator using the given unit, so that it will cancel.
3. a numerator using the unit wanted in the answer.

Example 1: 5 kg = __?__ g

$$5 \text{ kg} = 5 \text{ kg} \times 1$$
$$= 5 \text{ kg} \times \frac{1000 \text{ g}}{1 \text{ kg}}$$
$$= \frac{5 \times 1000 \text{ g}}{1}$$
$$= 5000 \text{ g}$$

(Note: 1 kg = 1000 g)

The unit-fraction should have kilograms (kg) on the bottom, so that kilograms (kg) cancel, leaving grams (g).

Example 2: 3450 mm = __?__ m

$$3450 \text{ mm} = 3450 \text{ mm} \times 1$$
$$= 3450 \text{ mm} \times \frac{.001 \text{ m}}{1 \text{ mm}}$$
$$= \frac{3450 \times .001 \text{ m}}{1}$$
$$= 3.45 \text{ m}$$

(Note: 1 mm = .001m)

The unit-fraction should have millimeters (mm) on the bottom, so that millimeters (mm) cancel, leaving meters (m).

Example 3: 65.4 cl = __?__ liters

$$65.4 \text{ cl} = 65.4 \text{ cl} \times 1$$
$$= 65.4 \text{ cl} \times \frac{.01 \text{ liter}}{1 \text{ cl}}$$
$$= \frac{65.4 \times .01 \text{ liter}}{1}$$
$$= .654 \text{ liters}$$

(Note: 1 cl = .01 liter)

The unit-fraction should have centiliters (cl) on the bottom, so that centiliters (cl) cancel, leaving liters.

A Short Cut

Look at the results of the last three examples.

$$5 \text{ kg} = 5000 \text{ g}$$

$$3450 \text{ mm} = 3.45 \text{ m}$$

$$65.4 \text{ cl} = .654 \text{ liters}$$

In each conversion, the difference between the two forms is merely the position of the decimal point. In fact, *that is the beauty of the metric system—units can be changed by a simple movement of the decimal point.*

5 kg = 5000 g	(decimal point moved 3 places right)
3450 mm = 3.450 m	(decimal point moved 3 places left)
65.4 cl = .654 liters	(decimal point moved 2 places left)

All you need to do to change units in the metric system is to learn which direction to move the decimal point and how many places it should be moved.

1. How do you determine which way to move the decimal point?

   If you study those examples you will see the following:

   a. When changing a larger unit to a smaller unit, move the decimal point to the right. (We need a larger number of the smaller units to express the same quantity; to make a number larger, we move the point right.)

   b. When changing a smaller unit to a larger unit, move the decimal point to the left. (We need a smaller number of the larger units to express the same quantity; to make a number smaller, we move the point left.)

2. How do you determine how many places to move the decimal point?

   The prefixes used tell you how many places to move the decimal point. If you consider the basic unit as the reference point, then:

   a. kilo represents a 3 place movement from the basic unit.
      (kilo = 1000, and 1000 has three zeros)

   b. centi represents a 2 place movement from the basic unit.
      (centi = 1/100, and 100 has two zeros)

   c. milli represents a 3 place movement from the basic unit.
      (milli = 1/1000, and 1000 has three zeros)

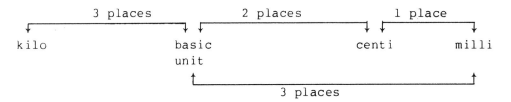

Using that short cut method just explained, try the following examples:

Example 4:   5.3 g = __?__ cg

> Centigrams are smaller than grams, so we need more of them to express the same quantity. To make the number larger, move the decimal point to the right.
>
> Gram to centigram is a 2 place movement.
>
> 5.3 g = 5.30 cg   (decimal moved 2 places right)
>
> = 530 cg

Example 5:   7500 ml = __?__ liters

> Liters are larger than milliliters, so we need fewer of them to express the same quantity. To make the number smaller, move the decimal point to the left.
>
> Milliliter to liter is a 3 place movement.
>
> 7500 ml = 7.500 liters   (decimal moved 3 places left)
>
> = 7.5 liters

Example 6:   642 mm = __?__ cm

> Centimeters are larger than millimeters, so we need fewer of them to express the same amount. To make the number smaller, move the decimal point to the left.
>
> Millimeter to centimeter is a 1 place movement.
> (.001 to .01 is a 1 decimal place difference.)
>
> 642 mm = 64.2 cm   (decimal moved 1 place left)
>
> = 64.2 cm

Example 7:   A bubbly punch recipe calls for 5 bottles of sparkling water and 2 bottles of punch mix. If the sparkling water comes in 750 ml bottles and the mix comes in 360 ml bottles, how many liters of punch will the recipe make?

Analyze:	Solve:
1. Find the total number of milliliters called for.	1. 5 x 750 ml = 3750 ml 2 x 360 ml = 720 ml total = 4470 ml
2. Change ml to liters.	2. 4470 ml = 4.470 liters

When making conversions in the metric system you can use either the unit-fraction method or the decimal movement short cut just explained. Try the following problems using whichever method you choose.

PROBLEM 1: 2.65 km = __?__ m

Answer: 2650 m

$$2.65 \text{ km} \times \frac{1000 \text{ m}}{1 \text{ km}} =$$

$$= \frac{2.65 \times 1000 \text{ m}}{1}$$

$$= 2650 \text{ m}$$

PROBLEM 2: 56 mg = __?__ g

Answer: 0.056 g

G is larger, so move the decimal to the left. Mg to g moves it 3 places.

56 mg = .056 g

PROBLEM 3: 35.6 cm = __?__ mm

Answer: 356 mm

Mm is smaller, so move the decimal to the right. Cm to mm moves it 1 place.

35.6 cm = 356 mm

PROBLEM 4: 7 kg = __?__ cg

Answer: 700,000 cg

Cg is smaller, so move the decimal to the right. Kg to cg is a 5 place movement (3 from kg to g and 2 from g to cg).

7 kg = 700000 cg

PROBLEM 5: A doctor's prescription calls for three 250 mg tablets four times a day. How many grams of medicine is taken each day?

Answer: 3 g

3 tablets = 3 × 250 mg
         = 750 mg
4 times a day = 4 × 750
              = 3000 mg

3000 mg = 3.000 g

EXERCISE 6-4   SET A

NAME _____   DATE _____

SOLVE FOR THE ? IN EACH PROBLEM: ANSWERS

1. 3 kg = ? g   2. 8 kl = ? liters   1. _____

2. _____

3. 5200 cl = ? liters   4. 7100 cm = ? m   3. _____

4. _____

5. 5 m = ? cm   6. 7 g = ? cg   5. _____

6. _____

7. 5000 ml = ? liters   8. 7000 mg = ? g   7. _____

8. _____

9. 4.6 liters = ? ml   10. 7.3 m = ? mm   9. _____

10. _____

11. 1697 mm = ? m   12. 1555 mg = ? g   11. _____

12. _____

13. 8.9 g = ? mg   14. 7.6 liters = ? ml   13. _____

14. _____

15. 570 liters = ? kl   16. 860 m = ? km   15. _____

16. _____

17. 40 cm = ? mm   18. 50 cl = ? ml   17. _____

18. _____

19. 7.6 ml = ? cl   20. 8.3 mg = ? cg   19. _____

20. _____

21. 4.2 kg = ? cg   22. 3.7 km = ? cm   21. _____

22. _____

23. 16.5 ml = ? liters   24. 180 cm = ? m   23. _____

24. _____

EXERCISE 6-4   SET B

NAME _____   DATE _____

SOLVE FOR THE ? IN EACH PROBLEM: ANSWERS

1. 4 kl = _?_ liters   2. 9 kg = _?_ g   1. _____

                                                          2. _____

3. 5300 cm = _?_ m   4. 4100 cl = _?_ liters   3. _____

                                                          4. _____

5. 9 g = _?_ cg   6. 2 m = _?_ cm   5. _____

                                                          6. _____

7. 3000 mg = _?_ g   8. 2000 ml = _?_ liters   7. _____

                                                          8. _____

9. 5.9 m = _?_ mm   10. 2.8 liters = _?_ ml   9. _____

                                                          10. _____

11. 3127 mg = _?_ g   12. 2095 mm = _?_ m   11. _____

                                                          12. _____

13. 4.3 liters = _?_ ml   14. 5.9 g = _?_ mg   13. _____

                                                          14. _____

15. 610 m = _?_ km   16. 230 liters = _?_ kl   15. _____

                                                          16. _____

17. 70 cl = _?_ ml   18. 60 cm = _?_ mm   17. _____

                                                          18. _____

19. 2.9 mg = _?_ cg   20. 3.2 ml = _?_ cl   19. _____

                                                          20. _____

21. 4.1 km = _?_ cm   22. 6.4 kg = _?_ cg   21. _____

                                                          22. _____

23. 280 mg = _?_ g   24. 140 mm = _?_ m   23. _____

                                                          24. _____

6-5 Conversions Between the Metric and U.S. Systems

CONVERSION: A PRISONER'S EXPLANATION OF A PROBLEM.

Since both the U.S. and the metric systems are used in America, it is important that you are able to find relationships between the two systems. In this section we will use the unit-fraction method to make conversions between the two systems. The following is a chart of the commonly used conversions between the two systems.

Length:	1 in.	$\doteq$ 2.54 cm
	39.4 in.	$\doteq$ 1 m
	0.621 mi	$\doteq$ 1 km
Weight:	1 oz	$\doteq$ 28.35 g
	1 lb	$\doteq$ 454 g
	2.2 lb	$\doteq$ 1 kg
Volume:	1.06 qt	$\doteq$ 1 liter

Applying the unit-fraction method of conversion and the values in that chart, we can convert metric measurements to U.S. units and vice versa.

Example 1: 12 in. = ___?___ cm

$$12 \text{ in.} = 12 \text{ in.} \times 1$$
$$= 12 \text{ in.} \times \frac{2.54 \text{ cm}}{1 \text{ in.}}$$
$$= \frac{12 \times 2.54 \text{ cm}}{1}$$
$$= 30.48 \text{ cm}$$

(Note: 1 in. $\doteq$ 2.54 cm)

The unit-fraction should have inches (in.) on the bottom, so that inches (in.) cancel, leaving centimeters (cm).

Example 2: 33 lb = ___?___ kg

$$33 \text{ lb} = 33 \text{ lb} \times 1$$
$$= 33 \text{ lb} \times \frac{1 \text{ kg}}{2.2 \text{ lb}}$$
$$= \frac{33 \times 1 \text{ kg}}{2.2}$$
$$= 15 \text{ kg}$$

(Note: 2.2 lb $\doteq$ 1 kg)

The unit-fraction should have pounds (lb) on the bottom, so that pounds (lb) cancel, leaving kilograms (kg).

Example 3: 13.25 qt = __?__ liters

$$13.25 \text{ qt} = 13.25 \text{ qt} \times 1 \quad \text{(Note: } 1.06 \text{ qt} \stackrel{\bullet}{=} 1 \text{ liter)}$$

$$= 13.25 \, \cancel{\text{qt}} \times \frac{1 \text{ liter}}{1.06 \, \cancel{\text{qt}}}$$

The unit-fraction should have quarts (qt) on the bottom, so that quarts (qt) cancel, leaving liters.

$$= \frac{13.25 \times 1 \text{ liter}}{1.06}$$

$$= 12.5 \text{ liters}$$

Example 4: 1589 g = __?__ lb

$$1589 \text{ g} = 1589 \text{ g} \times 1 \quad \text{(Note: } 1 \text{ lb} \stackrel{\bullet}{=} 454 \text{ g)}$$

$$= 1589 \, \cancel{\text{g}} \times \frac{1 \text{ lb}}{454 \, \cancel{\text{g}}}$$

The unit-fraction should have grams (g) on the bottom, so that grams (g) cancel, leaving pounds (lb).

$$= \frac{1589 \times 1 \text{ lb}}{454}$$

$$= 3.5 \text{ lb}$$

Example 5: 5 km = __?__ mi

$$5 \text{ km} = 1 \quad \text{(Note: } .621 \text{ mi} \stackrel{\bullet}{=} 1 \text{ km)}$$

$$= 5 \, \cancel{\text{km}} \times \frac{.621 \text{ mi}}{1 \, \cancel{\text{km}}}$$

The unit-fraction should have kilometers (km) on the bottom, so that kilometers (km) cancel, leaving miles (mi).

$$= \frac{5 \times .621 \text{ mi}}{1}$$

$$= 3.105 \text{ mi}$$

If you set up the unit-fraction correctly, there should be no confusion as to whether you multiply or divide to make the conversion from one system to the other. Let us now consider some problems that involve more than one conversion.

Example 6: 7 gal = __?__ liters

Analyze: The conversion chart has the conversion between quarts and liters, so we must first change the gallons to quarts and then change the quarts to liters.

Solve:

Convert gal to qt.
(1 gal = 4 qt)

$$7 \text{ gal} = 7 \text{ gal} \times 1$$

$$= 7 \, \cancel{\text{gal}} \times \frac{4 \text{ qt}}{1 \, \cancel{\text{gal}}}$$

$$= \frac{7 \times 4 \text{ qt}}{1}$$

$$= 28 \text{ qt}$$

Convert qt to liters.
(1.06 qt $\stackrel{\bullet}{=}$ 1 liter)

$$28 \text{ qt} = 28 \text{ qt} \times 1$$

$$= 28 \, \cancel{\text{qt}} \times \frac{1 \text{ liter}}{1.06 \, \cancel{\text{qt}}}$$

$$= \frac{28 \times 1 \text{ liter}}{1.06}$$

$$= 26.42 \text{ liters}$$

Example 7: 3000 kg = __?__ T

Analyze: Convert kg to lb.   Convert lb to T.
(2.2 lb ≟ 1 kg)   (1 T = 2000 lb)

Solve: 3000 kg = 3000 kg x 1   6600 lb = 6600 lb x 1

$= 3000 \text{ kg} \times \dfrac{2.2 \text{ lb}}{1 \text{ kg}}$ $= 6600 \text{ lb} \times \dfrac{1 \text{ T}}{2000 \text{ lb}}$

= 6600 lb   = 3.3 T

Example 8: Which is longer, the metric mile (1500 m) or the U.S. mile?

Analyze: The U.S. mile is 5280 ft, so convert the metric mile into feet and then compare.

Solve: Convert m to in.   Convert in. to ft.
(1 m ≟ 39.4 in.)   (12 in. = 1 ft)

1500 m = 1500 m x 1   59100 in = 59100 in. x 1

$= 1500 \text{ m} \times \dfrac{39.4 \text{ in.}}{1 \text{ m}}$ $= 59100 \text{ in.} \times \dfrac{1 \text{ ft}}{12 \text{ in.}}$

= 59100 in.   = 4925 ft

Answer: The U.S. mile (5280 ft) is longer than the metric mile (4925 ft).

Example 9: Which holds more: a fifth (4/5 of a quart), or a 750 ml bottle?

Analyze: 4/5 = .8, so a fifth is .8 qt. Convert the .8 qt to ml, and then compare.

Solve: Convert qt to liters.   Convert liters to ml.
(1.06 qt ≟ 1 liter)   (1 ml = .001 liter)

.8 qt = .8 qt x 1   .7547 liters x 1 =

$= .8 \text{ qt} \times \dfrac{1 \text{ liter}}{1.06 \text{ qt}}$ $.7547 \text{ liters} \times \dfrac{1 \text{ ml}}{.001 \text{ liter}} =$

= .7547 liters   754.7 ml

Answer: A fifth (754.7 ml) holds more than a 750 ml bottle.

After studying those examples on the last three pages, you should be ready to do some conversions by yourself. Refer to the conversion chart at the bottom of the page as you work through these problems.

PROBLEM 1:   36 in. = __?__ cm

Answer: 91.44 cm

$$36 \text{ in.} \times \frac{2.54 \text{ cm}}{1 \text{ in.}}$$

$$= \frac{36 \times 2.54 \text{ cm}}{1}$$

$$= 91.44 \text{ cm}$$

PROBLEM 2:   26.2 mi = __?__ km

Answer: 42.19 km

$$26.2 \text{ mi} \times \frac{1 \text{ km}}{.621 \text{ mi}}$$

$$= \frac{26.2 \times 1 \text{ km}}{.621}$$

$$\stackrel{.}{=} 42.19 \text{ km}$$

PROBLEM 3:   8 liters = __?__ gal

Answer: 2.12 gal

$$8 \text{ liters} \times \frac{1.06 \text{ qt}}{1 \text{ liter}}$$

$$= \frac{8 \times 1.06 \text{ qt}}{1} = 8.48 \text{ qt}$$

$$8.48 \text{ qt} \times \frac{1 \text{ gal}}{4 \text{ qt}}$$

$$= \frac{8.48 \text{ gal}}{4} = 2.12 \text{ gal}$$

U.S.--Metric Conversions

Length	Volume	Weight
1 in. $\stackrel{.}{=}$ 2.54 cm		1 oz $\stackrel{.}{=}$ 28.35 g
39.4 in. $\stackrel{.}{=}$ 1 m	1.06 qt $\stackrel{.}{=}$ 1 liter	1 lb $\stackrel{.}{=}$ 454 g
0.621 mi $\stackrel{.}{=}$ 1 km		2.2 lb $\stackrel{.}{=}$ 1 kg

EXERCISE 6-5   SET A

NAME _____   DATE _____

Answers may vary depending on conversions used: ANSWERS

1. 9 in. = __?__ cm   2. 8 liters = __?__ qt   1. _____

                                                                    2. _____

3. 33 lb = __?__ kg   4. 197 in. = __?__ m   3. _____

                                                                    4. _____

5. 4 m = __?__ in.   6. 7 lb = __?__ g   5. _____

                                                                    6. _____

7. 170.1 g = __?__ oz   8. 18.63 mi = __?__ km   7. _____

                                                                    8. _____

9. 24.75 m = __?__ in.   10. 58.95 kg = __?__ lb   9. _____

                                                                    10. _____

11. 147.4 lb = __?__ kg   12. 62.73 cm = __?__ in.   11. _____

                                                                    12. _____

13. .05 kg = __?__ oz   14. 10 ft = __?__ m   13. _____

                                                                    14. _____

15. 560 fl oz = __?__ ml   16. .001 oz = __?__ cg   15. _____

                                                                    16. _____

17. WHICH HOLDS MORE:
    A 50 gallon drum, or a 200 liter drum?   17. _____

18. WHICH IS HEAVIER:
    A man weighing 140 pounds, or a man weighing 64 kilograms?   18. _____

EXERCISE 6-5 SET B

NAME _____ DATE _____

Answers may vary depending on conversions used.   ANSWERS

1. 5 oz = _?_ g   2. 9 kg = _?_ lb   1. _____

                                                        2. _____

3. 591 in. = _?_ m   4. 66 lb = _?_ kg   3. _____

                                                        4. _____

5. 80 km = _?_ mi   6. 5 liters = _?_ qt   5. _____

                                                        6. _____

7. 1657.1 g = _?_ lb   8. 3.657 qt = _?_ liters   7. _____

                                                        8. _____

9. 53.25 in. = _?_ cm   10. 76.45 kg = _?_ lb   9. _____

                                                        10. _____

11. 84.6 in. = _?_ m   12. 319.6 g = _?_ oz   11. _____

                                                        12. _____

13. 60 yd = _?_ m   14. 15.5 c = _?_ liters   13. _____

                                                        14. _____

15. 32 liters = _?_ gal   16. $\frac{4}{5}$ pt = _?_ ml   15. _____

                                                        16. _____

17. WHICH IS TALLER:
    A woman who is 5' tall, or a woman who is 150 cm tall?   17. _____

18. WHICH MAKES A SMALLER HOLE:
    A 7/16 inch bit, or a 12 mm bit?   18. _____

6-6 U.S. and Metric Temperatures

In the last four sections we have discussed ways of measuring length, weight, and volume. Another quantity that is commonly measured is temperature. We are always concerned about how hot or cold the day is going to be. When we feel sick, we wonder if we are running a "temperature." In the U.S. system, we measure temperature using the Fahrenheit scale. In the metric system, temperature is measured using the Centigrade or Celsius scale. A comparison of the two scales is listed below:

Centigrade		Fahrenheit
100°	water boils	212°
37°	normal body temp	98.6°
30°	a hot day	86°
20°	room temperature	68°
10°	a cool day	50°
0°	water freezes	32°

This clever rhyme, courtesy of the Newhouse News Service, gives an easy way to relate to the temperatures on the Centigrade scale:

"Celsius: 30 is hot, 20 is nice, 10 is cool, and 0 is ice."

To convert a temperature in one scale to a temperature in the other scale, we will use one of the formulas below:

To Find the Centigrade Temperature

$$C = \frac{5}{9} \times (F - 32)$$

In other words, to find the Centigrade temperature, you should subtract 32 from the Fahrenheit temperature, then multiply that result by 5/9.

To Find the Fahrenheit Temperature

$$F = \frac{9}{5} \times C + 32$$

In other words, to find the Fahrenheit temperature, you should multiply the Centigrade temperature by 9/5, then add 32 to that result.

Example 1: 95°F = __?__ °C

Since we want to find the Centigrade temperature, we will use the first formula.

$C = \frac{5}{9} \times (F - 32)$

$= \frac{5}{9} \times (95 - 32)$ Replace the F in the formula with 95°.

$= \frac{5}{\cancel{9}_1} \times \cancel{63}^{7}$ Work inside the parentheses first, then cross cancel.

$= 35°$ Do the multiplication.

Example 2: 66°C = __?__ °F

Since we want to find the Fahrenheit temperature, we will use the second formula.

$F = \frac{9}{5} \times C + 32$

$= \frac{9}{5} \times 66 + 32$ Replace the C in the formula with 66°.

$= 118.8 + 32$ Do the multiplication first.

$= 150.8°$ Do the addition.

Example 3: On July 13, 1913, the temperature in Death Valley reached a temperature of 134°F in the shade. To the nearest tenth of a degree, what temperature would that be on the Celsius scale?

$C = \frac{5}{9} \times (F - 32)$

$= \frac{5}{9} \times (134 - 32) = \frac{5}{\cancel{9}_3} \times \cancel{102}^{34} = \frac{170}{3} \doteq 56.7°$

PROBLEM 1: In Browning, Montana, on January 23, 1916, the temperature changed 37.8°C in a 24 hour period. To the nearest degree, how much is that change on the Fahrenheit scale?

Answer: 100°F

$F = \frac{9}{5} \times 37.8 + 32$

$= 68.04 + 32$

$= 100.04$

$\doteq 100°$

EXERCISE 6-6   SET A

NAME _____   DATE _____

$$F = \frac{9}{5} \times C + 32 \qquad C = \frac{5}{9} \times (F - 32)$$

ANSWERS

1. 70°C = ___?___ °F   2. 45°C = ___?___ °F   1. _____

                                                    2. _____

3. 104°F = ___?___ °C   4. 50°F = ___?___ °C   3. _____

                                                    4. _____

5. 52°C = ___?___ °F   6. 82°C = ___?___ °F   5. _____

                                                    6. _____

7. 350°F = ___?___ °C   8. 100°F = ___?___ °C   7. _____

                                                    8. _____

9. 72.6°C = ___?___ °F   10. 82.7°C = ___?___ °F   9. _____

                                                    10. _____

WHICH IS HOTTER:

11. 200°C or 390°F?   11. _____

12. 25°C or 77°F?   12. _____

13. 39°F or 4°C?   13. _____

EXERCISE 6-6   SET B

NAME _____   DATE _____

$$F = \frac{9}{5} \times C + 32 \qquad\qquad C = \frac{5}{9} \times (F - 32)$$

ANSWERS

1. 60°C = _?_ °F   2. 85°C = _?_ °F   1. _____

                                                         2. _____

3. 68°F = _?_ °C   4. 95°F = _?_ °C   3. _____

                                                         4. _____

5. 43°C = _?_ °F   6. 73°C = _?_ °F   5. _____

                                                         6. _____

7. 400°F = _?_ °C   8. 275°F = _?_ °C   7. _____

                                                         8. _____

9. 93.4°C = _?_ °F   10. 75.3°C = _?_ °F   9. _____

                                                        10. _____

WHICH IS HOTTER:

11. 300°C or 570°F?   11. _____

12. 15°C or 59°F?   12. _____

13. 2°C or 36°F?   13. _____

6-7 Statistical Graphs

In order to present statistical information in a more visual and understandable manner, we often use bar, line, and circle graphs. These **graphs** are diagrams that display numerical facts in a way that aids us in interpreting and comparing those facts.

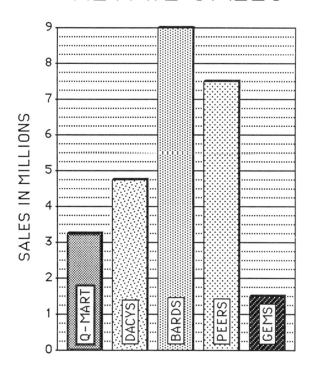

Bar Graphs

Bar graphs use parallel bars to represent numerical information. The bar graph on the left compares five retail stores and the amounts of sales each made during the year. Along the horizontal axis are the names of the stores. Along the vertical axis are the sales in millions of dollars. By studying the graph, you can quickly make comparisons and conclusions about the five stores.

Example 1: Which store had the most sales and how much did it sell?

Bards had the most sales, which was $9,000,000. (Its bar is the longest and reaches to the 9 on the vertical axis.)

Example 2: How much in sales did Dacys have?

Dacys sold $4,750,000, since the bar for Dacys reaches 3/4 of the way between 4 and 5 million, and 3/4 of $1,000,000 is $750,000.

Example 3: How much more did Q-Mart sell than Gems?

Q-Mart sold $1,750,000 more than Gems.

Q-Mart:	$3,250,000
Gems:	−$1,500,000
difference:	$1,750,000

Example 4: What is the average of the sales for the five stores?

To find the average, you add up the sales of the five stores and divide by five.

$26,000,000 ÷ 5 = $5,200,000

So the average is $5,200,000.

Q-Mary:	$3,250,000
Dacys:	$4,750,000
Bards:	$9,000,000
Peers:	$7,500,000
Gems:	$1,500,000
TOTAL:	$26,000,000

Line Graphs

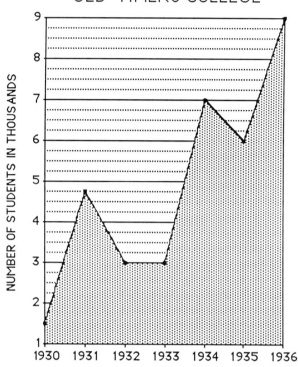

A graph that uses line segments to represent numerical information is called a **line graph**. Line graphs are used primarily when one of the axes measures time or distance. In the line graph on the left, you can quickly see that when the line rises, enrollment increases; when the line falls, enrollment decreases; when the line is level, enrollment is unchanged. By studying the graph, you can make various comparisons and conclusions about the enrollment at Old-Timers College.

Example 5: During what year was the enrollment lowest, and what was that enrollment?

The year of lowest enrollment was 1930, since in that year the graph reaches its lowest value of 1500.

Example 6: Between what two years was there the largest increase in enrollment, and what was that increase?

The largest increase in enrollment was between 1933 and 1934. The graph rises the most between those two years.

$$\begin{array}{rr} 1934: & 7000 \text{ students} \\ 1933: & -3000 \text{ students} \\ \hline \text{increase:} & 4000 \text{ students} \end{array}$$

Example 7: What percent did the enrollment decrease between 1931 and 1932?

$$\begin{array}{rr} 1931: & 4750 \text{ students} \\ 1932: & -3000 \text{ students} \\ \hline \text{decrease:} & 1750 \text{ students} \end{array}$$

To find the percent decrease, you can restate the problem as follows: "1750 is what percent of 4750?" Using the percent proportion, A = 1750, B = 4750, and P is unknown.

$$\frac{A}{B} = \frac{P}{100} \qquad \frac{1750}{4750} = \frac{P}{100}$$

$$\frac{4750 \times P}{4750} = \frac{175{,}000}{4750}$$

$$P \doteq 36.84$$

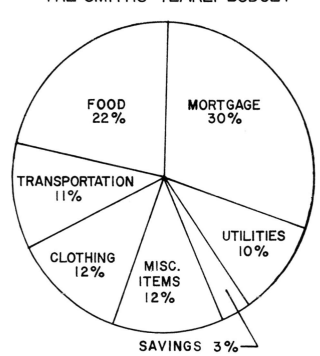

Circle Graphs

Circle graphs are used to show the division of a whole into parts. In the circle graph on the left, the whole represents the Smiths' yearly income and the parts represent the portion of that income spent on various items. By studying the size of the sectors of the circle, you can readily make conclusions about the Smiths' yearly budget.

Example 8: If the Smiths earn $19,500 per year, how much would they spend on food each year?

22% is spent on food each year

22% of $19,500 = .22 x $19,500

= $4290

Example 9: If the Smiths earn $24,000 per year, how much would they spend on clothing each month?

12% of the income is spent on clothing each year

12% of $24,000 = .12 x $24,000

= $2880 each year

To find the amount spent each month, divide by 12.

$2880 ÷ 12 = $240

Example 10: If the Smiths spend $1980 a year on transportation, what is their yearly income?

Since the Smiths spend 11% of their income on transportation, you can restate the problem as follows: "$1980 is 11% of what income?"

Using the percent proportion with A = 1980, P = 11, and B the unknown.

$$\frac{A}{B} = \frac{P}{100} \qquad\qquad \frac{1980}{B} = \frac{11}{100}$$

$$\frac{11 \times B}{11} = \frac{198{,}000}{11}$$

B = $18,000

When an object is dropped from above the earth, it accelerates to earth because of gravity. The graph on the right gives the distance an object falls after a given number of seconds.

DISTANCE TRAVELED BY FALLING OBJECTS

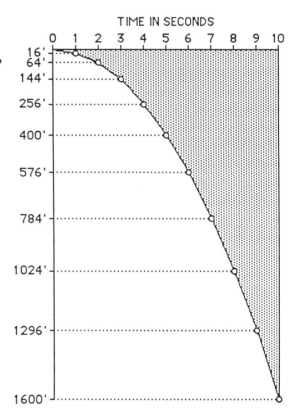

PROBLEM 1: How far did the object fall after 6 seconds.?

Answer: 576 ft

PROBLEM 2: How far did the object fall from the 5th to the 10th seconds?

Answer: 1200 ft

```
after 10 sec.:    1600
after  5 sec.:    -400
    difference:   1200
```

WIND-CHILL TEMPERATURES FOR 35° FAHRENHEIT

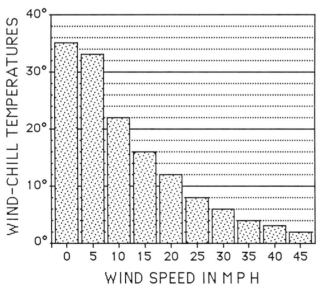

Wind-chill temperatures represent the combined effect of temperature and wind upon the body. That means the wind makes it feel colder than it really is. The graph on the left shows the effect the wind has when the temperature is actually 35° Fahrenheit.

PROBLEM 3: A wind of 15 mph makes a 35° temperature feel equivalent to what temperature?

Answer: 16°

PROBLEM 4: Between what two wind speeds is there the greatest decrease in wind-chill temperature?

Answer: Between 5 mph and 10 mph

EXERCISE 6-7   SET A

NAME _____   DATE _____

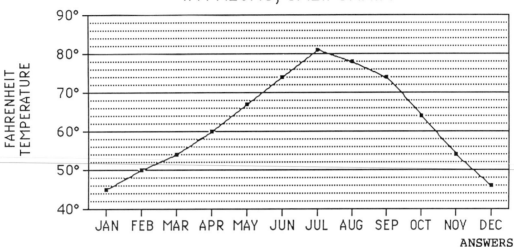

ANSWERS

1. What is the highest average monthly temperature in Fresno?   1. _____
2. What is the lowest average monthly temperature in Fresno?   2. _____
3. What is the average temperature for the year?   3. _____
4. What is the percent increase in temperature from Jan. to May?   4. _____
5. What is the percent decrease in temperature from May to Dec.?   5. _____

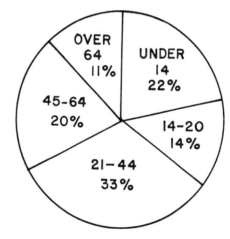

UNITED STATES POPULATION BY AGE GROUPS

IF THE POPULATION OF THE UNITED STATES IS 227,000,000,

6. What age group has the largest number of people?   6. _____
7. What age group has the smallest number of people?   7. _____
8. How many people live in the U.S. who are over 20?   8. _____
9. How many people live in the U.S. who are under 45?   9. _____
10. How many more people are under 14 than over 64?   10. _____

EXERCISE 6-7   SET B

NAME _____   DATE _____

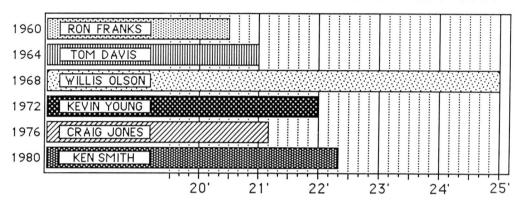

1. How long was the longest jump in feet and inches?   1. _____
2. How long was the shortest jump in feet and inches?   2. _____
3. What was the average distance jumped from 1960 to 1980? 3. _____
4. What is the percent decrease in distance from 1968 to 1972? 4. _____
5. What is the percent increase in distance from 1964 to 1968? 5. _____

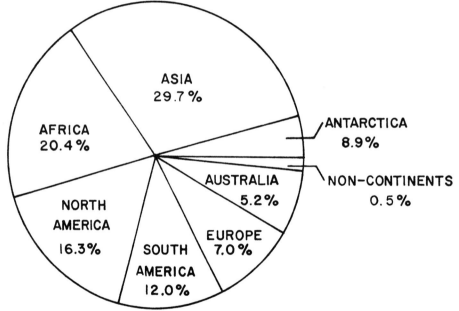

SINCE THE LAND AREA OF THE EARTH IS ABOUT 57,500,000 sq. miles,

6. What is the largest continent and what is its area?   6. _____
7. What is the smallest continent and what is its area? 7. _____
8. How much larger is Africa than Europe in sq. miles?   8. _____
9. How much larger is North America than South America? 9. _____
10. What is the total land area of North and South America? 10. _____

318

Chapter 6 Summary

Concepts

You may refer to the sections listed below to review how to:

1. read measuring devices. (6-1)

2. change units of measurement in the U.S. system using the unit-fraction method. (6-2)

3. represent the units of measurement in the metric system by using a combination of a prefix and a basic unit. (6-3)

4. change units of measurement in the metric system by using either the unit-fraction method or movement of the decimal point. (6-4)

5. make conversions between measurements in the U.S. and the metric systems utilizing the unit-fraction method. (6-5)

6. make conversions between Centigrade and Fahrenheit temperatures using the formulas:

$$C = \frac{5}{9} \times (F - 32) \quad\quad \text{or} \quad\quad F = \frac{9}{5} \times C + 32 \quad\quad (6\text{-}5)$$

7. analyze bar, line, and circle graphs. (6-6)

Terminology

This chapter's important terms and their corresponding page numbers are:

bar graph: a graph that uses parallel bars to represent numerical information. (313)

centi: prefix meaning hundredth. (293)

circle graph: a graph that uses sectors of a circle to show the division of a whole into parts. (315)

deca: prefix meaning ten. (293)

gram: the basic unit of weight in the metric system. (293)

hecto: prefix meaning hundred. (293)

kilo: prefix meaning thousand. (293)

line graph: a graph that uses line segments to represent numerical information. (314)

liter: the basic unit for volume in the metric system. (293)

meter: the basic unit for length in the metric system. (293)

milli: prefix meaning thousandth. (293)

unit-fraction: a fraction with equivalent measurements in its numerator and its denominator, giving it a value of 1. (289)

U.S. Conversions

Length

1 foot (ft) = 12 inches (in.)
1 yard (yd) = 3 feet (ft)
1 mile (mi) = 5280 feet (ft)

Weight

1 pound (lb) = 16 ounces (oz)
1 ton (T) = 2000 pounds (lb)

Volume

cup (c) = 8 fluid ounces (fl oz)
1 pint (pt) = 2 cups (c)
1 quart (qt) = 2 pints (pt)
1 gallon (gal) = 4 quarts (qt)

U.S.—Metric Conversions

Length

1 in. ≐ 2.54 cm
39.4 in. ≐ 1 m
0.621 mi ≐ 1 km

Volume

1.06 qt ≐ 1 liter

Weight

1 oz ≐ 28.35 g
1 lb ≐ 454 g
2.2 lb ≐ 1 kg

Fahrenheit—Centigrade Conversions

$$F = \frac{9}{5} \times C + 32 \qquad C = \frac{5}{9} \times (F - 32)$$

Student Notes

Chapter 6 Practice Test A

ANSWERS

1. What amount is indicated by the arrow below?

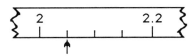

2. 8.7 tons = _?_ lb   3. 162 in. = _?_ ft

4. 24 pt = _?_ gal   5. 1 kg = _?_ g

6. 1 m = _?_ cm   7. 8 g = _?_ mg

8. 1697 ml = _?_ liters   9. 8.24 cm = _?_ mm

10. 5 km = _?_ mi   11. 101.6 cm = _?_ in.

12. 0.04 kg = _?_ oz   13. Which is longer: 400 m or 440 yd?

14. What Centigrade temperature corresponds to 104° Fahrenheit?

15. In the line graph on page 314, what percent did enrollment increase from 1930 to 1931?

1. _____
2. _____
3. _____
4. _____
5. _____
6. _____
7. _____
8. _____
9. _____
10. _____
11. _____
12. _____
13. _____
14. _____
15. _____

Chapter 6 Practice Test B

ANSWERS

1. What amount is indicated by the arrow below?

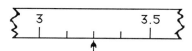

2. 12.5 ft = __?__ in. 3. 36 lb = __?__ oz

4. $8\frac{3}{4}$ pt = __?__ fl oz 5. 1 km = __?__ m

6. 1 liter = __?__ ml 7. 5 m = __?__ cm

8. 1555 mg = __?__ g 9. 6.23 mm = __?__ cm

10. 50 m = __?__ in. 11. 4.968 mi = __?__ km

12. 324 fl oz = __?__ liters 13. Which is heavier: 1 ton or 1000 kg?

14. What Fahrenheit temperature corresponds to 45° Centigrade?

15. In the line graph on page 317, what percent did the temperature increase in Fresno, California, from January to July?

1. _____
2. _____
3. _____
4. _____
5. _____
6. _____
7. _____
8. _____
9. _____
10. _____
11. _____
12. _____
13. _____
14. _____
15. _____

Chapter 6 Supplementary Exercises

Section 6-1

Determine the length in inches indicated by the letters:

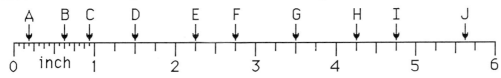

1. A
2. B
3. C
4. D
5. E
6. F
7. G
8. H
9. I
10. J

Determine the weight in kilograms (kg) indicated by the letters:

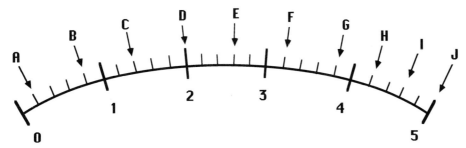

11. A
12. B
13. C
14. D
15. E
16. F
17. G
18. H
19. I
20. J

Section 6-2

1. 4 ft = __?__ in.
2. 18 ft = __?__ yd
3. 5 c = __?__ fl oz
4. 56 fl oz = __?__ c
5. 20 qt = __?__ gal
6. 50 gal = __?__ qt
7. 7920 ft = __?__ mi
8. 3.5 mi = __?__ ft
9. $4\frac{1}{2}$ yd = __?__ in.
10. 51 in. = __?__ ft
11. 7500 lb = __?__ T
12. 5.43 T = __?__ lb
13. 729.6 oz = __?__ lb
14. 3.125 lb = __?__ oz
15. 31,680 in. = __?__ mi
16. .62 mi = __?__ in.
17. 37,676 oz = __?__ T
18. 1.225 T = __?__ oz
19. 3.1 mi = __?__ yd
20. 1320 yd = __?__ mi

Section 6-3

Write in words:

1. cm 2. kl 3. mg 4. mm 5. kg 6. cl 7. km 8. ml

Replace the ? with the correct number:

9. 1 kg = _?_ g
10. 1 g = _?_ kg
11. 1 m = _?_ mm
12. 1 mm = _?_ m
13. 1 cl = _?_ liter
14. 1 liter = _?_ cl

Section 6-4

1. 6000 g = _?_ kg
2. 3 km = _?_ m
3. 7 liter = _?_ ml
4. 5 g = _?_ cg
5. 3250 mg = _?_ g
6. 5.6 kg = _?_ g
7. 6.85 m = _?_ cm
8. 12 ml = _?_ cl
9. 53 cl = _?_ ml
10. 16.25 g = _?_ mg
11. 328.8 cg = _?_ g
12. 72.6 mm = _?_ cm
13. 4.75 km = _?_ m
14. 0.85 m = _?_ mm
15. 12,500 mm = _?_ m
16. 1.06 liters = _?_ cl
17. 2.5 g = _?_ mg
18. 63.8 g = _?_ kg
19. 14 cl = _?_ ml
20. 9.4 liters = _?_ ml

Section 6-5

1. 17.78 cm = _?_ in.
2. 12 in. = _?_ cm
3. 13.2 lb = _?_ kg
4. 5 kg = _?_ lb
5. 8 m = _?_ in.
6. 157.6 in. = _?_ m
7. 170.1 g = _?_ oz
8. 5 oz = _?_ g
9. 15.9 qt = _?_ liters
10. 4.3 liters = _?_ qt
11. 5000 km = _?_ mi
12. 6.21 mi = _?_ km
13. 0.75 lb = _?_ g
14. 1021.5 g = _?_ lb
15. 3.5 ft = _?_ cm
16. 76.2 cm = _?_ ft

17. Which is heavier: 1000 pounds or 500 kilograms?
18. Which is longer: 3 miles or 5 kilometers?
19. John is 6 feet tall and Frank is 180 centimeters tall. Who is taller?
20. Hazel weighs 110 pounds and Susan weighs 55.5 kilograms. Who weighs more?
21. A road sign in Canada reads "Jasper 8.5 km." How far is that in miles?
22. A speed limit sign in Mexico, given in kilometers per hour, reads "90." What is that speed in miles per hour?

Section 6-6

1. 41°F = _?_ °C
2. 20°C = _?_ °F
3. 45°C = _?_ °F
4. 86°F = _?_ °C
5. 185°F = _?_ °C
6. 155°C = __ °F
7. 37.6°C = _?_ °F
8. 80.3°F = _?_ °C
9. 74.9°F = _?_ °C
10. 95°C = _?_ °F
11. 242.5°C = _?_ °F
12. 110.2°F = _?_ °C
13. 82.76°F = _?_ °C
14. 1000°C = _?_ °F
15. 88.25°C = _?_ °F
16. 91.85°F = _?_ °C
17. 212°F = _?_ °C
18. 7.75°C = _?_ °F
19. 64°C = _?_ °F
20. 109.76°F = _?_ °C

Section 6-7

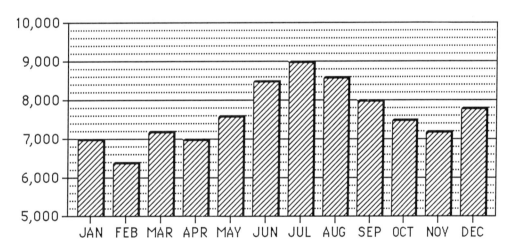

1. How many accidental deaths were recorded in March?
2. How many accidental deaths were recorded in June?
3. During what month were there 7500 accidental deaths?
4. During what month were there 7600 accidental deaths?
5. What is the percent increase in accidental deaths from February to July?
6. What is the percent decrease in accidental deaths from July to November?
7. What is the average number of accidental deaths occurring during the summer months of June, July, and August?
8. What is the average number of accidental deaths occurring during the winter months of December, January, and February?

Unit III Exam

<u>Chapters 5 and 6</u>   ANSWERS

Solve each proportion:

1. $\dfrac{7}{15} = \dfrac{N}{105}$ 2. $\dfrac{13.5}{7.6} = \dfrac{2.25}{Y}$

1. _____

2. _____

3. Express 45% as a decimal and reduced fraction.   3. _____

4. 54 is what percent of 72?   4. _____

5. 22% of what number is 19.03?   5. _____

6. 2.5 mi = _?_ in.   7. 5600 mg = _?_ g   6. _____

7. _____

8. 1.5 qt = _?_ ml   9. 60°C = _?_ °F   8. _____

9. _____

10. Which is heavier: 600 g or 30 oz?   10. _____

11. On a drafting ruler each inch is divided into 32 equal parts. What distance is indicated by the mark just before the mark at 5 inches?   11. _____

12. On a circle graph showing the budget for a company, the sector representing advertising is 12% of the graph. If the company's total budget is $256,000, how much is spent on advertising?   12. _____

13. The scale on a road map is 1.5 inches equals 90 miles. What is the actual distance between two cities that are 7 inches apart on that map?   13. _____

14. After one year the resale value of an $8000 truck was only $5,500. That drop in price represents what percent decrease in the value of the truck?   14. _____

15. How much simple interest is charged on a $876 loan for 2 years with a yearly interest rate of 11%?   15. _____

Unit IV

Introductions to Algebra and Geometry

In the first three units you covered basic arithmetic. The next step on the path of math development is algebra and geometry. You are probably thinking to yourself, "What are algebra and geometry? They sound very difficult." You shouldn't be too worried, because people seem to put an unwarranted mystique around these subjects.

To understand what algebra is all about, you must realize that, in algebra, letters are used to stand for numbers. Just as you operated with numbers in arithmetic, in algebra you simple replace those numbers with letters and work with them. In geometry you learn about the shapes you see in the world around you, and how the arithmetic you learned applies to those shapes.

In this unit you will see that algebra goes beyond the computations of arithmetic and gives you a system to solve problems of increased difficulty. You will also see how a knowledge of geometry will help you classify and measure the shapes you see each day. After finishing this unit you should be able to do the following:

1. work with algebraic variables.
2. operate with positive and negative numbers.
3. solve equations containing one variable.
4. define basic geometric shapes.
5. classify common geometric shapes.
6. find perimeters and areas of various plane figures.

Unit IV

Brain Buster

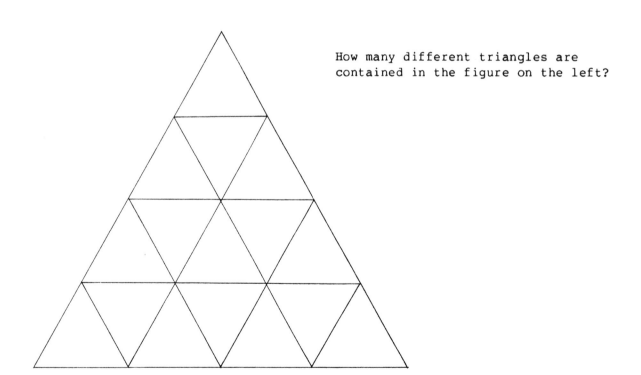

How many different triangles are contained in the figure on the left?

Chapter 7

Introduction to Algebra

After completing this chapter you should be able to do the following:

1. evaluate algebraic expressions.
2. compare the size of signed numbers.
3. operate with positive and negative numbers.
4. solve equations using the addition property.
5. solve equations using the multiplication and division properties.
6. solve word problems using equations.

On the next page you will find a pretest for this chapter. The purpose of the pretest is to help you determine which sections in this chapter you need to study in detail and which sections you can review quickly. By taking and correcting the pretest according to the instructions on the next page you can better plan your pace through this chapter.

Chapter 7 Pretest

Take and correct this test using the answer section at the end of the book. Those problems that give you difficulty indicate which sections need extra attention. Section numbers are in parentheses before each problem.

ANSWERS

(7-1) 1. If A = 5, B = 4, and C = 1, then 3A − 2B + C = ?   1. _____

(7-2) Replace the ? with > or <:

    2. 0 ? −9   3. −7 ? −5   2. _____

                                                                                                                              3. _____

Perform the indicated operations:

(7-3) 4. −16 + 9   (7-3) 5. −14 + (−17)   4. _____

                                                                                                                           5. _____

(7-4) 6. 5 − 17   (7-4) 7. −8 − (−6)   6. _____

                                                                                                                                    7. _____

(7-5) 8. 7(−6)   (7-5) 9. (−5)(−9)   8. _____

                                                                                                                                    9. _____

(7-6) 10. $\dfrac{-2 + 10(3 - 16)}{-5 - 7}$ 10. _____

Solve for X in each equation:

(7-7) 11. X + 6 = 15   (7-8) 12. X − 14 = −36   11. _____

                                                                                                                                12. _____

(7-9) 13. 152 = 4X   (7-10) 14. $\dfrac{X}{7} + 30 = 3$ 13. _____

                                                                                                                                14. _____

(7-11) 15. The difference between six times a number and 17 is −35. Find the number.   15. _____

330

7-1 From Numbers to Letters

IN ORDER TO UNDERSTAND "X" YOU MUST BE VARIABLE.

The basis of algebra is its use of letters to stand for numbers. The letters used are called **variables**. The first thing that you have to accept and understand in algebra is that letters can be used in the same ways as numbers. Consider the basic operations:

Addition

    A + B   means that two numbers are being added.
A and B represent the two addends.

Subtraction

    p - q   means that two numbers are being subtracted.
p and q represent those numbers.

Division

    $f \div R$ or $\frac{f}{R}$ means that two numbers are being divided. f and R represent those numbers.

Multiplication

    X·Y or (X)(Y) or XY   means that two numbers are being multiplied.
X and Y represent the two factors.

Note: We do not use the letter "x" to signify multiplication, since in algebra the "x" represents a number. Multiplication can be displayed in three other ways:

    1. using a dot: 5·K
    2. using parentheses: 5(k)
    3. using no sign between factors: 5k

Combined Operations

    4x + 3y   means 4 times a number (x) plus 3 times a number (y)

    $\frac{P}{RT}$ means a number (P) is divided by the product of two numbers (R and T).

Do not be confused by those letters in algebra. Just keep telling yourself that letters are used just like numbers, since they simply represent numbers.

The first type of algebra problem involves finding the value of **algebraic expressions**--expressions that have numbers, variables, and operations. We can find the value of an algebraic expression if we know the value of each variable and use the proper order of operations as discussed in Section 2-8.

Example 1: If $x = 5$, and $y = 2$, then $4x + 3y = ?$

$$4x + 3y = 4 \cdot 5 + 3 \cdot 2$$ Replace each variable with its value.

$$= 20 + 6$$

$$= 26$$ Do the computation using the proper order of operations.

Example 2: If $P = 10$ and $q = \frac{1}{2}$, then $3(P - 2q) = ?$

$$3(P - 2q) = 3(10 - 2 \cdot \tfrac{1}{2})$$ Replace each variable with its value.

$$= 3(10 - 1)$$

$$= 3(9)$$ Do the computation using the proper order of operations.

$$= 27$$

Example 3: If $S = P + PRT$ and $P = 5200$, $R = 12\%$, and $T = 3$, find the value of S.

$$S = P + PRT$$

$$= 5200 + 5200(.12)(3)$$ Replace the P, R, and T with the given values.

$$= 5200 + 1872$$

$$= 7072$$ Do the computation using the proper order of operations.

PROBLEM 1: If $H = .6$, $Z = 36$, and $W = 1$, then $ZH - W + \frac{Z}{H} = ?$

Answer: 80.6

$$= 36(.6) - 1 + \frac{36}{.6}$$

$$= 21.6 - 1 + 60$$

$$= 80.6$$

PROBLEM 2: If $A = \frac{3}{4}$ and $B = 6$, then $A(20 - 2B + 8) = ?$

Answer: 12

$$= \tfrac{3}{4}(20 - 2 \cdot 6 + 8)$$

$$= \tfrac{3}{4}(20 - 12 + 8)$$

$$= \tfrac{3}{4}(16) = 12$$

EXERCISE 7-1   SET A

NAME _____   DATE _____

IF A = 2, B = 7, AND C = 1, EVALUATE THE FOLLOWING: ANSWERS

1. A + B − C   2. B + C − A   1. _____

                                              2. _____

3. $\dfrac{4B}{A}$ 4. $\dfrac{7A}{B}$ 3. _____

                                              4. _____

5. 2BC − 7A   6. 14AC − 4B   5. _____

                                              6. _____

7. A(3B + 5A)   8. A(9C + 2B)   7. _____

                                              8. _____

EVALUATE THE FOLLOWING:

9. If x = 1.2, y = 5, and z = 24, then $xy + \dfrac{z}{x} - \dfrac{y}{2}$ = ?   9. _____

10. If p = 5.8, q = 6, and r = 29, then $pq - \dfrac{q}{5} + \dfrac{r}{p}$ = ?   10. _____

11. If M = $\dfrac{3}{4}$ and N = $5\dfrac{1}{2}$, then M(3N − 2M) = ?   11. _____

12. If R = $\dfrac{5}{8}$ and S = $3\dfrac{1}{5}$, then R(2S + 8) = ?   12. _____

13. If f = $\dfrac{2}{5}$, g = .65, and h = 15, then $3hf + h(\dfrac{h}{f} - 6g)$ = ?   13. _____

EXERCISE 7-1   SET B

NAME _____   DATE _____

IF A = 3, B = 6, AND C = 1, EVALUATE THE FOLLOWING: ANSWERS

1.   C + B - A   2.   A + B - C   1. _____

2. _____

3. $\dfrac{2B}{A}$ 4. $\dfrac{8A}{B}$ 3. _____

4. _____

5.   4AC - 2B   6.   2BC - 4A   5. _____

6. _____

7.   A(2C + 3B)   8.   B(2A + 4C)   7. _____

8. _____

EVALUATE THE FOLLOWING:

9.   If h = 3.5, j = 6, and k = 7, then $jk + \dfrac{h}{k} + \dfrac{h}{2}$ = ?   9. _____

10. If q = 2.8, r = 5, and s = 20, then $rq - \dfrac{r}{4} + \dfrac{r}{s}$ = ?   10. _____

11. If U = $\dfrac{2}{3}$ and V = $1\dfrac{1}{4}$, then U(3V - 3) = ?   11. _____

12. If p = $\dfrac{7}{8}$ and q = $2\dfrac{4}{5}$, then q(8p - 2) = ?   12. _____

13. If d = $\dfrac{4}{5}$, e = .45, and f = 25, then $2df + f(\dfrac{f}{5} - 8e)$ = ?   13. _____

7-2 Signed Numbers

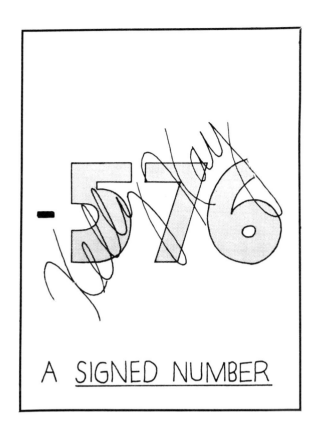

Through our study of arithmetic we have worked with numbers that were larger than zero. These numbers are called positive numbers, and are sometimes written with a "+" sign, such as +5 instead of just 5. Here the "+" sign is not used to signify the operation of addition. It is used to indicate that the number is larger than zero--a positive number.

Before we can proceed into an introduction of algebra, we must learn about another group of numbers called negative numbers. You have probably encountered these numbers before.

> If the temperature is 12° below zero, we say it is minus twelve degrees (-12°).

> If you write a check for $50.75, when you have only $20.00 in your checking account, you would be overdrawn and have a balance of -$30.75.

These numbers that have values less than zero are the negative numbers. They are the opposites of positive numbers. The "-" sign used with these numbers does not signify the operation of subtraction. It means you have a number that is less than zero. These two types of numbers, positive and negative numbers, are called **signed numbers,** since they are written with either a positive sign (+) or a negative sign (-).

The number zero separates the positive numbers and the negative numbers. It is neither positive nor negative. It is the only neutral number in our number system. We can show the relative values of signed numbers by placing them along a line called **a number line.**

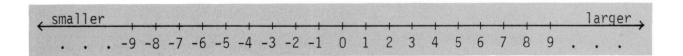

You can use the number line to compare signed numbers. *As you go to the right along the number line, the numbers get larger; and as you go to the left along the number line, the numbers get smaller.*

For example: 3 > -2   3 is greater than -2, since 3 is farther to the right on the number line than -2.

-4 < 0   -4 is less than 0, since -4 is farther to the left on the number line than 0.

-6 > -10   -6 is greater than -10, since -6 is farther to the right on the number line than -10.

-12 < -7   -12 is less than -7, since -12 is farther to the left on the number line than -7.

Fractional and decimal numbers can also be placed on the number line by placing them between the positive and negative whole numbers.

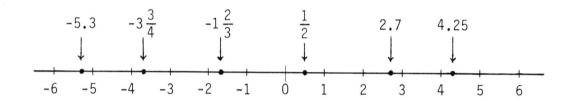

PROBLEM 1: Place the following numbers in the correct position on the number line:

2, -3, ¼, 3.6, -.2, -1½

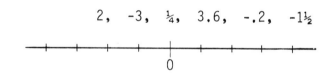

Answer:

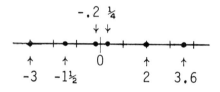

PROBLEMS: Replace the ? with > or <:

2. 0 ? -7

3. -400 ? -506

4. -4.5 ? $2\frac{1}{3}$

5. +45.67 ? +45.607

6. $-6\frac{2}{3}$? $-6\frac{1}{2}$

Answers:

2. >

3. >

4. <

5. >

6. <

EXERCISE 7-2   SET A

NAME _____   DATE _____

LOCATE THE FOLLOWING NUMBERS ON THE NUMBER LINE: ANSWERS

1. -2, 2, 3, -4

2. -3, 3, -1, 4

3. $\frac{3}{4}$, $-2\frac{1}{2}$, $3\frac{2}{3}$, $-3\frac{1}{3}$

4. $-\frac{1}{2}$, $1\frac{2}{3}$, $-3\frac{3}{4}$, $4\frac{1}{2}$

5. 3.5, -2.25, .6, -1.9

REPLACE THE ? WITH > OR <:

6. 5 ? 3   7. 4 ? 2   6. _____

                                          7. _____

8. 0 ? -3   9. 0 ? -2   8. _____

                                          9. _____

10. -6 ? 3   11. -7 ? 2   10. _____

                                          11. _____

12. -7 ? -4   13. -6 ? -2   12. _____

                                          13. _____

14. $-5\frac{1}{2}$? $-5\frac{3}{4}$ 15. $-7\frac{1}{2}$? $-7\frac{3}{4}$ 14. _____

                                          15. _____

16. -1.6 ? -1.5   17. -2.7 ? -2.6   16. _____

                                          17. _____

EXERCISE 7-2   SET B

NAME _____   DATE _____

LOCATE THE FOLLOWING NUMBERS ON THE NUMBER LINE: ANSWERS

1. -1, 1, -2, 4

2. -3, 3, 4, -4

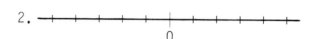

3. $\frac{1}{2}$, $-\frac{3}{4}$, $1\frac{2}{3}$, $-2\frac{1}{3}$

4. $-\frac{1}{3}$, $2\frac{1}{2}$, $-3\frac{3}{4}$, $3\frac{1}{4}$

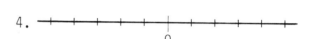

5. -2.6, .5, 3.5, -3.75

REPLACE THE ? WITH > OR <:

6. 7 ? 4   7. 8 ? 3   6. _____

                                                          7. _____

8. 0 ? -4   9. 0 ? -1   8. _____

                                                          9. _____

10. -7 ? 5   11. -8 ? 4   10. _____

                                                          11. _____

12. -9 ? -4   13. -7 ? -3   12. _____

                                                          13. _____

14. $-1\frac{1}{2}$? $-1\frac{2}{3}$ 15. $-3\frac{1}{2}$? $-3\frac{2}{3}$ 14. _____

                                                          15. _____

16. -4.7 ? -4.4   17. -5.5 ? -5.4   16. _____

                                                          17. _____

7-3 Adding Signed Numbers

Now that you understand what is meant by a signed number, you can learn to perform the basic operations on these numbers. The first operation is the addition of signed numbers. To help explain how this is done, we will use the number line. **Adding signed numbers** can be illustrated on the number line as follows:

> 1. Adding a positive number is the same as moving that many units in the positive direction.
> 2. Adding a negative number is the same as moving that many units in the negative direction.

Example 1: $5 + (-3) = ?$

That problem suggests that you start at 5 on the number line and go 3 units in the negative direction.

So $5 + (-3) = 2$

Example 2: $-7 + 4 = ?$

Start at -7 on the number line and go 4 units in the positive direction.

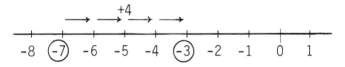

So $-7 + 4 = -3$

Example 3: $-2 + (-6) = ?$

Start at -2 on the number line and go 6 units in the negative direction.

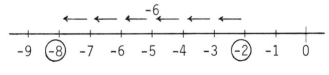

So $-2 + (-6) = -8$

If the numbers to be added had many digits, such as $-576 + 329$, using a number line to find the answer would really be impractical. Because of that difficulty, let us consider a method that will enable us to add signed numbers more quickly than using a number line.

Every number has two parts: a "+" or "-" sign, and a number part.

```
      number                number
sign┐ ┌part          sign┐ ┌part
    ↓ ↓                  ↓ ↓
    -8                   +9
```

If a number has no sign, it is understood to be a positive number.

$$16 = +16$$

> **Rules for Adding Signed Numbers**
>
> 1. If the numbers have the same signs, add their number parts and use that same sign as the sign of the answer.
>
> 2. If the numbers have different signs, subtract their number parts and use the sign of the larger number part.

Example 4: -75 + (-46) = ?

    The numbers have the same signs.

    Add the number parts.   75
                                               46
                                           121

    Since both numbers are negative, the answer is negative.

    So -75 + (-46) = -121

Example 5: 27 + (-52) = ?

    The numbers have different signs.

    Subtract their number parts.   52
                                                  27
                                               25

    Since the larger number part (52) is negative, the answer is negative.

    So -52 + 27 = -25

Even if you use the above rules to get an answer, you can picture movement on a number line to check your results.

PROBLEM 1: 167 + (-88) = ?   Answer: + 79   167
                                                                            (subtract)   88
                                                                                           79

                                                                                Use the sign of the larger number part--the "+" of 167.

EXERCISE 7-3   SET A

NAME _____ DATE _____

ADD THE FOLLOWING: ANSWERS

1.   7 + 5   2.   6 + 8   1. _____

                                                2. _____

3.   8 + (-5)   4.   9 + (-7)   3. _____

                                                4. _____

5.   -12 + 8   6.   -13 + 7   5. _____

                                                6. _____

7.   -6 + (-4)   8.   -7 + (-5)   7. _____

                                                8. _____

9.   -7 + 9   10. -9 + 12   9. _____

                                                10. _____

11. -3 + (-15)   12. -2 + (-16)   11. _____

                                                12. _____

13. 12 + (-17) + (-4) 14. 16 + (-3) + (-9)   13. _____

                                                14. _____

15. -20 + 9 + 11   16. -30 + 8 + 22   15. _____

                                                16. _____

17. 75 + (-37)   18. 84 + (-49)   17. _____

                                                18. _____

19. -124 + 76   20. -145 + 95   19. _____

                                                20. _____

21. -83 + (-156)   22. -92 + (-189)   21. _____

                                                22. _____

23. 274 + (-406)   24. 316 + (-578)   23. _____

                                                24. _____

EXERCISE 7-3   SET B

NAME _____   DATE _____

ADD THE FOLLOWING: ANSWERS

1. 8 + 4   2. 5 + 9   1. _____

                                                      2. _____

3. 7 + (-4)   4. 6 + (-5)   3. _____

                                                      4. _____

5. -11 + 5   6. -17 + 9   5. _____

                                                      6. _____

7. -7 + (-8)   8. -6 + (-8)   7. _____

                                                      8. _____

9. -5 + 12   10. -7 + 18   9. _____

                                                     10. _____

11. -4 + (-12)   12. -1 + (-16)   11. _____

                                                     12. _____

13. 13 + (-15) + (-9)   14. 15 + (-17) + (-8)   13. _____

                                                     14. _____

15. -24 + 11 + 13   16. -34 + 9 + 25   15. _____

                                                     16. _____

17. 87 + (-49)   18. 92 + (-38)   17. _____

                                                     18. _____

19. -150 + 77   20. -127 + 89   19. _____

                                                     20. _____

21. -65 + (-146)   22. -98 + (-216)   21. _____

                                                     22. _____

23. 329 + (-417)   24. 546 + (-679)   23. _____

                                                     24. _____

7-4 Subtracting Signed Numbers

In the previous section, we learned to add signed numbers by using either a number line or the rules stated on page 340. In order to perform subtraction on signed numbers, we will learn to change a subtraction problem into an addition problem. Consider the problems $5 - 2 = ?$ and $5 + (-2) = ?$.

We know that $5 - 2 = 3$ and $5 + (-2) = 3$.

Those examples are different ways to represent the same problem. The first one, $5 - 2$, is a subtraction problem, and the second one, $5 + (-2)$, is an addition problem. That discussion points out how to subtract signed numbers.

> **Rules for Subtracting Signed Numbers**
>
> 1. Change the sign of the number being subtracted.
> 2. Change the subtraction to addition.

That is, for any numbers represented by A and B:

$$A - B = A + (-B)$$

For example:

$$13 - 7 = 13 + (-7) = 6$$

$$6 - 9 = 6 + (-9) = -3$$

$$-4 - 6 = -4 + (-6) = -10$$

$$75 - 97 = 75 + (-97) = -22$$

$$-37 - 21 = -37 + (-21) = -58$$

Notice that each subtraction was changed into addition of signed numbers, then the rules for addition were applied.

PROBLEM 1: $15 - 23 = ?$ Answer: -8

$$15 - 23 = 15 + (-23) = -8$$

PROBLEM 2: $-30 - 12 = ?$ Answer: -42

$$-30 - 12 = -30 + (-12) = -42$$

From the above explanations, you will notice that the "-" sign is used in different ways.

1. To represent subtraction.
   $12 - 8$ means 12 subtract 8.

2. To signify a negative number.
   -7 indicates a number that is less than zero--a negative number.

There is yet another way to use the "−" sign. It is sometimes used to indicate the inverse or opposite of a number.

$$-(+8) \text{ indicates the opposite of a } +8, \text{ which is } -8.$$

$$-(+8) = -8$$

$$-(-3) \text{ indicates the opposite of a } -3, \text{ which is } +3.$$

$$-(-3) = +3 = 3$$

This usage of the "−" sign gives us a way to determine the result of a number that has two negative signs in front of it.

$$-(-7) = +7 = 7 \qquad -(-35) = +35 = 35 \qquad -(-103) = +103 = 103$$

That property of signed numbers should be familiar to us, since, even in everyday English, two negatives give a positive statement.

For example: "I'm *not mis*behaving!" implies the positive meaning that you are behaving.

So $5 - (-2)$ becomes $5 + 2$, which equals 7.

Similarly,
$$8 - (-7) = 8 + 7 = 15$$

$$-17 - (-8) = -17 + 8 = -9$$

$$-67 - (-67) = -67 + 67 = 0$$

You will notice that this is consistent with the rules for subtracting signed numbers as stated on page 343.

PROBLEM 3: $28 - (-12) = ?$ Answer: 40

$$28 - (-12) = 28 + 12 = 40$$

PROBLEM 4: $-8 - (-15) = ?$ Answer: 7

$$-8 - (-15) = -8 + 15 = 7$$

PROBLEM 5: $-20 - 11 = ?$ Answer: −31

$$-20 - 11 = -20 + (-11) = -31$$

PROBLEM 6: $-43 - (-31) = ?$ Answer: −12

$$-43 - (-31) = -43 + 31 = -12$$

EXERCISE 7-4   SET A

NAME _____ DATE _____

PERFORM THE INDICATED SUBTRACTIONS: ANSWERS

1. 9 - 6   2. 10 - 5   1. _____

                                                             2. _____

3. 6 - 10   4. 4 - 9   3. _____

                                                             4. _____

5. -7 - 3   6. -4 - 8   5. _____

                                                             6. _____

7. 5 - (-3)   8. 6 - (-4)   7. _____

                                                             8. _____

9. -7 - (-5)   10. -8 - (-3)   9. _____

                                                             10. _____

11. -7 - (-7)   12. -4 - (-4)   11. _____

                                                             12. _____

13. 52 - 6 - 19   14. 34 - 5 - 15   13. _____

                                                             14. _____

15. 20 - 11 - 17   16. 24 - 9 - 17   15. _____

                                                             16. _____

17. 53 - 143   18. 72 - 165   17. _____

                                                             18. _____

19. 72 - (-94)   20. 83 - (-77)   19. _____

                                                             20. _____

21. -51 - (-38)   22. -27 - (-96)   21. _____

                                                             22. _____

23. -125 - (-72)   24. -146 - (-61)   23. _____

                                                             24. _____

EXERCISE 7-4   SET B

NAME _____   DATE _____

PERFORM THE INDICATED SUBTRACTIONS: ANSWERS

1. 8 - 5   2. 9 - 4   1. _____

                                                    2. _____

3. 3 - 9   4. 5 - 11   3. _____

                                                    4. _____

5. -4 - 9   6. -6 - 8   5. _____

                                                    6. _____

7. 7 - (-2)   8. 8 - (-3)   7. _____

                                                    8. _____

9. -5 - (-3)   10. -7 - (-5)   9. _____

                                                    10. _____

11. -3 - (-3)   12. -5 - (-5)   11. _____

                                                    12. _____

13. 42 - 6 - 20   14. 41 - 8 - 19   13. _____

                                                    14. _____

15. 27 - 16 - 15   16. 25 - 18 - 15   15. _____

                                                    16. _____

17. 84 - 176   18. 93 - 157   17. _____

                                                    18. _____

19. 63 - (-82)   20. 75 - (-56)   19. _____

                                                    20. _____

21. -43 - (-39)   22. -27 - (-95)   21. _____

                                                    22. _____

23. -175 - (-82)   24. -163 - (-17)   23. _____

                                                    24. _____

7-5 Multiplying and Dividing Signed Numbers

Multiplying Signed Numbers

To arrive at a method for multiplying signed numbers, we must consider four possibilities.

1. **A positive number times a positive number.**

$$(+3) \cdot (+4) = ?$$

We know that the answer will be a positive number, 12.

2. **A positive number times a negative number.**

$$(3) \cdot (-4) = ?$$

In our first discussion of multiplication in Section 1-5, we saw that multiplication is actually a way to represent repeated addition. $3 \cdot (-4)$ means that you have three -4's; that is, $(-4) + (-4) + (-4)$. The answer to that is -12.

$$\text{So } 3 \cdot (-4) = -12$$

The result of multiplying a positive times a negative number is a negative number.

3. **A negative number times a positive number.**

$$-3 \cdot 4 = ?$$

By similar reasoning, the result of a negative number times a positive number is a negative number.

$$-3 \cdot 4 = -12$$

4. **A negative number times a negative number.**

$$-3 \cdot (-4) = ?$$

Consider the list below:

```
-3·3    = -9  ) +3      As we proceed down the list,
-3·2    = -6  ) +3      each new answer can be obtained
-3·1    = -3  ) +3      by adding 3 to a previous answer.
-3·0    = 0   ) +3
-3·(-1) = ?   ) +3      If you continue the pattern,
-3·(-2) = ?   ) +3      you will find that:
-3·(-3) = ?   ) +3
-3·(-4) = ?              -3·(-4) = 12
```

Thus a negative number times a negative number is a positive number.

The results of the discussion on the previous page can be summarized as follows:

$$(+) \cdot (+) = (+)$$
$$(-) \cdot (-) = (+)$$
$$(+) \cdot (-) = (-)$$
$$(-) \cdot (+) = (-)$$

Notice, if the numbers being multiplied have the same signs, the answer will be positive. If they have different signs, the answer will be negative.

Dividing Signed Numbers

You will be pleased to know that there are no new rules for dividing signed numbers. As we showed in Section 3-5, dividing by a number will give the same result as multiplying by an appropriate fraction. For example,

$$8 \div 2 = 8 \cdot \frac{1}{2} = 4 \qquad \text{and} \qquad 12 \div 3 = 12 \cdot \frac{1}{3} = 4$$

Since division can be changed into multiplication, the rules to obtain the sign of the answer in division are the same as those in multiplication.

> **Rules for Multiplying and Dividing Signed Numbers**
>
> 1. Multiply or divide the number parts.
> 2. If the numbers have the same signs, the answer will be positive.
> 3. If the numbers have different signs, the answer will be negative.

PROBLEMS: Perform the indicated operation: Answers:

1. $-7 \cdot 5 = ?$ 1. -35
2. $-9 \cdot (-6) = ?$ 2. 54
3. $40 \div (-8) = ?$ 3. -5
4. $+27 \div 3 = ?$ 4. 9
5. $4 \cdot (-7) = ?$ 5. -28
6. $-45 \div (-15) = ?$ 6. 3
7. $-12 \cdot 1 = ?$ 7. -12
8. $24 \div (-4) = ?$ 8. -6

EXERCISE 7-5   SET A

NAME _____   DATE _____

MULTIPLY THE FOLLOWING: ANSWERS

1. $5 \cdot (+3)$ 2. $6 \cdot (+4)$ 1. _____

                                                             2. _____

3. $3(-7)$ 4. $2(-9)$ 3. _____

                                                             4. _____

5. $-4 \cdot 8$ 6. $-5 \cdot 6$ 5. _____

                                                             6. _____

7. $(-8)(-6)$ 8. $(-7)(-5)$ 7. _____

                                                             8. _____

9. $-9 \cdot (-8)$ 10. $-7 \cdot (-9)$ 9. _____

                                                             10. _____

DIVIDE THE FOLLOWING:

11. $\dfrac{+36}{9}$ 12. $\dfrac{+27}{3}$ 11. _____

                                                             12. _____

13. $\dfrac{-40}{5}$ 14. $\dfrac{-42}{7}$ 13. _____

                                                             14. _____

15. $\dfrac{56}{-8}$ 16. $\dfrac{63}{-9}$ 15. _____

                                                             16. _____

17. $\dfrac{-72}{-9}$ 18. $\dfrac{-32}{-8}$ 17. _____

                                                             18. _____

19. $\dfrac{-25}{-5}$ 20. $\dfrac{-36}{-6}$ 19. _____

                                                             20. _____

21. $\dfrac{143}{-11}$ 22. $\dfrac{144}{-12}$ 21. _____

                                                             22. _____

EXERCISE 7-5   SET B

NAME _____   DATE _____

MULTIPLY THE FOLLOWING: ANSWERS

1. 3·(+7)   2. 8·(+3)   1. _____

                                                        2. _____

3. 5(-8)   4. 3(-9)   3. _____

                                                        4. _____

5. -6·7   6. -9·7   5. _____

                                                        6. _____

7. (-8)(-7)   8. (-6)(-6)   7. _____

                                                        8. _____

9. -5·(-9)   10. -8·(-4)   9. _____

                                                        10. _____

DIVIDE THE FOLLOWING:

11. $\frac{+45}{9}$ 12. $\frac{+32}{8}$ 11. _____

                                                        12. _____

13. $\frac{-24}{6}$ 14. $\frac{-35}{7}$ 13. _____

                                                        14. _____

15. $\frac{63}{-9}$ 16. $\frac{48}{-8}$ 15. _____

                                                        16. _____

17. $\frac{-36}{-4}$ 18. $\frac{-30}{-5}$ 17. _____

                                                        18. _____

19. $\frac{-81}{-9}$ 20. $\frac{-49}{-7}$ 19. _____

                                                        20. _____

21. $\frac{156}{-13}$ 22. $\frac{180}{-12}$ 21. _____

                                                        22. _____

7-6 The Order of Operations—Signed Numbers

In Section 2-8 you encountered problems that had a combination of operations with whole numbers. In this section you will combine operations with signed numbers. The order in which the operations are done is still the same.

> **The Order of Operations**
>
> First: Do operations inside parentheses ( ) or brackets [ ].
> Second: Do powers or square roots.
> Third: Do multiplications or divisions from left to right.
> Fourth: Do additions or subtractions from left to right.

Note: If a problem contains a fraction line, do the operations above and below the fraction line separately before simplifying the fraction. Inside the parentheses and brackets you must again do multiplications and divisions before you do additions and subtractions.

Example 1: $6^2 + (5)(-2) = ?$

$= 36 + (5)(-2)$	Do the powers.
$= 36 + (-10)$	Do the multiplication.
$= 26$	Do the addition.

Example 2: $4 - \dfrac{15}{3} - 10(2) = ?$

$= 4 - 5 - 20$	Do the multiplication and division.
$= 4 + (-5) + (-20)$	Change subtractions to adding the opposite of the number.
$= -21$	Do the additions.

Example 3: $\dfrac{5(-3 + 7)}{4 + (-2)(-3)} = ?$

$= \dfrac{5(4)}{4 + 6}$	Do the addition inside the parentheses above the fraction line. Do the multiplication below the fraction line.
$= \dfrac{20}{10}$	Do the multiplication above the fraction line. Do the addition below the fraction line.
$= 2$	Simplify the fraction (divide 20 by 10).

Example 4: $-7[5 + 4(-2)] = ?$

$\quad = -7[5 + (-8)]$ Do the multiplication inside the brackets.

$\quad = -7[-3]$ Do the addition inside the brackets.

$\quad = 21$ Do the multiplication.

Example 5: $-8 - (4 - 11) + 5 = ?$

$\quad = -8 - (-7) + 5$ Do the subtraction inside the brackets.

$\quad = -8 + 7 + 5$ Change subtraction to adding the opposite of the number.

$\quad = 4$ Do the additions.

Example 6: $\dfrac{15[-4 + (-2)]}{(-4 - 1)(-9 + 8)} = ?$

$\quad = \dfrac{15[-6]}{(-5)(-1)}$ Do the operations inside the parentheses and brackets above and below the fraction line.

$\quad = \dfrac{-90}{5}$ Do the multiplication above the fraction line. Do the multiplication below the fraction line.

$\quad = -18$ Simplify the fraction (divide -90 by 5).

PROBLEM 1: $3 - 7 + 4(-2) = ?$

Answer: -12
$= 3 - 7 + (-8)$
$= -4 + (-8)$
$= -12$

PROBLEM 2: $3^2 + [-5 + (4 - 9)] = ?$

Answer: -1
$= 9 + [-5 + (-5)]$
$= 9 + [-10]$
$= -1$

PROBLEM 3: $\dfrac{2(7-4)}{5 - 9} = ?$

Answer: $-3/2$
$= \dfrac{2(3)}{-4}$
$= \dfrac{6}{-4} = -\dfrac{3}{2}$

EXERCISE 7-6 SET A

NAME _____ DATE _____

PERFORM THE INDICATED OPERATIONS: ANSWERS

1. $4^2 + (2)(-3)$ 2. $6^2 + (5)(-2)$

1. _____

2. _____

3. $(-2)(6) - 3 \cdot 5$ 4. $(4)(-3) - 2 \cdot 7$

3. _____

4. _____

5. $-5(-3 + 7)$ 6. $6(-5+8)$

5. _____

6. _____

7. $8 - \frac{8}{2} - 5 \cdot 3$ 8. $-4 + \frac{9}{3} - 6 \cdot 2$

7. _____

8. _____

9. $-3 - (4 - 6)$ 10. $5 - (6 - 7)$

9. _____

10. _____

11. $(-5 + 2 \cdot 2)(7 - 9)$ 12. $(3 \cdot 2 - 9)(-4 + 1)$

11. _____

12. _____

13. $5 + 2[3 - (-4)]$ 14. $17 - 3[2 - (-3)]$

13. _____

14. _____

15. $\dfrac{-4 - 2(8 - 5)}{5}$ 16. $\dfrac{6 + 3(5 - 3)}{3}$

15. _____

16. _____

17. $\dfrac{8 + (-6)}{5 - (-3)}$ 18. $\dfrac{3 - (-2)}{-7 + (-3)}$

17. _____

18. _____

19. $\dfrac{-9[3 + 2(-4)]}{3(4 - 1)}$ 20. $\dfrac{2(7 - 1)}{3[5 + (-3)]}$

19. _____

20. _____

EXERCISE 7-6   SET B

NAME _____   DATE _____

PERFORM THE INDICATED OPERATIONS: ANSWERS

1. $3^2 + (3)(-5)$ 2. $5^2 + (-4)(6)$ 1. _____

                                                                2. _____

3. $(-2)(5) - 3 \cdot 6$ 4. $(4)(-2) - 5 \cdot 2$ 3. _____

                                                                4. _____

5. $7(-8 + 10)$ 6. $-4(-2 + 8)$ 5. _____

                                                                6. _____

7. $-6 + \dfrac{6}{3} - 3 \cdot 2$ 8. $4 - \dfrac{8}{4} - 2 \cdot 5$ 7. _____

                                                                8. _____

9. $4 - (3 - 7)$ 10. $-6 - (2 - 9)$ 9. _____

                                                                10. _____

11. $(2 - 6)(-6 + 2 \cdot 3)$ 12. $(-8 + 3 \cdot 4)(3 - 8)$ 11. _____

                                                                12. _____

13. $[2 - (-5)] \cdot 3 - 4$ 14. $[6 - (-8)] \cdot 2 + 9$ 13. _____

                                                                14. _____

15. $\dfrac{-8 + 2(7 - 6)}{2}$ 16. $\dfrac{15 - 3(6 - 5)}{3}$ 15. _____

                                                                16. _____

17. $\dfrac{2 - (-2)}{-6 - 4}$ 18. $\dfrac{1 - (-3)}{-2 - 4}$ 17. _____

                                                                18. _____

19. $\dfrac{3(5 - 1)}{2[8 + 2(-3)]}$ 20. $\dfrac{4[-5 + 2(-1)]}{-2(3 - 4)}$ 19. _____

                                                                20. _____

7-7 What Are Equations?

Now that we have covered the operations with signed numbers, we can return to the use of variables in algebra. In Section 7-1 you learned what a variable is and how to find the value of an algebraic expression. The next and probably the most important process in elementary algebra is learning how to solve algebraic equations.

An **equation** is a statement that two quantities are equal. The value of the terms on the left of the equals sign (the left side) is equal to the value of the terms on the right of the equals sign (the right side). The following are examples of equations:

$$y - 5 = 2$$

$$X + 7 = -3$$

$$21 = 4P + 1$$

$$-\frac{1}{2}R = 8$$

In each of the above equations, a variable is used. We know that a variable is simply a letter that represents a number. In an equation, our objective is to determine what number the variable represents.

Let's try to determine what number the variable represents in each of the above equations.

1. $y - 5 = 2$ What value for y will make y - 5 equal 2?
   $y = 7$ since 7 - 5 gives an answer of 2.

2. $X + 7 = -3$ What value for X will make X + 7 equal -3?
   $X = -10$ since -10 + 7 gives an answer of -3.

3. $21 = 4p + 1$ What value for p will make 4p + 1 equal 21?
   $5 = p$ since 4·5 + 1 gives an answer of 21.

4. $-\frac{1}{2}R = 8$ What value for R will make - 1/2 times R equal 8?
   $R = -16$ since - 1/2 times -16 gives an answer of 8.

What we have done in the previous examples is solve for the variable in each equation. We determined the value for the variable that made the left side equal to the right side of each equation, without using any specific method to find the answers. We looked at each equation and mentally figured out the answers. We solved them by inspection. See if you can determine the solutions to the following equations in the same manner.

PROBLEM 1: Solve for X:

$X + 6 = 17$

Answer: X = 11

since 11 + 6 = 17

PROBLEM 2: Solve for S:

$-4 = S - 7$

Answer: S = 3

since −4 = 3 − 7

PROBLEM 3: Solve for z:

$6z = -42$

Answer: z = −7

since $6 \cdot (-7) = -42$

PROBLEM 4: Solve for T:

$3T - 4 = 5$

Answer: T = 3

since $3 \cdot 3 - 4 = 5$

PROBLEM 5: Is X = −1 a solution of this equation?

$\frac{3}{4}X + \frac{1}{2} = \frac{1}{4}$

Answer: No

$\frac{3}{4} \cdot (-1) + \frac{1}{2} = -\frac{3}{4} + \frac{2}{4}$

$= -\frac{1}{4}$

PROBLEM 6: Is Y = 9 a solution of this equation?

$0.5(Y + 3) = 6$

Answer: Yes

$0.5(9 + 3) = 0.5(12)$

$= 6$

EXERCISE 7-7   SET A

NAME _____   DATE _____

SOLVE BY INSPECTION: ANSWERS

1. A + 3 = 5   2. B + 2 = 6   1. _____

                                                       2. _____

3. X − 4 = 7   4. Y − 3 = 8   3. _____

                                                       4. _____

5. 3P = 12   6. 5z = 15   5. _____

                                                       6. _____

7. $\frac{r}{5} = 4$ 8. $\frac{S}{6} = 3$ 7. _____

                                                       8. _____

9. −5 = t + 4   10. −3 = v + 5   9. _____

                                                       10. _____

11. −16 = −8T   12. −15 = −3N   11. _____

                                                       12. _____

13. 3y + 2 = 8   14. 2k + 1 = 7   13. _____

                                                       14. _____

15. 5 = 5 − 2m   16. 7 = 7 − 3n   15. _____

                                                       16. _____

IS X = −2 A SOLUTION OF THE FOLLOWING EQUATIONS?

17. $\frac{-3X}{2} - 6 = -9$ 17. _____

18. $\frac{6X}{9} - 1 = -1$ 18. _____

19. 5X + (−7) = −17   19. _____

20. 8 + 7X = −6   20. _____

EXERCISE 7-7   SET B

NAME _____   DATE _____

SOLVE BY INSPECTION: ANSWERS

1. X + 2 = 6   2. Y + 3 = 8   1. _____

                                                       2. _____

3. C - 5 = 2   4. D - 7 = 3   3. _____

                                                       4. _____

5. 4r = 24   6. 3s = 18   5. _____

                                                       6. _____

7. $\frac{t}{5} = 3$ 8. $\frac{v}{4} = 2$ 7. _____

                                                       8. _____

9. -1 = A + 3   10. -2 = B + 4   9. _____

                                                       10. _____

11. -8 = -2R   12. -24 = -8T   11. _____

                                                       12. _____

13. 2n + 6 = 10   14. 3m + 2 = 11   13. _____

                                                       14. _____

15. 8 = 8 - 5Z   16. 6 = 6 - 3r   15. _____

                                                       16. _____

IS X = -3 A SOLUTION OF THE FOLLOWING EQUATIONS?

17. $\frac{-6X}{9} - 1 = 3$ 17. _____

18. $\frac{4X}{3} - 1 = -3$ 18. _____

19. 2X + 5 = -1   19. _____

20. 6 + 3X = -9   20. _____

7-8 Solving Equations Using the Addition Property

PROPERTY: A SOPHISTICATED ENGLISH DRINK.

The inspection method for solving equations as seen in the previous section works as long as you can mentally determine the correct value for the variable in the equation. As the problems get more involved, it will be very difficult to solve the equation just by mental inspection. In this and in the next two sections we will develop a method for solving equations that have one variable.

An equation is like a balance. If you add the same amount to both sides of the balance, it will still be in balance.

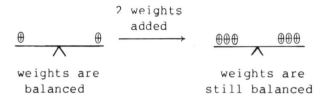

weights are balanced   weights are still balanced

What this means is that in an equation you can add the same amount to both sides of the equation and the results will be equal. That fact is known as:

> **The Addition Property of Equality**
>
> For any numbers A, B, and C, if A = B, then A + C = B + C.

You can use this property to help solve equations.

Example 1: Solve for P:

$$P - 7 = 19$$
$$\underline{+ 7 \quad\quad +7}$$
$$P + 0 = 26$$

If you add 7 to both sides of the equation, you get P on the left hand side, and the answer for the variable (26) on the right.

$$P = 26$$

(Note: $-7 + 7 = 0$ and $P + 0 = P$ on the left side of the equation.)

Example 2: Solve for X:

$$X + 5 = -3$$
$$\underline{-5 \quad\quad -5}$$
$$X = -8$$

If you add -5 to both sides, you will have X on the left side, and the answer (-8) on the right.

The method, then, for solving equations is to go through a series of steps so that the variable is alone on one side of the equals sign, and a number is on the other side. That number will be the solution for the variable in the equation. Adding the same quantity to both sides of an equation is sometimes all you need to do to find the number which the variable represents.

Example 3: Solve for g:

$$-5.3 + g = 12.8$$
$$\underline{+5.3 \quad\quad\quad +5.3}$$
$$g = 18.1$$

Add 5.3 to both sides, and get g on the left, and the answer (18.1) on the right.

Example 4: Solve for R:

$$-37 = R + 16$$
$$\underline{-16 \quad\quad -16}$$
$$-53 = R$$

Add -16 to both sides, and get R on the right, and the answer (-53) on the left.

PROBLEM 1: Solve for B:

$$B + \tfrac{1}{2} = \tfrac{1}{4}$$

Answer: $B = -\tfrac{1}{4}$

$$B + \tfrac{1}{2} = \tfrac{1}{4}$$
$$\underline{-\tfrac{1}{2} \quad -\tfrac{1}{2}}$$
$$B = -\tfrac{1}{4}$$

PROBLEM 2: Solve for k:

$$187 = 79 + k$$

Answer: K = 108

$$187 = 79 + k$$
$$\underline{-79 \quad -79}$$
$$108 = k$$

PROBLEM 3: Solve for T:

$$-67 + T = -19$$

Answer: T = 48

$$-67 + T = -19$$
$$\underline{+67 \quad\quad\quad +67}$$
$$T = 48$$

Note: You can always check your answer by replacing the variable with the answer you obtained to see if it, in fact, makes the left side equal to the right side of the equation.

EXERCISE 7-8   SET A

NAME _____   DATE _____

SOLVE THE FOLLOWING EQUATIONS: ANSWERS

1. S + 7 = 9   2. S + 4 = 10   1. _____

                                                   2. _____

3. t - 7 = 5   4. t - 2 = 7   3. _____

                                                   4. _____

5. 1 = y + 5   6. 2 = y + 7   5. _____

                                                   6. _____

7. -6 = X - 4   8. -7 = X - 6   7. _____

                                                   8. _____

9. X + 9 = 22   10. X + 7 = 36   9. _____

                                                   10. _____

11. K - 15 = 17   12. K - 17 = 25   11. _____

                                                   12. _____

13. 12 = Z + 15   14. 13 = Z + 17   13. _____

                                                   14. _____

15. -2.5 + p = -1.7   16. -4.6 + p = -2.5   15. _____

                                                   16. _____

17. R - (-8) = 20   18. R - (-9) = 30   17. _____

                                                   18. _____

19. $A + \dfrac{2}{3} = \dfrac{4}{5}$ 19. _____

EXERCISE 7-8   SET B

NAME _____   DATE _____

SOLVE THE FOLLOWING EQUATIONS: ANSWERS

1. $Z + 3 = 12$ 2. $X + 7 = 15$ 1. _____

                                                     2. _____

3. $K - 5 = 7$ 4. $K - 8 = 2$ 3. _____

                                                     4. _____

5. $-3 = X + 1$ 6. $-2 = Z + 2$ 5. _____

                                                     6. _____

7. $-8 = R - 5$ 8. $-9 = R - 3$ 7. _____

                                                     8. _____

9. $S + 7 = 26$ 10. $S + 8 = 34$ 9. _____

                                                     10. _____

11. $t - 17 = 14$ 12. $t - 12 = 29$ 11. _____

                                                     12. _____

13. $10 = y + 13$ 14. $7 = y + 16$ 13. _____

                                                     14. _____

15. $-4.6 + N = -2.7$ 16. $-3.9 + N = -1.8$ 15. _____

                                                     16. _____

17. $M - (-9) = 17$ 18. $M - (-4) = 19$ 17. _____

                                                     18. _____

19. $B + \frac{3}{4} = \frac{2}{3}$ 19. _____

7-9 Solving Equations Using Multiplication/Division Properties

Not all equations can be solved by adding a number to both sides of the equals sign. The Addition Property only works in an equation that has a number added to or subtracted from the variable. There are other equations where you have to multiply or divide both sides of the equation by a number in order to obtain the variable on one side and the answer on the other side.

If you multiply or divide both sides of an equation by the same number, both sides are still equal. These properties can be stated as follows:

The Multiplication Property of Equality

For any numbers A, B, and C, if $A = B$, then $A \cdot C = B \cdot C$.

The Division Property of Equality

For any numbers A, B, and C, if $A = B$, then $\frac{A}{C} = \frac{B}{C}$

(Note: $\frac{A}{C}$ is another way to write $A \div C$. C can not be equal to zero since we can not divide by zero.)

Here is how those properties are used to solve equations:

Example 1: Solve for Y:

$$6Y = -30$$

$$\frac{6Y}{6} = \frac{-30}{6}$$ Divide both sides by 6, leaving Y on the left side, and the answer (-5) on the right.

$$Y = -5$$ (Note: $\frac{6Y}{6} = 1Y = Y$)

Example 2: Solve for X:

$$\frac{1}{3}X = 9$$

$$3 \cdot \frac{1}{3}X = 9 \cdot 3$$ Multiply both sides by 3, leaving the X on the left, and the answer (27) on the right.

$$X = 27$$ (Note: $3 \cdot \frac{1}{3}X = 1X = X$)

Example 3: Solve for V:

$$-24 = -4V$$

$$\frac{-24}{-4} = \frac{-4V}{-4}$$ Divide both sides by −4, leaving V on the right, and the answer (6) on the left.

$$6 = V$$

Example 4: Solve for m:

$$\frac{m}{7} = 8$$

$$7 \cdot \frac{m}{7} = 8 \cdot 7$$ Multiply both sides by 7, leaving m on the left, and the answer (56) on the right.

$$m = 56$$

The Multiplication and Division Properties enable you to solve equations where the variable is divided or multiplied by a number. You must remember that you want to get the variable alone on one side of the equation. If you have a number *times* the variable, you must *divide* both sides by the number. If you have the variable *divided* by a number, you must *multiply* both sides by that divisor.

PROBLEM 1: Solve for k:

$$-8k = 19$$

Answer: $k = -2\frac{3}{8}$

$$\frac{-8k}{-8} = \frac{19}{-8}$$

$$k = -2\frac{3}{8}$$

PROBLEM 2: Solve for d:

$$\frac{d}{6} = -3$$

Answer: $d = -18$

$$6 \cdot \frac{d}{6} = -3 \cdot 6$$

$$d = -18$$

PROBLEM 3: Solve for R:

$$-9 = -2R$$

Answer: $R = 4\frac{1}{2}$

$$\frac{-9}{-2} = \frac{-2R}{-2}$$

$$4\frac{1}{2} = R$$

PROBLEM 4: Solve for H:

$$\frac{1}{5}H = 17$$

Answer: $H = 85$

$$5 \cdot \frac{1}{5}H = 17 \cdot 5$$

$$H = 85$$

EXERCISE 7-9   SET A

NAME _____   DATE _____

SOLVE THE FOLLOWING EQUATIONS: ANSWERS

1. $5m = 45$ 2. $7m = 49$ 1. _____

                                                       2. _____

3. $-8Z = 64$ 4. $-6X = 72$ 3. _____

                                                       4. _____

5. $\dfrac{N}{5} = 7$ 6. $\dfrac{N}{6} = 4$ 5. _____

                                                       6. _____

7. $\dfrac{R}{-9} = -72$ 8. $\dfrac{R}{-4} = -52$ 7. _____

                                                       8. _____

9. $-48 = 6K$ 10. $-32 = 4K$ 9. _____

                                                       10. _____

11. $-8 = \dfrac{1}{4}t$ 12. $-12 = \dfrac{1}{3}t$ 11. _____

                                                       12. _____

13. $-90 = -6y$ 14. $-98 = -7y$ 13. _____

                                                       14. _____

15. $85v = 595$ 16. $76v = 456$ 15. _____

                                                       16. _____

17. $-3.2m = 20.8$ 18. $-4.4m = 15.4$ 17. _____

                                                       18. _____

19. $78 = \dfrac{T}{-12}$ 19. _____

EXERCISE 7-9   SET B

NAME _____   DATE _____

SOLVE THE FOLLOWING EQUATIONS: ANSWERS

1. $3m = 24$ 2. $6m = 42$ 1. _____

                                                      2. _____

3. $-9X = 72$ 4. $-8X = 56$ 3. _____

                                                      4. _____

5. $\dfrac{N}{3} = 7$ 6. $\dfrac{N}{4} = 9$ 5. _____

                                                      6. _____

7. $\dfrac{R}{-7} = -9$ 8. $\dfrac{R}{-6} = -8$ 7. _____

                                                      8. _____

9. $-54 = 9K$ 10. $-72 = 8K$ 9. _____

                                                      10. _____

11. $-9 = \dfrac{1}{3}t$ 12. $-11 = \dfrac{1}{5}t$ 11. _____

                                                      12. _____

13. $-104 = -8y$ 14. $-126 = -9y$ 13. _____

                                                      14. _____

15. $57r = 513$ 16. $69r = 552$ 15. _____

                                                      16. _____

17. $-5.8m = 14.5$ 18. $-6.2m = 27.9$ 17. _____

                                                      18. _____

19. $69 = \dfrac{T}{-13}$ 19. _____

7-10 More on Solving Equations

In the last two sections we covered the basic properties that are needed to solve equations with one variable. In those sections we used just one of the properties to obtain the answer for each equation. Some equations, however, require more than one property to get its solution. The most efficient order to apply those properties is:

First: If a number is added to or subtracted from the variable term, use the Addition Property to get that variable term alone on one side of the equation.

Second: If the variable is divided or multiplied by a number, use the Multiplication or Division Property to get the variable equal to its answer.

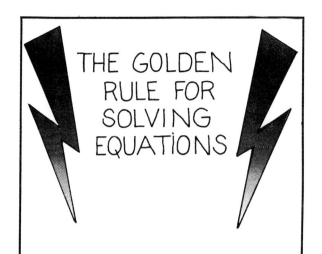

THE GOLDEN RULE FOR SOLVING EQUATIONS

"DO UNTO ONE SIDE AS YOU DO UNTO THE OTHER."

Example 1: Solve for X:

$$5X - 4 = 31$$

$$\begin{array}{r} 5X - 4 = 31 \\ +4 +4 \\ \hline 5X = 35 \end{array}$$

Add 4 to both sides, leaving the variable term (5X) on the left side.

$$\frac{5X}{5} = \frac{35}{5}$$

Divide both sides by 5, leaving X on the left, and the answer (7) on the right.

$$X = 7$$

Example 2: Solve for N:

$$6 - 3N = 18$$

$$\begin{array}{r} 6 - 3N = 18 \\ -6 -6 \\ \hline -3N = 12 \end{array}$$

Add −6 to both sides, to get the variable term (−3N) on one side.

$$\frac{-3N}{-3} = \frac{12}{-3}$$

Divide both sides by −3, leaving N on the left, and the answer (−4) on the right.

$$N = -4$$

Example 3:   Solve for S:

$$\frac{S}{4} + 7 = 10$$

$$\begin{array}{r} \frac{S}{4} + 7 = 10 \\ -7 \quad -7 \end{array}$$ Add -7 to both sides, leaving the variable term (S/4) on the left side.

$$\frac{S}{4} = 3$$

$$4 \cdot \frac{S}{4} = 3 \cdot 4$$ Multiply both sides by 4, leaving S on the left, and the answer (12) on the right.

$$S = 12$$

Example 4:   Solve for n:

$$-4 = \frac{n}{2} - 7$$

$$\begin{array}{r} -4 = \frac{n}{2} - 7 \\ +7 \quad\quad +7 \end{array}$$ Add 7 to both sides, to get the variable term (n/2) on one side.

$$3 = \frac{n}{2}$$

$$2 \cdot 3 = \frac{n}{2} \cdot 2$$ Multiply both sides by 2, leaving n on the right, and the answer (6) on the left.

$$6 = n$$

PROBLEM 1:   Solve for z:

$$2z - 2 = 13$$

Answer: $z = 7\frac{1}{2}$

$$\begin{array}{r} 2z - 2 = 13 \\ +2 \quad +2 \end{array}$$

$$\frac{2z}{2} = \frac{15}{2}$$

$$z = 7\frac{1}{2}$$

PROBLEM 2:   Solve for N:

$$-1 = \frac{N}{-3} + 5$$

Answer: $N = 18$

$$\begin{array}{r} -1 = \frac{N}{-3} + 5 \\ -5 \quad\quad -5 \end{array}$$

$$-6 = \frac{N}{-3}$$

$$(-3)(-6) = \frac{N}{-3}(-3)$$

$$18 = N$$

EXERCISE 7-10   SET A

NAME _____   DATE _____

SOLVE THE FOLLOWING EQUATIONS: ANSWERS

1. $2X + 5 = 9$ 2. $3X + 1 = 13$ 1. _____

                                                    2. _____

3. $40 = -8y + 8$ 4. $27 = -3y + 9$ 3. _____

                                                    4. _____

5. $-3 + 5z = 12$ 6. $-6 + 4z = 14$ 5. _____

                                                    6. _____

7. $-18 = -3K - 6$ 8. $-7 = -2K - 1$ 7. _____

                                                    8. _____

9. $\dfrac{N}{3} + 6 = 10$ 10. $\dfrac{N}{2} + 5 = 13$ 9. _____

                                                    10. _____

11. $\dfrac{1}{6}m - 4 = 3$ 12. $\dfrac{1}{4}m - 5 = 2$ 11. _____

                                                    12. _____

13. $-12 = \dfrac{R}{5} - 8$ 14. $-16 = \dfrac{R}{6} - 7$ 13. _____

                                                    14. _____

15. $4 + \dfrac{t}{-7} = -8$ 16. $5 + \dfrac{t}{-3} = -7$ 15. _____

                                                    16. _____

17. $\dfrac{3X}{4} = 27$ 18. $\dfrac{2X}{3} = 18$ 17. _____

                                                    18. _____

19. $5.6 = \dfrac{1}{7}N + 4.5$ 19. _____

EXERCISE 7-10   SET B

NAME _____   DATE _____

SOLVE THE FOLLOWING EQUATIONS   ANSWERS

1. 2X + 1 = 13   2. 3X + 2 = 17   1. _____

                                                       2. _____

3. 33 = −5y + 3   4. 37 = −4y + 5   3. _____

                                                       4. _____

5. −5 + 7z = 9   6. −3 + 8z = 21   5. _____

                                                       6. _____

7. −10 = −2k − 6   8. −8 = −3k − 5   7. _____

                                                       8. _____

9. $\frac{N}{4} + 5 = 7$ 10. $\frac{N}{5} + 3 = 8$ 9. _____

                                                       10. _____

11. $\frac{1}{2}m - 7 = 5$ 12. $\frac{1}{3}m - 6 = 2$ 11. _____

                                                       12. _____

13. $-15 = \frac{R}{6} - 12$ 14. $-18 = \frac{R}{7} - 13$ 13. _____

                                                       14. _____

15. $3 + \frac{t}{-8} = -7$ 16. $9 + \frac{t}{-3} = -6$ 15. _____

                                                       16. _____

17. $\frac{5X}{8} = 30$ 18. $\frac{4X}{5} = 16$ 17. _____

                                                       18. _____

19. $8.7 = \frac{1}{5}N + 7.4$ 19. _____

7-11 Word Problems

The final task in our brief introduction to algebra is to apply our knowledge of equations to some word problems. Following the steps listed below will help in solving word problems.

1. **Read:** Read the problem slowly and carefully.

2. **Represent the Unknown** Use a variable to represent the unknown quantity in the problem.

3. **Translate:** Translate the words of the problem into an equation using that variable.

4. **Solve:** Solve the equation.

5. **Check:** Check your answer using the words of the original problem.

COMMUTERS HATE IT WHEN THE <u>TRANSLATE</u>.

Besides those steps, there are some words that indicate the various operations of mathematics. Knowing these will help in the translating phase of solving word problems.

	Word(s)	Statement	Algebraic Translation
Addition:	*plus*	6 *plus* a number	6 + X
	add	*add* 7 to a number	N + 7
	sum	the *sum* of a number and 2	y + 2
	added to	13 *added to* a number	k + 13
	more than	5 *more than* a number	P + 5
	increased by	a quantity *increased by* 9	M + 9
	total of	the *total of* 40 and a number	40 + R
Subtraction:	*minus*	a number *minus* 6	X − 6
	subtract from	*subtract* 3 *from* a number	n − 3
	difference	the *difference* between Y and 2	Y − 2
	fewer than	8 *fewer than* a number	k − 8
	less than	7 *less than* a number	P − 7
	decreased by	a quantity *decreased by* 4	J − 4
Division:	*divided by*	a number *divided by* 7	
	quotient	the *quotient* of P and 7	$\frac{P}{7}$
	ratio	a *ratio* of a number and 7	

	Word(s)	Statement	Algebraic Translation
Multiplication:	times	6 *times* a number	6N
	multiplied by	a number *multiplied by* 3	3k
	product	the *product* of 8 and a number	8y
	twice	*twice* a number	2X
	doubled	a number is *doubled*	2A
	tripled	a number is *tripled*	3R
	of	three-fourths *of* a number	$\frac{3}{4}B$
Equality:	is equal to	twice a number *is equal to* 13	$2X = 13$
	is	the sum of a number and 8 *is* -3	$X + 8 = -3$
	result is	the *result of* Y tripled *is* 6	$3Y = 6$
	is the same as	the quotient of a number and 5 *is the same as* 18 minus 8	$\frac{k}{5} = 18 - 8$

Before we proceed into the solution of word problems, you should first practice translating word statements into algebraic ones.

For example: Answers:

1. the sum of 7 and a number   1. $7 + X$

2. the product of -5 and a number   2. $-5X$

3. the difference between twice a number and 6   3. $2X - 6$

4. the quotient of a number and 4, increased by 2   4. $\frac{X}{4} + 2$

5. 3 less than 8 times a number   5. $8X - 3$

PROBLEMS: Express the following using X as the variable: Answers:

1. the total of a number and seven   1. $X + 7$

2. 12 fewer than twice a number   2. $2X - 12$

3. 7 more than five times a number   3. $5X + 7$

4. the ratio of a number and -7   4. $\frac{X}{-7}$

5. a quantity is tripled and decreased by 10   5. $3X - 10$

Now that we have learned to translate word statements into algebraic ones, let's attempt some word problems that require the use of algebra.

Example 1: The total of a number and 9 is 28. Find the number.

Represent the unknown: X = the number

Translate: $X + 9 = 28$

Solve:
$$\begin{array}{r} -9 \quad -9 \\ \hline X \quad = 19 \end{array}$$

Check:
$$\begin{array}{r} 19 \\ + 9 \\ \hline 28 \end{array}$$

Example 2: The quotient of a number and -6 is 7. Find the number.

Represent the unknown: X = the number

Translate: $\dfrac{X}{-6} = 7$

Solve: $(-6) \cdot \dfrac{X}{-6} = 7 \cdot (-6)$

$X = -42$

Check: $\dfrac{-42}{-6} = 7$

Example 3: The cost of two pencils minus 7¢ is 29¢. Find the cost of a pencil.

Represent the unknown: X = the cost of a pencil

Translate: $2X - 7 = 29$

Solve:
$$\begin{array}{r} +7 \quad +7 \\ \hline \dfrac{2X}{2} = \dfrac{36}{2} \end{array}$$

$x = 18$

Check: The two pencils cost 18¢ + 18¢, which is 36¢. 36¢ minus 7¢ is 29¢.

Example 4: Helen's age is four more than twice John's age. If Helen is 20 years old, how old is John?

Represent the unknown: X = John's age

Translate: $2X + 4 = 20$

Solve:
$$\begin{array}{r} -4 \quad -4 \\ \dfrac{2X}{2} = \dfrac{16}{2} \end{array}$$

$$X = 8$$

Check: Twice John's age, $2 \cdot 8 = 16$; four more $(16 + 4)$ does equal 20, Helen's age.

PROBLEM 1: Three times a number, decreased by 7, is 12. What is the number?

Answer: $6\dfrac{1}{3}$

X = the number

$$3X - 7 = 12$$
$$+7 \quad +7$$
$$\dfrac{3X}{3} = \dfrac{19}{3}$$
$$x = 6\dfrac{1}{3}$$

PROBLEM 2: Three cokes and one milk shake cost $1.45. If the milk shake costs $.55, what does a coke cost?

Answer: 30¢

X = cost of a coke

$$3X + 55 = 145$$
$$-55 \quad -55$$
$$\dfrac{3X}{3} = \dfrac{90}{3}$$
$$X = 30$$

PROBLEM 3: The difference between three times Don's age and Ginny's age is 76. If Ginny is 29, how old is Don?

Answer: 35

X = Don's age
$$3X - 29 = 76$$
$$+29 \quad +29$$
$$\dfrac{3X}{3} = \dfrac{105}{3}$$
$$X = 35$$

EXERCISE 7-11   SET A

NAME _____   DATE _____

EXPRESS THE FOLLOWING USING X AS THE VARIABLE: ANSWERS

1. The sum of a number and six.   1. _____

2. The sum of a number and three.   2. _____

3. Five fewer than twice a number.   3. _____

4. Six fewer than twice a number.   4. _____

5. The product of a number and six.   5. _____

6. The product of a number and nine.   6. _____

7. Four more than three times a number.   7. _____

8. Two more than five times a number.   8. _____

9. The quotient of a number and seven.   9. _____

10. The quotient of a number and four.   10. _____

SOLVE THE FOLLOWING WORD PROBLEMS:

11. The sum of a number and 8 is 45. Find the number.   11. _____

12. The difference between a number and 3 is 19. Find the
    number.   12. _____

13. Five times a number, decreased by 6, is 29. Find the
    number.   13. _____

14. Twice a number, increased by 9, is 37. Find the number. 14. _____

15. The product of -7 and a number is 112. Find the number. 15. _____

16. The quotient of a number and -9 is 8. Find the number. 16. _____

17. One pen and three pencils cost 95¢. If the pen costs 50¢,
    what is the cost of a pencil?   17. _____

18. The sum of Joe's age and twice Sue's age is 52. If Joe
    is 18 years old, how is Sue?   18. _____

19. The difference between three times Kevin's weight and twice
    Craig's weight is 280 pounds. If Craig weighs 100 pounds,
    how much does Kevin weigh?   19. _____

20. The quotient of a number and -4, decreased by 28, gives the
    same result as the product of 3 and -11. What is the
    number?   20. _____

EXERCISE 7-11   SET B

NAME _____   DATE _____

EXPRESS THE FOLLOWING USING X AS THE VARIABLE   ANSWERS

1. The sum of a number and 12.   1. _____

2. The sum of a number and -9.   2. _____

3. Three fewer than a number tripled in value.   3. _____

4. Seven fewer than a number doubled in value.   4. _____

5. The product of a number and -7.   5. _____

6. The product of a number and 39.   6. _____

7. Ten more than twice a number.   7. _____

8. Nine more than five times a number.   8. _____

9. The quotient of a number and 68.   9. _____

10. The quotient of a number and -4.   10. _____

SOLVE THE FOLLOWING WORD PROBLEMS:

11. The sum of a number and 17 is 56. Find the number.   11. _____

12. The difference between a number and 5 is 13. Find the
    number.   12. _____

13. Six times a number, increased by 5, is 29. Find the
    number.   13. _____

14. One-half of a number, decreased by 5, is 17. Find the
    number.   14. _____

15. The quotient of a number and -4 is -9. Find the number.   15. _____

16. The product of 8 and a number is -272. Find the number.   16. _____

17. The cost of one candy bar and two packs of gum is 86¢.
    If the candy bar costs 30¢, what does one pack of gum cost? 17. _____

18. The sum of three times Ann's age and twice Jill's age is
    96 years. If Ann is 20, how old is Jill?   18. _____

19. The difference between four times Dick's height and three
    times Tom's height is 30 inches. If Tom is 70 inches tall,
    how tall is Dick?   19. _____

20. Two-thirds of a number, increased by eight, gives the same
    result as the product of -2 and 11. What is the number?   20. _____

Chapter 7 Summary

Concepts

You may refer to the sections listed below to review how to:

1. evaluate algebraic expressions by replacing each variable with the number that it represents. (7-1)

2. compare signed numbers by understanding that numbers get larger proceeding to the right along a number line and smaller proceeding to the left. (7-2)

3. add signed numbers using either movement on a number line or the following rule: if the numbers have the same signs, add their number parts and use that same sign as the sign of the answer; if the numbers have different signs, subtract their number parts and use the sign of the larger number part. (7-3)

4. subtract signed numbers by changing the sign of the number being subtracted and converting the subtraction to addition. (7-4)

5. multiply and divide signed numbers by multiplying or dividing their number parts and using the fact that same signs give positive answers and different signs give negative answers. (7-5)

6. Use the order of operations on signed numbers. (7-6)

7. solve simple equations by mentally determining what number the variable represents. (7-7)

8. solve more complex equations using the Addition, Multiplication, and Division Properties of Equality. (7-8), (7-9), (7-10)

9. solve word problems by reading, representing the unknown, translating, solving, and checking. (7-11)

Terminology

This chapter's important terms and their corresponding page numbers are:

algebraic expression: a quantity containing numbers, variables, and operations. (332)

equation: a statement that two quantities are equal. (355)

number line: a straight line whose points are associated with numbers. (335)

signed number: a positive or negative number. (335)

variable: a letter used to represent a number. (331)

Student Notes

Chapter 7 Practice Test A

ANSWERS

1. If $x = 6$, $y = 2$, and $z = 5$, then $\frac{x}{y} + 2y - z = ?$ 1. _____

Replace the ? with > or <:

2. $-7 \; ? \; 0$ 3. $-17 \; ? \; -18$ 2. _____

3. _____

Perform the indicated operations:

4. $-12 + 8$ 5. $-24 + (-18)$ 4. _____

5. _____

6. $16 - 27$ 7. $-4 - (-10)$ 6. _____

7. _____

8. $5(-8)$ 9. $\frac{63}{-9}$ 8. _____

9. _____

10. $\frac{6[3 - (-8)]}{-5 + 2(-3)}$ 10. _____

Solve the following equations:

11. $X + 37 = 61$ 12. $28 = P - 8$ 11. _____

12. _____

13. $342 = 9N$ 14. $\frac{r}{-3} + 22 = 3$ 13. _____

15. The sum of seven times a number and 8 is -27. What is the number?   15. _____

379

Chapter 7 Practice Test B

ANSWERS

1. If r = 6, s = 1, and t = 2, then $\frac{r}{t}$ + 3r − 2s = ?

1. _____

Replace the ? with > or <:

2. 0 ? −6

3. −24 ? −25

2. _____

3. _____

Perform the indicated operations:

4. 17 + (−14)

5. −36 + (−17)

4. _____

5. _____

6. 14 − 29

7. 8 − (−6)

6. _____

7. _____

8. (−6)·9

9. (−4)(−8)

8. _____

9. _____

10. $\frac{(7 - 11)(2 - 3 \cdot 6)}{-5 - (-3)}$

10. _____

Solve the following equations:

11. 70 = X + 44

12. P − 9 = 27

11. _____

12. _____

13. 273 = 7n

14. −2n + 7 = 23

13. _____

15. The quotient of a number and −4 gives the same result as the product of 6 and −2. What is the number?

15. _____

Chapter 7 Supplementary Exercises

Section 7-1
If $P = 3$, $Q = 5$, and $R = 1$, evaluate the following:

1. $Q + P - R$
2. $4Q + 2P$
3. $3PQR$
4. $5(2P - 3R)$
5. $RQ - RP$
6. $4(Q + 6 - R)$
7. $\frac{1}{2}Q + \frac{1}{4}R$
8. $\frac{1}{4}(Q - \frac{1}{2}P)$
9. $2Q + 2P - 2(R + P)$
10. $\frac{6Q}{P}$
11. $\frac{5P + R}{2}$
12. $\frac{9Q - P}{R + 2}$

Section 7-2
1. $8 \ ? \ 0$
2. $\frac{1}{2} \ ? \ \frac{1}{4}$
3. $-41 \ ? \ -81$
4. $-8 \ ? \ 0$
5. $-\frac{1}{2} \ ? \ -\frac{1}{4}$
6. $0 \ ? \ -0.5$
7. $-5 \ ? \ 4$
8. $-5 \ ? \ -56$
9. $6\frac{1}{2} \ ? \ -6\frac{1}{4}$
10. $-5 \ ? \ -4$
11. $2.7 \ ? \ -2.6$
12. $-3.8 \ ? \ -3.9$

Section 7-3
1. $7 + 5$
2. $-7 + 5$
3. $7 + (-5)$
4. $-7 + (-5)$
5. $-15 + 18$
6. $15 + (-18)$
7. $-15 + (-18)$
8. $-6 + (-35)$
9. $-40 + 18$
10. $14 + (-33)$
11. $56 + (-87)$
12. $44 + (-29)$
13. $-4 + 6 + (-7)$
14. $8 + (-9) + (-13)$
15. $-15 + 9 + (-7)$
16. $-77 + 18 + 9$
17. $14 + (-32) + (-12)$
18. $54 + (-77) + 5$

Section 7-4
1. $12 - 7$
2. $7 - 12$
3. $13 - 25$
4. $-7 - 9$
5. $-9 - 7$
6. $-6 - 4$
7. $-38 - 56$
8. $18 - 6$
9. $-18 - 6$
10. $9 - (-3)$
11. $5 - (-2)$
12. $12 - (-7)$
13. $-18 - (-6)$
14. $-7 - (-14)$
15. $-3 - (-17)$
16. $25 - 10 - (-6)$
17. $35 - 6 - 27$
18. $-5 - 5 - (-5)$
19. $-12 + 32 - (-7)$
20. $-56 - 7 + 23$
21. $-22 + 17 - (-9)$
22. $-23 - (-55) + 17$
23. $34 - (-21) + 9$
24. $1 - 9 + 8 - 6$

Section 7-5

1. $5 \cdot (7)$
2. $6 \cdot (-3)$
3. $-4(7)$
4. $-5 \cdot 3$
5. $(-9)(-6)$
6. $-3(-5)$
7. $(-1)(-27)$
8. $-8(12)$
9. $5(-31)$
10. $8 \cdot (-3)(-2)$
11. $-7(-3) \cdot 9$
12. $-6 \cdot 8 \cdot (-3)$
13. $-2(-4)(-7)$
14. $-3(-3)(-3)$
15. $(-2)(-8)(-1)(5)$
16. $\dfrac{-45}{9}$
17. $\dfrac{-36}{4}$
18. $\dfrac{-72}{-8}$
19. $\dfrac{-52}{-4}$
20. $\dfrac{-81}{-9}$
21. $\dfrac{52}{-4}$
22. $\dfrac{-225}{5}$
23. $\dfrac{441}{-3}$
24. $\dfrac{-700}{-25}$

Section 7-6

1. $5 - 2 \cdot 7$
2. $-6 + 3 \cdot 2$
3. $-6 - (3+8)$
4. $9 - (8 - 5)$
5. $3(-2) + 2(-3)$
6. $(-4)(6) + 8 \cdot 3$
7. $(-5)(-2) + 3(-6)$
8. $4(-9) + (-5)(-4)$
9. $3[-2 + 4(8-1)]$
10. $-2[7 - 5(3-9)]$
11. $25 - [4 - (3-10)]$
12. $-34 + [2(4-12) + (-5)]$
13. $4^2 - (2)(-5)$
14. $2^3 - (-5)(6)$
15. $(-7 + 10)^3 - (-4)$
16. $(-6 + 9)^2 - (-8)$
17. $\dfrac{4 - 2 \cdot 7}{-3 + (-7)}$
18. $\dfrac{-8 - 1}{5 + 3(-2)}$
19. $\dfrac{-5 + 13}{2(7-9)}$
20. $\dfrac{-4(6+8)}{-7 + 5}$
21. $\dfrac{-6(4) - (8)(-3)}{(-7 - 5)(8 - 3)}$
22. $\dfrac{(-6)(-6) + 4(-9)}{3(2 - 8) - (-4)}$
23. $\dfrac{7 - 3(-4) - 4}{3[8 - (-5)]}$
24. $\dfrac{8 + 2(-3) - (-4)}{-5[-6 + (-4)]}$

Section 7-7

Solve for X:

1. $X + 3 = 5$
2. $2 + X = 8$
3. $-1 = X + 9$
4. $X - 4 = 10$
5. $7 - X = -2$
6. $3 = X - 5$
7. $3X = 12$
8. $-5X = -30$
9. $56 = -7X$
10. $\frac{X}{5} = 2$
11. $-3 = \frac{X}{7}$
12. $3X + 1 = 13$

Section 7-8

Solve for N:

1. $N + 7 = 19$
2. $-4 = N + 3$
3. $18 + N = 47$
4. $N - 23 = 14$
5. $25 = N - 7$
6. $12 = N - 12$
7. $N - 43 = -43$
8. $5.6 + N = 8.7$
9. $-5.3 + N = -2.4$
10. $-(-9) + N = 6$
11. $N - (-8) = -4$
12. $-16 = N - (-16)$

Section 7-9

Solve for P:

1. $4P = 32$
2. $-5P = 35$
3. $-36 = 9P$
4. $-120 = -6P$
5. $15P = -45$
6. $-16P = -64$
7. $\frac{1}{2}P = -18$
8. $9 = -\frac{1}{4}P$
9. $-\frac{1}{2}P = -7$
10. $\frac{P}{7} = 4$
11. $\frac{P}{-3} = 12$
12. $-6 = \frac{P}{8}$

Section 7-10

Solve for X:

1. $3X + 1 = 16$
2. $2X - 3 = 15$
3. $7 - 2X = -11$
4. $-18 = -5X - 3$
5. $5X + 8 = 49$
6. $81 = 6 + 3X$
7. $6X - 8 = 34$
8. $-2X + 3 = 7$
9. $46 + 8X = 6$
10. $3.5X + 6 = 13$
11. $50 = -4X + 6$
12. $-7X - 3 = -24$
13. $\frac{1}{2}X + 6 = 8$
14. $\frac{1}{4}X - 4 = 15$
15. $-15 = \frac{1}{2}X - 6$

Section 7-11

1. Twice a number, decreased by 7, is 49. Find the number.
2. The product of -4 and a number is 52. Find the number.
3. The sum of a number and 7 is -7. Find the number.
4. Half a number, increased by 3.7, is -4.9. Find the number.
5. The quotient of three times a number and 7 is 45. Find the number.

Chapter 8

Introduction to Geometry

After completing this chapter you should be able to do the following:

1. define basic geometric terms.
2. distinguish between different kinds of angles.
3. classify different types of triangles.
4. identify various kinds of plane figures.
5. draw common geometric shapes.
6. find perimeters and areas of plane figures.

On the next page you will find a pretest for this chapter. The purpose of the pretest is to help you determine which sections in this chapter you need to study in detail and which sections you can review quickly. By taking and correcting the pretest according to the instructions on the next page you can better plan your pace through this chapter.

Note: Unless otherwise indicated in this chapter, when the answers have more than three decimal digits, they will be rounded off to the nearest hundredth.

Chapter 8 Pretest

Take and correct this test using the answer section at the end of the book. Those problems that give you difficulty indicate which sections of this chapter need extra attention. Section numbers are in parentheses before each problem.

ANSWERS

(8-1) 1. List all the line segments in the figure on the right.

1. _____

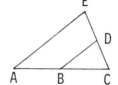

(8-1) 2. In the figure on the right, if B is the midpoint of segment $\overline{AC}$, and AC = 25, then AB = ?

2. _____

(8-2) 3. What kind of angle is <G, if it measures 75°?

3. _____

(8-2) 4. Find the supplement and complement of a 46° angle.

4. _____

(8-3) 5. Draw an isosceles triangle, △ DEF.

5. _____

(8-3) 6. Draw a right triangle, △ ABC, with <A a right angle.

6. _____

(8-4) 7. Draw parallelogram MNOP, with <N an obtuse angle.

7. _____

(8-4) 8. What is the name of the plane figure on the right?

8. _____

(8-5) 9. Find the circumference and area of a circle with a 25 cm diameter.

9. _____

(8-5) 10. Find the area and perimeter of the figure on the right.

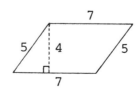

10. _____

8-1 Basic Geometric Objects

As you look around the world we live in, you see many different objects, each having its own size and shape. Since early times, man has measured the earth and classified the shapes around him. This study we call geometry. To begin an introduction to geometry, we will consider the most basic objects -- the point, line, and plane.

Point

A point is a location. It has no size; it has only position. For example, there is a point on the upper right-hand corner of this page. Even though you can not see it, the location is there. A point is represented by a dot and named by a letter placed next to it. Shown below are three points -- A, B, and C.

WHAT DO LITTLE ACORNS SAY WHEN THEY GROW UP? GEOMETRY!

```
            B .
                             C.
   A.
```

Line

A line is made up of points. It is straight and extends forever in opposite directions. Only one line can be drawn connecting two points. A line is represented by a double-headed arrow and is named by a single lower-case letter or by two points that are on the line. Shown below is line w or line $\overleftrightarrow{AB}$.

Plane

A plane is a flat surface that has no thickness and extends indefinitely in every direction. A thin plate of glass that stretches forever in all directions is a good model of a plane. A plane is represented by a four-sided figure as pictured below and is named by a single letter.

plane E (The letter E may be placed in any corner.)

PROBLEM 1: Draw and label plane K containing point P and $\overleftrightarrow{AD}$.

Answer: (One of many possible sketches.)

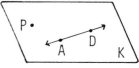

387

Intersecting and Parallel Lines

Lines that cross each other at one point are called *intersecting lines*. Lines in the same plane that never intersect are called *parallel lines*.

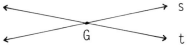

Line s intersects line t at point G.

Line p is parallel to line q, written p ∥ q.

Line Segment

A **line segment** is the part of a line that consists of two points and all the points in between those two points.

The line segment connecting points D and F is written $\overline{DF}$ or $\overline{FD}$.

Note: The order of the letters does not make a difference.

Length of a Line Segment

Every line segment has a length, a distance from one end point to the other end point. The line segment above is 2 inches long. We write DF = 2" or FD = 2". In the length of a line segment the dash is not written above the letters. $\overline{DF}$ is the line segment while DF represents the length of the line segment.

Midpoint of a Line Segment

A point that divides a line segment into two equal parts is called the **midpoint** of the line segment.

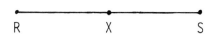

X is the midpoint of the line segment $\overline{RS}$ since RX = XS.

PROBLEM 2: List all the different line segments on $\overleftrightarrow{DB}$.

Answer: $\overline{DA}$, $\overline{AB}$, $\overline{DB}$

PROBLEM 3: In the line above, if DB = 12 and A is the midpoint of $\overline{DB}$, how long is $\overline{DA}$?

Answer: 6

PROBLEM 4: Draw and label k ∥ j with $\overleftrightarrow{MN}$ intersecting both k and j.

Answer:

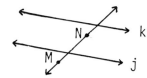

EXERCISE 8-1   SET A

NAME _____   DATE _____

DRAW AND LABEL THE FOLLOWING:

1. line w containing points Y, P, and F

2. $\overleftrightarrow{JK}$ passing through points C and X

3. plane G containing points J and V

4. plane W containing line d

5. lines g and h with g ∥ h

6. $\overleftrightarrow{FE} \parallel \overleftrightarrow{GH}$

7. $\overleftrightarrow{QP}$ intersecting line segment $\overline{CD}$ at point Y

8. line r intersecting line segment $\overline{AB}$ at point F

LIST ALL THE LINE SEGMENTS IN EACH FIGURE: ANSWERS

9. 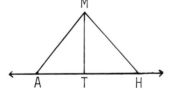   10. 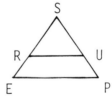   9. _____

10. _____

DETERMINE THE FOLLOWING LENGTHS:

11. If ES = 10, A is the midpoint of $\overline{ES}$, and S is the midpoint of $\overline{AY}$, find the lengths of $\overline{EA}$, $\overline{AS}$, and $\overline{AY}$.   11. _____

        E   A   S   Y

12. If IE = 24, C is the midpoint of $\overline{IE}$, and NI = IE, find the lengths of $\overline{NI}$, $\overline{IC}$, and $\overline{CE}$.   12. _____

        N   I   C   E

EXERCISE 8-1   SET B

NAME _____   DATE _____

DRAW AND LABEL THE FOLLOWING:

1. line v passing through points A, R, T, and S.

2. $\overleftrightarrow{MN}$ containing points B and E

3. plane R containing intersecting lines p and q

4. plane Q containing point A and line segment $\overline{LT}$

5. lines a, b, c with a ∥ b ∥ c

6. $\overleftrightarrow{GH}$ ∥ $\overleftrightarrow{LD}$

7. line segment $\overline{AB}$ intersecting line segment $\overline{XZ}$ at point T

8. line segment $\overline{TQ}$ intersecting line b at point R

LIST ALL THE LINE SEGMENTS IN EACH FIGURE: ANSWERS

9. 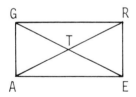   10.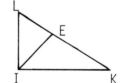

9. _____

10. _____

DETERMINE THE FOLLOWING LENGTHS:

11. If HP = 6, O is the midpoint of $\overline{HP}$, and HP = PE, find the lengths of $\overline{HO}$, $\overline{OP}$, and $\overline{PE}$.

•———•———•————————•
H   O   P   E

11. _____

12. If LC = 14, U is the midpoint of $\overline{LC}$, and C is the midpoint of $\overline{UK}$, find the lengths of $\overline{LU}$, $\overline{UC}$, and $\overline{CK}$.

•————•————•————•
L   U   C   K

12. _____

8-2 Angles

A **ray** is part of a line that has one end point and extends indefinitely in only one direction. When two rays meet at the same end point, they form an **angle**. The two rays are called the sides of the angle and the common end point is the vertex of the angle. In the angle below, ray $\overrightarrow{XZ}$ and ray $\overrightarrow{XT}$ are the sides of the angle, and point X is its vertex. An angle is named using the angle symbol <.

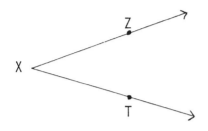

This angle can be named:

<X or <ZXT or <TXZ.

Note: When using a three-letter name for an angle, the vertex is always the middle letter.

Example 1: Name the three angles at point D in the figure below:

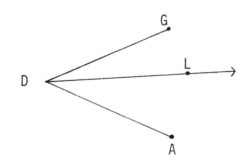

The three angles at point D are:

<GDL (or <LDG)
<LDA (or <ADL)
<GDA (or <ADG).

Note: None of these angles may be name <D because there is more than one angle having a vertex at D.

Measuring Angles

Angles have a measurement determined by the amount of rotation needed to swing one side of the angle around to the other side of the angle. That amount of rotation is measured in degrees. A complete rotation is 360 degrees (360°) and 1/360 of a full rotation is one degree (1°).

360° ⟲ ———   1° ———

PROBLEM 1: What are the five different angles in the figure below?

Answers: <A, <D, <ABD, <DBE, and <ABE

Types of Angles

Half of a complete rotation forms a 180° angle, called a **straight angle**. One-fourth of a complete rotation forms a 90° angle, called a **right angle**.

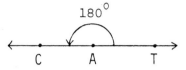

<CAT is a straight angle.

<RAT is a right angle.
(The box in the angle means 90°.)

If an angle measures between 0° and 90°, it is called an **acute angle**. If an angle measures between 90° and 180°, it is called an **obtuse angle**.

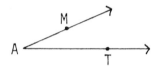
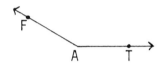

<MAT is an acute angle.

<FAT is an obtuse angle.

If the measures of two angles add up to 90°, they are called **complementary angles**. If the measures of two angles add up to 180°, they are called **supplementary angles**.

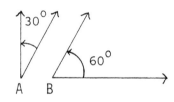
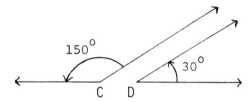

<A and <B are complementary
(<A is the complement of <B).

<C and <D are supplementary
(<C is the supplement of <D).

PROBLEMS: Referring to <A below:

2. What type of angle is <A?

3. What is the complement of <A?

4. What is the supplement of <A?

5. Draw line f through point B, making a right angle with ray $\overrightarrow{AB}$.

Answers:

2. acute angle
   (less than 90°)

3. 50°
   (40 + 50 = 90)

4. 140°
   (40 + 140 = 180)

5.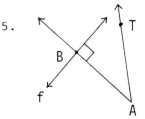

392

EXERCISE 8-2   SET A

NAME _____   DATE _____

DRAW AND LABEL THE FOLLOWING:

1. right angle, <BIT.

2. acute angle, <FIT.

3. obtuse angle, <HIT.

4. straight angle, <KIT.

5. plane F containing an acute angle, <LIT.

6. plane G containing a right angle, <MIT.

7. plane H containing a straight angle, <NIT.

8. plane J containing an obtuse angle, <PIT.

CLASSIFY <SIT AS A RIGHT ANGLE, STRAIGHT ANGLE, ACUTE ANGLE, OR OBTUSE ANGLE, IF:

ANSWERS

9. <SIT = 65°

10. <SIT = 117°

9. _____

10. _____

11. <SIT = 180°

12. <SIT = 90°

11. _____

12. _____

FIND THE NUMBER OF DEGREES IN BOTH THE COMPLEMENT AND SUPPLEMENT OF <WIT, IF:

13. <WIT = 45°

14. <WIT = 60°

13. _____

14. _____

15. <WIT = $22\frac{1}{2}$°

16. <WIT = 15.5°

15. _____

16. _____

EXERCISE 8-2   SET B

NAME _____   DATE _____

DRAW AND LABEL THE FOLLOWING:

1. right angle, <BAD.

2. acute angle, <CAD.

3. obtuse angle, <FAD.

4. straight angle, <HAD.

5. plane H containing an acute angle, <LAD.

6. plane I containing a right angle, <MAD.

7. plane J containing a straight angle, <PAD.

8. plane K containing an obtuse angle, <RAD.

CLASSIFY <SAD AS A RIGHT ANGLE, STRAIGHT ANGLE, ACUTE ANGLE, OR OBTUSE ANGLE, IF:

ANSWERS

9. <SAD = 43°

10. <SAD = 105°

9. _____

10. _____

11. <SAD = 90°

12. <SAD = 180°

11. _____

12. _____

FIND THE NUMBER OF DEGREES IN BOTH THE COMPLEMENT AND SUPPLEMENT OF <TAD, IF:

13. <TAD = 15°

14. <TAD = 30°

13. _____

14. _____

15. <TAD = $17\frac{1}{2}$°

16. <TAD = 30.5°

15. _____

16. _____

8-3 Triangles

A **triangle** is a figure that has three sides. A triangle is named by using the triangle symbol, △, followed by the letters of its three vertex points written in any order.

For example: △ABC or △BAC

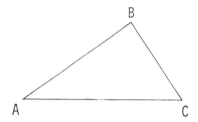

Its three vertices are A, B, C.
Its three sides are $\overline{AB}$, $\overline{BC}$, $\overline{CA}$.
Its three angles are <A, <B, <C.

Types of Triangles

One way to classify triangles is by the lengths of the sides, as shown below.

A PRETTY TRIANGLE: AN <u>ACUTE</u> TRIANGLE

A **scalene triangle** is a triangle in which each side has a different length.

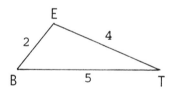

In △BET, no two sides are equal.

△BET is a scalene triangle.

An **isosceles triangle** is a triangle that has two sides of equal length.

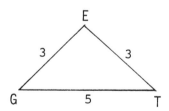

In △GET, two sides are equal:
EG = ET.

△GET is an isosceles triangle.

An **equilateral triangle** is a triangle with all three sides equal

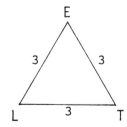

In △LET, all sides are equal:
LE = ET = LT.

△LET is an equilateral triangle.

Triangles are also classified by the size of their angles. A **right triangle** is a triangle that has one right angle.

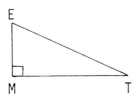

In △ MET, <M = 90°

△ MET is a right triangle.

An **acute triangle** is a triangle in which each angle measures less than 90°.

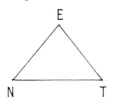

In △ NET, <N, <E, and <T are all less than 90°

△ NET is an acute triangle.

An **obtuse triangle** is a triangle in which one angle measures more than 90°

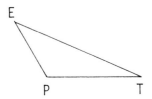

In △ PET, <P is larger than 90°

△ PET is an obtuse triangle.

PROBLEM 1: Name the three triangles in the figure below.

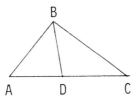

Answers:
△ BAD, △ BDC, △ BAC

PROBLEM 2: Classify the triangles below as acute, right, or obtuse

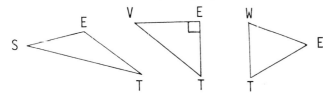

Answers:
△ SET - obtuse
△ VET - right
△ WET - acute

PROBLEM 3: Classify the triangles below as scalene, isosceles, or equilateral.

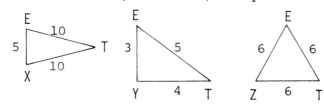

Answers:
△ XET - isosceles
△ YET - scalene
△ ZET - equilateral

EXERCISE 8-3 SET A

NAME _____ DATE _____

DRAW AND LABEL THE FOLLOWING:

1. right triangle, △LAW, with <W the right angle.

2. obtuse triangle, △SON, with <S the obtuse angle.

3. isosceles triangle, △FUN, with side $\overline{UN}$ the unequal side.

4. equilateral triangle, △MUT.

NAME ALL THE DIFFERENT TRIANGLES IN EACH FIGURE BELOW: ANSWERS

5.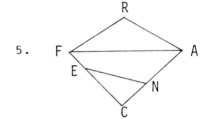

6.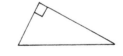

5. _____

6. _____

CLASSIFY THE FOLLOWING TRIANGLES AS ACUTE, RIGHT, OR OBTUSE:

7.

8.

9.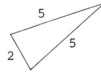

10.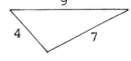

7. _____

8. _____

9. _____

10. _____

CLASSIFY THE FOLLOWING TRIANGLES AS SCALENE, ISOSCELES, OR EQUILATERAL:

11.

12.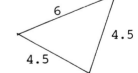

13.

14.

11. _____

12. _____

13. _____

14. _____

EXERCISE 8-3 SET B

NAME _____ DATE _____

DRAW AND LABEL THE FOLLOWING:

1. acute triangle, △POT.

2. equilateral triangle, △CUP.

3. right triangle, △PAN, with <P the right angle.

4. isosceles triangle, △LID, with side $\overline{ID}$ the unequal side.

NAME ALL THE DIFFERENT TRIANGLES IN EACH FIGURE BELOW.   ANSWERS

5.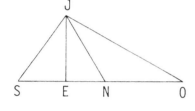

6.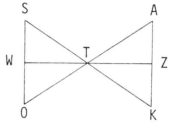

5. _____

6. _____

CLASSIFY THE FOLLOWING TRIANGLES AS ACUTE, RIGHT, OR OBTUSE:

7.

8.

9.

10.

7. _____

8. _____

9. _____

10. _____

CLASSIFY THE FOLLOWING TRIANGLES AS SCALENE, ISOSCELES, OR EQUILATERAL:

11.

12.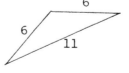

13.

14.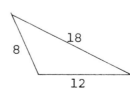

11. _____

12. _____

13. _____

14. _____

8-4 Other Common Geometric Shapes

There are many other geometric shapes besides those discussed in the first three sections of this chapter. In an introduction to geometry we could not cover all of them, so let us consider only some of the most common **plane figures,** those that can be contained in a plane.

Quadrilaterals

Quadrilaterals are four-sided figures. A **rectangle** is a quadrilateral with four right angles. A **square** is a quadrilateral with four right angles and four equal sides. Quadrilaterals are named by placing letters at their vertex points and writing these in either a clockwise or counter clockwise direction.

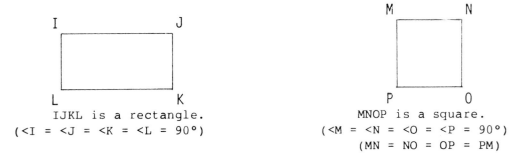

IJKL is a rectangle.
(<I = <J = <K = <L = 90°)

MNOP is a square.
(<M = <N = <O = <P = 90°)
(MN = NO = OP = PM)

A **trapezoid** is a quadrilateral that has one pair of sides that are parallel. A **parallelogram** is a quadrilateral that has both pairs of sides parallel.

QRST is a trapezoid
($\overline{QR} \parallel \overline{TS}$)

UVWX is a parallelogram.
($\overline{UV} \parallel \overline{XW}$ and $\overline{UX} \parallel \overline{VW}$)

Other Plane Geometric Figures

Other common plane figures are classified by the number of sides in the figure. A **pentagon** has five sides, a **hexagon** has six sides, and an **octagon** has eight sides.

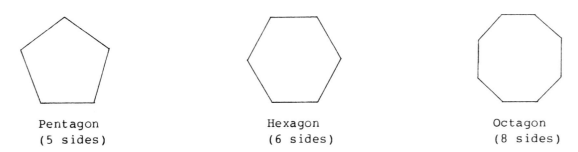

Pentagon
(5 sides)

Hexagon
(6 sides)

Octagon
(8 sides)

Circles

Another very common geometric shape is the circle. A **circle** is a plane figure that consists of all points that are the same distance away from a center point. Any line segment that joins the center and a point on the circle is a **radius** of the circle. Any line segment that joins two points of the circle and also passes through the center is a **diameter** of the circle. Since a diameter (D) is twice as long as a radius (r), we can say that D = 2r.

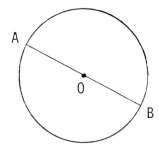

In the circle on the left:

O is the center.

$\overline{OA}$ and $\overline{OB}$ are radii,

$\overline{AB}$ is a diameter.

PROBLEMS: Name the shapes below:

1.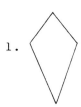
2.
3.

4. 5. 6.

Answers:

1. quadrilateral
2. triangle
3. pentagon
4. hexagon
5. quadrilateral
6. octagon

PROBLEMS: In the circle below:

7. List all radii and diameters.

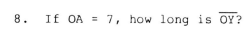

8. If OA = 7, how long is $\overline{OY}$?

9. If PA = 17, how long is $\overline{OP}$?

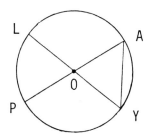

Answers:

7. radii: $\overline{OP}, \overline{OL}, \overline{OA}, \overline{OY}$
   diameters: $\overline{PA}, \overline{LY}$

8. OY = 7
   All radii of a circle are equal.

9. OP = 8½
   A diameter is twice the radius.

EXERCISE 8-4   SET A

NAME _____   DATE _____

DRAW AND LABEL THE FOLLOWING:

1. rectangle FDIC.

2. square FICA.

3. trapezoid USMC, with $\overline{US} \parallel \overline{CM}$.

4. quadrilateral USAF, with no two sides of equal length.

5. parallelogram YMCA, with <Y acute.

6. parallelogram YWCA, with <Y obtuse.

7. a seven-sided plane figure.

8. an octagon.

9. a circle with $\overline{TV}$ a radius.

10. a circle with $\overline{MX}$ a diameter.

REFER TO THE CIRCLE BELOW FOR THE FOLLOWING PROBLEMS: ANSWERS

11. List all radii.

12. List all diameters.

13. If ED = 4, EA = ?

14. If EN = 5, DN = ?

15. If DN = 13, EI = ?

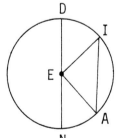

11. _____

12. _____

13. _____

14. _____

15. _____

401

EXERCISE 8-4   SET B

NAME _____   DATE _____

DRAW AND LABEL THE FOLLOWING:

1. square OPEC.

2. rectangle NATO.

3. trapezoid UCLA, with <U a right angle.

4. a trapezoid that has its nonparallel sides equal in length.

5. a quadrilateral that has no two sides parallel.

6. a nine-sided plane figure.

7. a pentagon.

8. a hexagon.

9. a circle with diameter $\overline{YD}$.

10. a circle with radius $\overline{FT}$.

REFER TO THE CIRCLE BELOW FOR THE FOLLOWING PROBLEMS: ANSWERS

11. List all radii.   11. _____

12. List all diameters.   12. _____

13. If AE = 5, AD = ?   13. _____

14. If AE = 4, FE = ?   14. _____

15. If FE = 25, AD = ?   15. _____

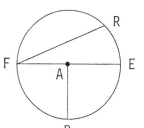

8-5 Perimeters and Areas

In Section 6-4, we encountered perimeters and areas. Remember, a perimeter is the distance around a figure and an area is the measure of the amount of space inside the figure. In this section you will learn more about perimeters and areas.

A FAMOUS ENGLISH KNIGHT: <u>CIRCUMFERENCE</u>

Perimeter

To find the perimeter of a plane figure you need to find the distancce around it. Therefore, if a figure has line segments as sides, you simply add up the sides to find the perimeter (P).

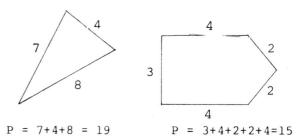

P = 7+4+8 = 19   P = 3+4+2+2+4=15

Area

To find the area (A) of a plane figure, you use a specific formula. Let me state some of the formulas and show you how to use them.

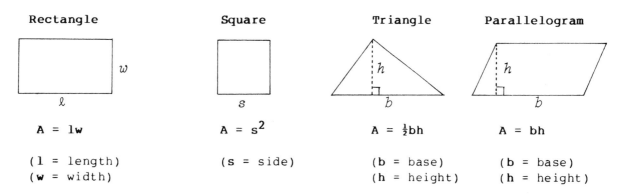

Rectangle	Square	Triangle	Parallelogram
$A = lw$	$A = s^2$	$A = \tfrac{1}{2}bh$	$A = bh$
(l = length) (w = width)	(s = side)	(b = base) (h = height)	(b = base) (h = height)

Example 1: Find the perimeter and area of the square below.

Perimeter:

P = 3+3+3+3

= 12 inches

Area:

$A = s^2$

$= 3^2 = 9$ square inches

Example 2: Find the perimeter and area of the triangle below.

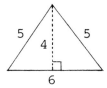

Perimeter:

P = 5+5+6

= 16

Area:

A = ½bh

= ½(6)(4) = 12

Circles

Since a circle is not made up of line segments, you can not find the distance around the circle, the **circumference,** by simply adding up the sides. You also can not multiply lengths of sides to find its area. To find these we must introduce the number pi, π. Pi is approximately equal to 3.14 and is obtained by dividing the circumference (C) of any circle by its diameter (D).

$$\frac{C}{D} = \pi \doteq 3.14$$

Using pi, the formulas for the circumference and the area of a circle with radius r are as folows:

Circumference:

C = 2 π r

Area:

A = πr^2

Example 3: Find the circumference and area of the circle below.

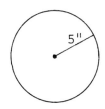

Circumference:

C = 2πr

= 2(3.14)(5)

= 31.4 in.

Area:

A = πr^2

= 3.14 $(5)^2$

= 3.14(25) = 78.5 sq in.

PROBLEM 1: The figure below consists of a half-circle and a parallelogram What is its area?

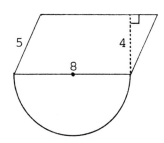

Answer: 57.12

parallelogram:
A = bh
= 8(4) = 32

half-circle:

A = ½πr^2

= ½(3.14)$(4)^2$

= ½(3.14)(16)

= 25.12

total: 57.12

404

EXERCISE 8-5   SET A

NAME _____   DATE _____

FIND THE PERIMETERS AND AREAS OF THE FOLLOWING: ANSWERS

1. a rectangle with 5' and 9' sides.
2. a rectangle with 4' and 7' sides.

1. _____
2. _____

3. a square with 8½" sides.
4. a square with 11½" sides.

3. _____
4. _____

5. a circle with a 6.7 cm radius.
6. a circle with an 8.3 cm radius.

5. _____
6. _____

7. the triangle below.
8. the triangle below.

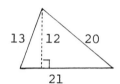
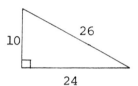

7. _____
8. _____

9. the parallelogram below.
10. the parallelogram below.

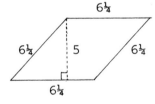
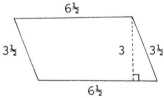

9. _____
10. _____

11. the half-circle below.
12. the half-circle below.

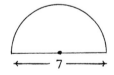
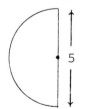

11. _____
12. _____

13. the trapezoid below.
14. the trapezoid below.

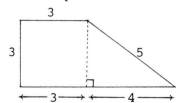
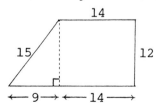

13. _____
14. _____

EXERCISE 8-5   SET B

NAME _____   DATE _____

FIND THE PERIMETERS AND AREAS OF THE FOLLOWING: ANSWERS

1. a rectangle with   2. a rectangle with   1. _____
   5 cm and 8 cm sides.   6 cm and 9 cm sides.
                                                                  2. _____

3. a square with   4. a square with   3. _____
   6¼ inch sides.   11¼ inch sides.
                                                                  4. _____

5. a circle with   6. a circle with   5. _____
   a 12.7 m radius.   a 15.6 m radius.
                                                                  6. _____

7. the triangle below.   8. the triangle below.   7. _____

                                                                  8. _____
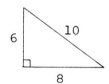   

9. the parallelogram below.   10. the parallelogram below.   9. _____

                                                                 10. _____
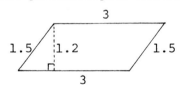   

11. the quarter-circle below.   12. the quarter-circle below. 11. _____

                                                                 12. _____

13. the pentagon below.   14. the pentagon below.   13. _____

                                                                 14. _____
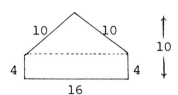   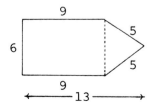

Chapter 8 Summary

Concepts

You may refer to the sections listed below to review how to:

1. illustrate the basic geometric objects -- points, lines, planes, and line segments. (8-1)

2. distinguish between different types of angles. (8-2)

3. identify different kinds of triangles. (8-3)

4. classify common plane figures. (8-4)

5. find areas and perimeters of plane figures. (8-5)

6. find areas and circumferences of circles. (8-5)

Terminology

This chapter's important terms and their corresponding page numbers are:

acute angle: an angle that measures between 0° and 90°. (392)

acute triangle: a triangle with three acute angles. (396)

angle: the figure formed by two rays with the same end point. (391)

circle: the set of points the same distance from a center point. (400)

circumference: the distance around a circle. (404)

complementary angles: two angles that add up to 90°. (392)

diameter: a line segment that joins two points of a circle and also passes through the center. (400)

equilateral triangle: a triangle that has all sides equal. (395)

hexagon: a six-sided plane figure. (396)

intersecting lines: two lines that cross each other at one point. (388)

isosceles triangle: a triangle that has two equal sides. (395)

line segment: a part of a line consisting of two points and all the points between them. (388)

midpoint: a point that divides a line segment into two equal parts. (388)

obtuse angle: an angle that measures between 90° and 180°. (392)

obtuse triangle: a triangle with one obtuse angle. (396)

octagon: a eight-sided plane figure. (399)

parallel lines: lines in the same plane that never intersect. (388)

parallelogram: a quadrilateral with both pairs of sides parallel. (399)

pentagon: a five-sided plane figure. (399)

pi: (π) the number obtained by dividing the circumference of a circle by its diameter, approximately equal to 3.14. (404)

plane figure: a figure that is contained in a plane. (399)

quadrilateral: a four-sided plane figure. (399)

radius: the line segment joining the center of a circle and a point on the circle. (400)

ray: part of a line that has one end point and extends indefinitely in only one direction. (391)

rectangle: a quadrilateral with four right angles. (399)

right angle: an angle that measures 90°. (392)

right triangle: a triangle with one right angle. (396)

scalene triangle: a triangle with each side having different length. (395)

square: a quadrilateral with equal sides and four right angles. (399)

straight angle: an angle that measures 180°. (392)

supplementary angles: two angles that add up to 180°. (392)

trapezoid: a quadrilateral with one pair of parallel sides. (399)

triangle: a three-sided figure. (395)

Student Notes

Chapter 8 Practice Test A

ANSWERS

1. List all the line segments in the figure on the right.

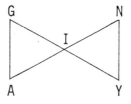

2. In the figure on the right, if I is the midpoint of segment $\overline{GY}$, and GY = 17, then GI = ?

3. What kind of angle is <C, if it measures 136°?

4. Find the supplement and complement of a 37° angle.

5. Draw an equilateral triangle, △ KLS.

6. Draw an obtuse triangle, △ CRS, with <S an obtuse angle.

7. Draw parallelogram QFUZ, with <F an acute angle.

8. What is the name of the plane figure on the right.

9. Find the circumference and area of a circle with a 37 cm diameter.

10. Find the area and perimeter of the figure on the right.

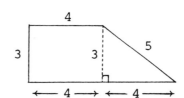

1. _____

2. _____

3. _____

4. _____

5. _____

6. _____

7. _____

8. _____

9. _____

10. _____

Chapter 8 Practice Test B

ANSWERS

1. List all the line segments in the figure on the right.

2. In the figure on the right, if X is the midpoint of segment $\overline{ST}$, and SX = 19, then ST = ?

3. What kind of angle is <R, if it measures 90°?

4. Find the supplement and complement of a 29° angle.

5. Draw a scalene triangle, △ MQT, with MQ > QT.

6. Draw an acute triangle, △ SLT.

7. Draw trapezoid ACFG, with <C a right angle.

8. What is the name of the plane figure on the right?

9. Find the circumference and area of a circle with a 43 cm diameter.

10. Find the area and perimeter of the figure on the right.

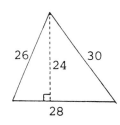

1. _____

2. _____

3. _____

4. _____

5. _____

6. _____

7. _____

8. _____

9. _____

10. _____

Chapter 8 Supplementary Exercises

Section 8-1

Draw and label the following:

1. $\overleftrightarrow{PR}$ containing point S.
2. line n containing points A and B.
3. plane E containing $\overleftrightarrow{AB}$.
4. plane D containing segment $\overline{AB}$.
5. $\overleftrightarrow{TQ} \parallel \overleftrightarrow{SP}$.
6. lines t, s, r, with t ∥ s ∥ r.
7. $\overleftrightarrow{RT}$ intersecting $\overleftrightarrow{TQ}$.
8. line m intersecting line n.
9. line segment $\overline{FD}$ with midpoint X.
10. line segment $\overline{GH}$ with midpoint K.

List all the line segments in each figure:

11.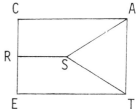
12.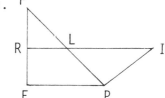
13.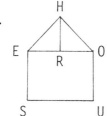

Section 8-2

Draw and label the following:

1. acute <COT.
2. straight <GOT.
3. right <DOT.
4. obtuse <HOT.
5. <JOT with $\overleftrightarrow{PQ} \parallel \overrightarrow{OT}$.
6. <KOT with line m ∥ $\overleftrightarrow{OK}$.
7. plane F containing <LOT.
8. plane G containing <MOT and <NOT.
9. <POT with line t passing through point P.
10. <ROT with line t passing through point R.

Find the complement and supplement of <SOT if it is:

11. 35°
12. 65°
13. 70°
14. 45°
15. 17.5°
16. 52.3°
17. 78¼°
18. 80.7°

Section 8-3

Draw and label the following:

1. acute △ABC.
2. obtuse △ABD.
3. right △ABE.
4. scalene △ABF.
5. equilateral △ABG.
6. isosceles △ABH.
7. line t intersecting side $\overline{AB}$ of isosceles △ABI.
8. plane E containing right △ABJ with <J the right angle.

Section 8-4

Draw and label the following:

1. trapezoid FTBL.
2. parallelogram BASE.
3. rectangle BSKT.
4. square GOLF.
5. a hexagon.
6. a nine-sided plane figure.
7. a circle with $\overline{UN}$ a radius.
8. a circle with $\overline{US}$ a diameter.
9. rectangle BELT, with $\overline{BE}$ longer than $\overline{EL}$.
10. parallelogram SALT, with $\overline{SA}$ shorter than $\overline{AL}$.

Refer to the circle below for the following problems:

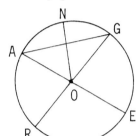

11. List all radii.
12. List all diameters.
13. If RG = 18, OA = ?
14. If ON = $6\frac{1}{2}$, AE = ?
15. If AE = 75.6, RG = ?
16. If OA = 7.9, RG = ?

Section 8-5

Find the perimeters and areas of the following:

1. a square with 10" sides.
2. a square with 6" sides.
3. a rectangle with 5' and 7' sides.
4. a rectangle with 3' and 8' sides.
5. a circle with a 4.6 cm radius.
6. a circle with a 7.8 cm diameter.
7. the triangle below.
8. the parallelogram below.

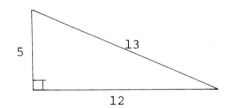

9. the circle below
10. the square below.

Unit IV Exam

Chapters 7 and 8   ANSWERS

1. If x = 3 and y = 5, then 3x − y + 16 = ?   1. _____

2. 8 + (−17)   3. 16 − 24   2. _____

                                                                           3. _____

4. −3 − (−7)   5. (−5)(−9)   4. _____

                                                                           5. _____

6. $\dfrac{-125}{25}$ 7. 5(−8 − 2) + 7 − 2(3)   6. _____

                                                                           7. _____

Solve for x in each equation:

8. −13 = x − 3   9. 3x + 5 = 26   8. _____

                                                                           9. _____

10. One-half a number, decreased by 4, is −10
    What is the number?   10. _____

11. What type of angle measures less than 90°?   11. _____

12. What type of triangle has three sides of equal length?   12. _____

13. Find the complement and supplement of a 42° angle.   13. _____
                                                                               _____

14. Find the area and perimeter of a square that has
    8 inch sides.   14. _____
                                                                               _____

15. Find the area and circumference of a circle with
    a diameter of 6 cm.   15. _____
                                                                               _____

Final Exam

Chapters 1 to 8   ANSWERS

Perform the indicated operations:

1. 4076 x 5000   2. 35,000 - 7,509   1. _____

    2. _____

3. 49,329 ÷ 87   4. $9^2 - \sqrt{49} + 3 \times 6$ 3. _____

    4. _____

5. $6\frac{3}{8} \div 2\frac{1}{3}$ 6. $\frac{3}{4} + \frac{5}{14} - \frac{1}{21}$ 5. _____

    6. _____

7. $85\frac{1}{2} - 31\frac{2}{3}$ 8. 74.05 x 3.6   7. _____

    8. _____

9. 3.6 + 0.185 + 91 + 3.26   10. 19.723 ÷ 3.26   9. _____

    10. _____

11. 45% of 982   12. 3(-7) + (-8)   11. _____

    12. _____

13. Express $\frac{18}{37}$ as a decimal rounded to the thousandths place.   13. _____

14. 16.52 is what percent of 472?   14. _____

15. 5800 g = ? kg   16. 14.5 m = ? in.   15. _____

    16. _____

Final Exam

Chapters 1 to 8   ANSWERS

Solve for N in each of the following:

17. $\dfrac{5}{8} = \dfrac{N}{140}$ 18. 4N + 3 = 51   17. _____

18. _____

19. Your scores out of 100 on chapter tests in the semester were 89, 95, 78, 82, 86, 93, 92, and 97. What is your chapter test average for the semester?   19. _____

20. You started the week with an $85.06 balance in your checking account. During the week you deposited $195.87 and wrote checks for $12.74, $38.56, $107.50, $20.00, and $7.65. What was the balance in your account at the end of the week?   20. _____

21. What would it cost to rent a truck, if you drive it 485 miles in two days and are charged $37.50 a day plus 19.6¢ a mile?   21. _____

22. What would the monthly payments be to pay off a loan of $775 in two years with a simple interest rate of 12% per year?   22. _____

23. A 15% discount is given on the purchase of a $784.00 item. If 5% sales tax is calculated on the discount price, what is the final cost of the item?   23. _____

24. Find the total cost of carpeting a 9' by 12' room, if the carpet with pad costs $24.95 a square yard, and a 6.5% sales tax and $75.00 installation fee are added.   24. _____

25. Find the area and circumference of a circle with a diameter of 20 cm.   25. _____

Answers

UNIT I

Page 2 — Brain Buster

a. $524,288
b. $1,048,575

CHAPTER 1

Page 4 — Pretest

1. 602,067,005
2. forty-two thousand, one hundred seven
3. 543,000
4. 540,000
5. 1325
6. 97,763
7. 2492
8. 3,090,474
9. 23,560,000,000
10. 3,224,007,254
11. 8
12. 7
13. <
14. >
15. $30

Exercise 1-1

Page 7 — Set A

1. eight hundred fifty-two
2. four thousand, two hundred fifty-six
3. seventeen thousand, one hundred nine
4. three million, fifty-seven thousand, ten
5. fourteen million, one hundred thousand, seven hundred
6. nine hundred forty-six thousand, three
7. one billion, three hundred fifty-seven million, nine hundred twenty-six thousand, one hundred eighty-three
8. 745
9. 50,068
10. 105,006
11. 40,000,036
12. 5,007,238
13. 12,015,000,000
14. 89,000,000,089
15. 313,710,000
16. 712,000,422
17. 5 millions
18. 9 tens
19. 8 hundred thousands

Page 8 — SET B

1. nine hundred twenty-five
2. six thousand, one hundred seventy-two
3. thirteen thousand, four hundred two
4. six million, twenty-six thousand, fifty
5. seventeen million, two hundred thousand, four hundred
6. one hundred ninety-six thousand, two
7. four billion, three hundred seven million, nine hundred sixteen thousand, four hundred fifty-two
8. 916
9. 30,043
10. 207,004
11. 50,072,040
12. 4,006,385
13. 18,012,000,000
14. 79,000,000,035
15. 144,730,000
16. 913,000,553
17. 2 ten thousands
18. 8 hundred thousands
19. 1 million

Exercise 1-2

Page 11 SET A   Page 12 SET B

* 1. 50   1. 70
  2. 80   2. 50
  3. 30   3. 70
  4. 510   4. 310
  5. 1470   5. 2580
  6. 5900   6. 4800
* 7. 7000   7. 4000
  8. 500   8. 700
  9. 900   9. 600
  10. 700   10. 900
  11. 14,700   11. 25,600
  12. 27,900   12. 37,600
*13. 180,000   13. 80,000
  14. 1,538,300   14. 4,356,500
  15. 215,749,540   15. 927,563,740
  16. 215,749,500   16. 927,563,700
*17. 215,750,000   17. 927,564,000
  18. 215,750,000   18. 927,560,000
  19. 215,700,000   19. 927,600,000
  20. 216,000,000   20. 928,000,000
  21. 220,000,000   21. 930,000,000

*Selected Solutions from Set A

1. $53 \stackrel{\bullet}{=} 50$

   tens place ↑ ↳digit to the right
                   is less than 5

Round off to the lower multiple, 50.

7. $6997 \stackrel{\bullet}{=} 7000$

   tens place ↑ ↳digit to the right
                   is more than 5

Round off to the higher multiple, 7000.

13. $179,950 \stackrel{\bullet}{=} 180,000$

   hundreds ↑   ↳digit to the
   place   right is 5

Round off to the higher multiple by
increasing the hundreds place by 1.

17. $215,749,538 \stackrel{\bullet}{=} 215,750,000$

   thousands ↑   ↳digit to the
   place   right is 5

Round off to the higher multiple by
increasing the thousands place by 1.

Exercise 1-4

Page 17 SET A   Page 18 SET B

* 1. 68   1. 77
  2. 88   2. 59
* 3. 380   3. 773
  4. 775   4. 292
* 5. 6018   5. 5091
  6. 9304   6. 10,180
* 7. 5898   7. 4596
  8. 2853   8. 2095
* 9. 467   9. 4555
  10. 5484   10. 1637
*11. 204   11. 341
  12. 433   12. 846
*13. 1746   13. 4866
  14. 584,757   14. 327,702
*15. 9975   15. 7469
  16. 5,856,296   16. 6,767,575
*17. 3574   17. 5203

*Selected Solutions from Set A

1.   45   3.   356
   + 23   + 24
     68   380

      11   1
5. 5072   7. 426
   + 946   5382
     6018   +   90
                     5898

      11   11
9.   5   11.   76
     76   + 128
    130   204
   + 256
    467

      1
13. 1400   15.   11
       7   52
     322   9000
   + 17   876
    1746   43
                   +   4
                     9975

    133
17. 18
    196
     45
   2463
    757
   + 95
   3574

Exercise 1-6

Page 23 SET A   Page 24 SET B

* 1. 212   1. 243
  2. 522   2. 148
* 3. 702   3. 603
  4. 776   4. 696
* 5. 1626   5. 1692
  6. 2860   6. 2456
* 7. 965   7. 1470
  8. 3682   8. 6132
* 9. 11,412   9. 11,216
  10. 23,424   10. 52,236
*11. 107,282   11. 170,632
  12. 193,480   12. 601,192
*13. 10,304   13. 7752
  14. 975   14. 103,780
*15. 140,400   15. 919,692
  16. 529,718   16. 28,005,306
*17. 9,907,800   17. 454,384
  18. 141,913,863 18. 38,155,284
*19. 28,608   19. 285,768

*Selected Solutions from Set A

```
        1                    7
1.    53            3.    78
     x 4                  x 9
     212                  702

        1                   41
5.   542            7.   193
     x 3                  x 5
    1626                  965

     1 1                 3214
9.  5706           11.  15326
     x 2                  x 7
   11412                107282

      232                  22
13. 2576           15. 23400
     x 4                  x 6
   10304                140400

    136364
17. 1238475
     x     8
    9907800

      222
19.  4768           14304
     x  3            x  2
    14304           28608
```

Exercise 1-7

Page 27 SET A   Page 28 SET B

* 1. 943   1. 946
  2. 408   2. 1333
* 3. 65,928   3. 54,768
  4. 50,997   4. 40,964
* 5. 21,571   5. 59,585
  6. 42,252   6. 39,494
* 7. 145,040   7. 112,320
  8. 430,430   8. 691,840
* 9. 8,097,408   9. 8,390,928
  10. 11,337,844   10. 34,194,638
*11. 290,928   11. 291,729
  12. 232,468   12. 295,106
*13. 10,524,046   13. 12,064,035
  14. 31,316,615   14. 26,052,422
*15. 5,361,845   15. 4,880,925
  16. 252,793,536   16. 53,348,040

*Selected Solutions from Set A

```
1.     41           3.     984
      x 23                x  67
       123                6888
        82                5904
       943               65928

5.     407          7.    518
      x  53              x 280
      1221                000
      2035               4144
     21571               1036
                       145040

9.    14256        11.   5016
     x  568             x   58
     114048            40128
      85536            25080
      71280           290928
     8097408

13.   74113        15.    6515
     x   142             x  823
     148226              19545
     296452              13030
      74113              52120
    10524046           5361845
```

Exercise 1-8

Page 31 SET A

* 1. 4270
  2. 4350
* 3. 32,576
  4. 60,525
* 5. 9,821,000
  6. 12,046,000
* 7. 280,000
  8. 300,000
* 9. 592,000,000
 10. 318,000,000
*11. 126,210
 12. 189,378
*13. 10,031,024
 14. 18,971,745
*15. 9,805,000
 16. 3,264,000,000
*17. 65,858,780
 18. 8,036,000,000

Page 32 SET B

1. 3640
2. 2450
3. 21,939
4. 67,260
5. 6,766,000
6. 19,909,000
7. 300,000
8. 630,000
9. 135,000,000
10. 532,000,000
11. 92,115
12. 216,108
13. 13,667,416
14. 23,443,812
15. 30,832,000
16. 2,001,000,000
17. 76,845,763
18. 14,049,000,000

*Selected Solutions from Set A

```
1.    61|
    x  7|0
    427 |0

3.      64
    x  509
       576
      3200
     32576

5.   427|
   x  23|000
    1281|000
     854 |
    9821|000

7.    700|
    x   4|00
     2800|00

9.  74000|
   x    8|000
   592000|000

11.     42
    x 3005
       210
     12600
    126210

13.    2504
     x 4006
       15024
     1001600
    10031024

15.    185|
     x  53|000
       555|000
       925 |
      9805|000

17.     2195
      x 30004
         8780
      6585000
     65858780
```

Exercise 1-9

Page 35 SET A

* 1. 9
  2. 49
  3. 144
  4. 225
* 5. 64
  6. 8
* 7. 0
  8. 0
* 9. 1
 10. 1
 11. 32
 12. 243
 13. 1,000,000
 14. 10,000
 15. 1156
 16. 2025
 17. 2
 18. 3
*19. 5
 20. 4
 21. 6
 22. 7
 23. 1
 24. 8
 25. 0
 26. 15
*27. 10
 28. 9
*29. =
 30. <
 31. =
 32. >
*33. <
 34. >
*35. <

Page 36 SET B

1. 16
2. 36
3. 169
4. 324
5. 125
6. 27
7. 0
8. 0
9. 1
10. 1
11. 243
12. 32
13. 100,000
14. 1000
15. 1849
16. 576
17. 4
18. 1
19. 6
20. 5
21. 7
22. 3
23. 0
24. 10
25. 14
26. 11
27. 9
28. 2
29. =
30. <
31. =
32. >
33. <
34. >
35. <

*Selected Solutions from Set A

1. $3^2 = 3 \times 3 = 9$
5. $4^3 = 4 \times 4 \times 4 = 64$
7. $0^5 = 0 \times 0 \times 0 \times 0 \times 0 = 0$
9. $1^4 = 1 \times 1 \times 1 \times 1 = 1$
19. 5; ($5 \times 5 = 25$)
27. 10; ($10 \times 10 = 100$)
29. $24 = 24$
33. $5 < 25$
35. $0 < 2$

Exercise 1-10

Page 39 SET A   Page 40 SET B

* 1. $52   1. $174
  2. $58,500   2. 435 cal
* 3. 585 mi   3. 10,048 invoices
  4. 12,000 sheets   4. 168 lb
* 5. 2190 cubes   5. 336 gal
  6. 336 pieces   6. 1152 peanuts
* 7. 1649 sq ft   7. $427
  8. 4248 pts.   8. 338 pts.
* 9. 430 ft   9. $470
  10. 5400 eggs   10. 144 pens
*11. $4875   11. $25

*Selected Solutions from Set A

1. Add the amounts.

   $15 + $28 + $9 = $52

3. Multiply the number of gallons by number of miles per gallon.

   15 x 39 = 585 mi

5. Cubes per day: 2 x 3 = 6 cubes
   For 365 days: 365 x 6 = 2190 cubes

7. bedrooms: 3 x 144 = 432 sq ft
   master bdrm: = 300 sq ft
   bathrooms: 2 x 56 = 112 sq ft
   kitchen: = 240 sq ft
   living rm: = 460 sq ft
   hallway: = 105 sq ft
                total: 1649 sq ft

9. pipe: 24 x 16 = 384 ft
   joints: 23 x 2 = 46 ft
                total: 430 ft

11. general: 575 x $5 = $2875
    reserved: 250 x $8 = $2000
                 total: $4875

Chapter 1
Practice Test A
Page 43

1. 254,030,107
2. three million, four hundred seven thousand, one hundred twenty-three
3. 753,000
4. 800,000
5. 4116
6. 86,385
7. 4614
8. 4,730,778
9. 39,032,000,000
10. 255,502,190
11. 81
12. 3
13. <
14. =
15. 1048 lb

Chapter 1
Practice Test B
Page 44

1. 1,025,306,517
2. six hundred seventy-eight thousand, three hundred nine
3. 1,257,380
4. 1,260,000
5. 1437
6. 94,041
7. 3824
8. 1,659,860
9. 19,040,000
10. 5,436,000,906
11. 64
12. 2
13. =
14. >
15. 85°

Chapter 2
Pretest
Page 50

1. 331   9. 295
2. 66,604   10. 606 R20
3. 5053   11. 600
4. 6   12. 9008
5. 5 R6   13. 97
6. impossible   14. 2x2x5x7
7. 53   15. 91
8. 2063 R3

Exercise 2-2

Page 55 SET A   Page 56 SET B

```
 *1. 64              1. 33
  2. 34              2. 23
 *3. 501             3. 805
  4. 802             4. 702
 *5. 8842            5. 4764
  6. 4464            6. 2762
 *7. 56              7. 35
  8. 28              8. 18
 *9. 234             9. 369
 10. 375            10. 469
*11. 1578           11. 2387
 12. 1788           12. 3089
*13. 548            13. 438
 14. 729            14. 828
*15. 3245           15. 2377
 16. 4365           16. 4267
*17. 362            17. 4439
 18. 1703           18. 5725
*19. 9552           19. 9191
 20. 8008           20. 9009
*21. 749,257        21. 2,111,795
```

*Selected Solutions from Set A

```
1.    9 6         3.    5 1 4
    - 3 2             -   1 3
    -----             -------
      6 4               5 0 1


5.    9 8 7 6     7.    7 13
    - 1 0 3 4           ⁄8 ⁄3
    ---------         -   2 7
      8 8 4 2         -------
                        5 6

      2 12              1415
9.    ⁄3 ⁄2 9    11.  3 ⁄4 ⁄5 13
    -     9 5         ⁄4 ⁄5 ⁄6 ⁄3
      -------       -   2 9 8 5
        2 3 4         -----------
                        1 5 7 8

      5 9 17            3 9 9 13
13.   ⁄6 ⁄0 ⁄7    15. ⁄4 ⁄0 ⁄0 ⁄3
    -       5 9      -       7 5 8
      ---------       -----------
        5 4 8            3 2 4 5

      4 9 9 10                5 12
17.   ⁄5 ⁄0 ⁄0 ⁄0  19.  1 4 ⁄6 ⁄2 5
    -   4 6 3 8         -   5 0 7 3
      -----------       -----------
            3 6 2         9 5 5 2

            8 16 7 15
21.    1 5 ⁄9 ⁄6 3 ⁄8 ⁄5
     -     8 4 7 1 2 8
       -----------------
           7 4 9 2 5 7
```

Exercise 2-3

Page 59 SET A   Page 60 SET B

```
 1. 1               1. 4
 2. 1               2. 2
 3. 1               3. 5
 4. 5               4. 4
 5. 4               5. 9
 6. 4               6. 6
 7. 3               7. 9
 8. 3               8. 2
 9. 4               9. 6
10. 5              10. 5
11. 5              11. 6
12. 7              12. 6
13. 5              13. 5
14. 2              14. 4
15. 8              15. 8
16. 5              16. 3
17. 2              17. 8
18. 9              18. 4
19. 8              19. 7
20. 6              20. 2
21. 4              21. 8
22. 6              22. 8
23. 9              23. 2
24. 5              24. 6
25. 8              25. 7
26. 7              26. 9
27. 6              27. 5
28. 8              28. 6
29. 9              29. 8
30. 9              30. 7
31. 4              31. 9
32. 5              32. 7
33. 3              33. 7
34. 5              34. 5
35. 7              35. 7
36. 6              36. 1
37. 7              37. 6
38. 6              38. 9
39. 3              39. 6
```

Selected Solutions from Set A

```
40.   20   ↗16   ↗12   ↗8   ↗4
     - 4  / - 4 / - 4 /- 4 /- 4
     ----    ----    ---   ---   ---
      16     12      8     4     0
```

Thus 20 ÷ 4 = 5

```
41.   18   ↗15  ↗12  ↗9   ↗6   ↗3
     - 3 / - 3/ - 3/- 3 /- 3 /- 3
     ----   ----  ---   ---  ---  ---
      15    12    9     6    3    0
```

Thus 3⟌18 = 6

Exercise 2-4

Page 63   SET A   Page 64 SET B

```
 * 1.  6 R1            1.  4 R1
   2.  5 R2            2.  6 R1
 * 3.  9 R2            3.  9 R2
   4.  7 R1            4.  7 R4
 * 5.  impossible      5.  impossible
   6.  8 R3            6.  8 R7
   7.  6 R1            7.  8 R1
   8.  impossible      8.  impossible
 * 9.  7               9.  6
  10.  9              10.  16
 *11.  3 R6           11.  7 R3
  12.  5 R5           12.  6 R6
 *13.  6 R3           13.  9 R3
  14.  8 R8           14.  7 R7
  15.  2 R5           15.  4 R3
  16.  8 R1           16.  5 R3
 *17.  8 R2           17.  9 R3
  18.  7 R1           18.  8 R3
 *19.  7 R4           19.  9 R2
  20.  2 R3           20.  7 R1
 *21.  8 R3           21.  8 R6
  22.  7 R4           22.  9 R1
 *23.  4              23.  5
  24.  8              24.  8
```

*Selected Solutions from Set A

```
         6                    9
 1.  4⟌25           3.  5⟌47
        24                   45
         1                    2

 5. impossible       9.      7
    to divide           7⟌49
    by zero                49

         3                    6
11.  8⟌30          13.  5⟌33
        24                   30
         6                    3

         8                    7
17.  7⟌58          19.  7⟌53
        56                   49
         2                    4

         8                    4
21.  9⟌75          23.  9⟌36
        72                   36
         3
```

Exercise 2-5

Page 67   SET A   Page 68   SET B

```
 * 1.  243            1.  341
   2.  332            2.  232
   3.  24             3.  13
   4.  24             4.  23
 * 5.  63             5.  54
   6.  57             6.  84
   7.  86 R3          7.  61 R2
   8.  84 R3          8.  73 R3
   9.  257            9.  542 R2
  10.  653           10.  197 R2
 *11.  2457 R1      11.  243 R2
  12.  174 R5       12.  657 R1
 *13.  4772 R1      13.  3216 R1
  14.  3909 R2      14.  2783 R1
 *15.  6853         15.  2496
  16.  2457         16.  2468
 *17.  51,249       17.  24,583
  18.  68,792       18.  57,842
```

*Selected Solutions from Set A

```
         243                   63
 1.  2⟌486         5.  7⟌441
        4                    42
        8                    21
        8                    21
         6
         6

        2457                  4772
11.  3⟌7372       13.  4⟌19089
        6                    16
       13                    30
       12                    28
        17                    28
        15                    28
        22                     9
        21                     8
         1                     1

        6853                 51249
15.  7⟌47971      17.  9⟌461241
        42                   45
        59                   11
        56                    9
        37                   22
        35                   18
        21                   44
        21                   36
                             81
                             81
```

Exercise 2-6

Page 73 SET A   Page 74 SET B

* 1. 17   1. 26
  2. 23   2. 67
* 3. 53   3. 35
  4. 36   4. 62
* 5. 47 R6   5. 53 R11
  6. 61 R20   6. 26 R42
* 7. 65   7. 65
  8. 65   8. 75
* 9. 135 R12   9. 47 R205
 10. 86 R101   10. 74 R223
*11. 453   11. 347
 12. 357   12. 428
*13. 86 R218   13. 48 R324
 14. 47 R123   14. 72 R825
*15. 175   15. 255
 16. 195   16. 275

*Selected Solutions from Set A

```
          17                    53
1. 23)391           3. 47)2491
       23                   235
       161                  141
       161                  141

          47                    65
5. 69)3249          7. 232)15080
       276                  1392
       489                  1160
       483                  1160
         6

         135                   453
9. 247)33357       11. 706)319818
       247                  2824
       865                  3741
       741                  3530
      1247                  2118
      1235                  2118
        12

          86                    175
13. 4125)354968    15. 5806)1016050
        33000                 5806
        24968                43545
        24750                40642
          218                29030
                            29030
```

Exercise 2-7

Page 77 SET A   Page 78 SET B

* 1. 400   1. 500
  2. 700   2. 800
* 3. 250   3. 250
  4. 250   4. 250
* 5. 400   5. 600
  6. 500   6. 900
* 7. 503   7. 307
  8. 206   8. 408
* 9. 406 R40   9. 302 R37
 10. 304 R50   10. 406 R25
*11. 1002   11. 1007
 12. 1003   12. 1009
*13. 602 R50   13. 307 R20
 14. 703 R40   14. 205 R20
*15. 2030   15. 2004
 16. 3020   16. 3006

*Selected Solutions from Set A

```
         400                   250
1. 6)2400           3. 24)6000
      24                    48
       0                   120
       0                   120
       0                     0
       0                     0

         400                   503
5. 70)28000         7. 43)21629
      280                  215
        0                  129
        0                  129
        0
        0

         406                  1002
9. 68)27648        11. 23)23046
      272                   23
      448                   046
      408                    46
       40

         602                  2030
13. 503)302856     15. 345)700350
        3018                 690
        1056                1035
        1006                1035
          50                   0
                               0
```

Exercise 2-8

Page 81 SET A   Page 82 SET B

* 1. 17   1. 20
  2. 19   2. 44
* 3. 32   3. 56
  4. 54   4. 72
* 5. 16   5. 16
  6. 16   6. 49
* 7. 36   7. 81
  8. 25   8. 64
  9. 1   9. 62
  10. 12   10. 78
*11. 162   11. 50
  12. 75   12. 567
*13. 31   13. 19
  14. 25   14. 16
  15. 16   15. 14
  16. 19   16. 17
*17. 597,000   17. 39,800
  18. 4,970,000   18. 7,995,000
*19. 23   19. 28
  20. 30   20. 62

*Selected Solutions from Set A

```
1.  5 + 3 x 4 =        3. (5 + 3) x 4 =
    5 + 12   =                 8  x 4 =
         17                        32

5. 24 x 4 ÷ 6 =        7. 24 ÷ 4 x 6 =
   96   ÷ 6 =                6   x 6 =
        16                       36

11. (2 + 1)^4 x (12 - 2 x 5) =
      3^4    x (12 - 10) =
      81     x    2    =
                162

13. √100 + 3 x √49 =
     10  + 3 x  7  =
     10  +   21    =
             31

17. √36 x 10^5 - 3 x 10^3 =
     6 x 100,000 - 3 x 1000 =
       600,000   -  3000    =
              597,000

19. 16 - 12 ÷ 4 x 3 + 8 x 2 =
    16 -    3    x 3 + 8 x 2 =
    16 -        9     + 16   =
               7       + 16  =
                      23
```

Exercise 2-9

Page 85 SET A   Page 86 SET B

* 1. prime   1. composite
  2. prime   2. prime
* 3. composite   3. prime
  4. composite   4. composite
* 5. composite   5. composite
  6. composite   6. composite
* 7. 2x3x3   7. 2x2x2x2
  8. 2x2x2x3   8. 3x3x3
* 9. 2x23   9. 2x29
  10. 2x17   10. 2x31
*11. 2x3x7   11. 2x2x3x3
  12. 2x3x11   12. 2x2x2x5
  13. 5x11   13. 2x5x5
  14. 5x7   14. 5x5x5
*15. 3x5x5   15. 2x7x7
  16. 5x7x7   16. 2x3x3x3
*17. 5x5x5x11   17. 5x5x7x11
  18. 5x5x5x13   18. 3x5x5x7
  19. 2x2x3x5x11   19. 2x2x2x5x13
  20. 2x5x5x13   20. 2x2x3x5x7
*21. 2x3x37   21. 2x3x3x13
  22. 3x3x43   22. 2x3x7x41
  23. 7x19   23. 11x19
  24. 11x13   24. 7x23

*Selected Solutions from Set A

1. No divisors except for 1 and itself.
3. 3 is a divisor.
5. 5 is a divisor.

```
7.  2|18            9.  2|46
    3| 9               23|23
    3| 3                  1
       1              46 = 2x23
   18 = 2x3x3

11. 2|42           15. 5|75
    3|21               5|15
    7| 7               3| 3
       1                  1
   42 = 2x3x7         75 = 3x5x5

17. 5|1375         21. 2|222
    5| 275            3|111
    5|  55           37| 37
   11|  11               1
        1
   1375 = 5x5x5x11   222 = 2x3x37
```

Exercise 2-10

Page 89 SET A

* 1. $949
2. 1148 mi
* 3. $150
4. 40 min
* 5. 168
6. 71 lb
* 7. $232
8. $85
* 9. $1516
10. 9 yrs; 13 yrs
*11. $188

Page 90 SET B

1. $61
2. 19 tickets
3. $204,600,000
4. $589
5. 23 pts.
6. 29 students
7. $75
8. $55,450
9. 443 mi
10. $604
11. 27 people

*Selected Solutions from Set A

1. Subtract the discount from the original price.

   $1199 - 250 = $949

3. Divide total to be saved by the number of months.

   2 years = 24 months

   $3600 ÷ 24 = $150

5. Add the scores and divide by 3.

   ```
   171
   149          168
   184        3)504
   504
   ```

7. Number of overtime hrs:
   52 - 40 = 12

   Regular pay: $4 x 40 = $160
   Overtime pay: $6 x 12 =   72
                  total: $232

9. Divide the yearly earnings by 12.

   $18,192 ÷ 12 = $1516 per month

11. Subtract costs from amount received and divide that amount into four parts.

    $787 - 35 = $752
    $752 ÷ 4 = $188

Chapter 2 Practice Test A

Page 93

1. 225
2. 28,824
3. 30,416
4. 8
5. 8 R3
6. impossible
7. 69 R5
8. 2972
9. 657
10. 309 R5
11. 800
12. 30,030
13. 49
14. 2x3x5x5x5
15. 163

Chapter 2 Practice Test B

Page 94

1. 521
2. 67,655
3. 5214
4. 5
5. 8 R1
6. impossible
7. 78 R1
8. 2063
9. 356
10. 409 R20
11. 800
12. 370
13. 56
14. 2x2x3x5x11
15. 106 boxes

Unit 1 Exam

Page 98

1. five billion, twelve thousand, five hundred seventy-six
2. 3717
3. 83,259
4. 3,502,625
5. 174,320,000
6. 73
7. 4636
8. 45,643
9. 398
10. 82 R3
11. 3006
12. 76
13. 2x2x2x5x17
14. $140
15. $164

UNIT II

Page 100 **Brain Buster**

approximately 31.7 years

CHAPTER 3

Page 102 **Pretest**

1. 49
2. 3 2/3
3. 27/50
4. 7/10
5. 5 16/21
6. 185 people
7. 2/3
8. 7/16
9. 1 29/240
10. 7/72
11. 8/9
12. 6 7/15
13. 1 13/22
14. 8/9, 7/8, 2/3
15. 12¢

Exercise 3-1

Page 105 SET A   Page 106 SET B

* 1. 9/16   1. 3/5
  2. 7/16   2. 2/5
* 3. 4/7   3. 5/9
  4. 3/7   4. 4/9
* 5. 3/8   5. 3/8
  6. 5/8   6. 5/8
* 7. 17/20   7. 19/25
  8. 3/20   8. 6/25
* 9. 3/11   9. 5/12
  10. 8/11   10. 7/12
*11. 17/30   11. 23/40
  12. 13/30   12. 17/40
*13. 1/2, 3/5, 15/16 13. 1/5, 2/3, 5/8
  14. 9/6, 8/1, 7/7   14. 9/4, 7/1, 3/3
*15. 7/8, 3/4   15. 17/32, 3/5, 1/10
  16. 12/7, 7/3, 10/1, 19/19   16. 11/8, 5/1, 5/5

*Selected Solutions from Set A

1. 9 out of 16 sections are shaded. It is 9/16 shaded.

3. 4 out of 7 sections are shaded. It is 4/7 shaded.

5. 3 out of 8 sections are shaded. It is 3/8 shaded.

7. 17 out of 20 = 17/20

9. 3 out of 11 = 3/11

11. 17 out of 30 = 17/30

13. Proper fractions have numerators that are smaller than their **denominators. Thus 1/2, 3/5, and 15/16 are proper fractions.**

15. 7/8 and 3/4 have numerators that are smaller than their denominators, so they are proper fractions.

Exercise 3-2

Page 111 SET A	Page 112 SET B
* 1. 3	1. 5
2. 4	2. 2
3. 6	3. 12
4. 9	4. 8
* 5. 28	5. 28
6. 32	6. 24
* 7. 60	7. 24
8. 36	8. 35
9. 30	9. 63
10. 12	10. 60
*11. 1/5	11. 1/4
12. 1/3	12. 1/2
13. 5/9	13. 2/3
14. 2/3	14. 3/5
15. 7/11	15. 9/11
16. 9/13	16. 7/13
*17. 2/3	17. 3/4
18. 3/8	18. 2/5
19. 5/12	19. 25/36
20. 11/15	20. 22/45
21. 26/147	21. 32/105
22. 28/117	22. 29/138
*23. 4/3	23. 59/47
24. 17/13	24. 47/37
*25. 2/5	25. 5/6
26. 3/4	26. 3/4
*27. 4/7	27. 3/4
28. 6/7	28. 36/29

*Selected Solutions from Set A

1. $\dfrac{1}{2} \xrightarrow{\times 3} \dfrac{3}{6}$ (×3 top and bottom)

5. $\dfrac{8}{7} \xrightarrow{\times 4} \dfrac{32}{28}$ (×4 top and bottom)

7. $\dfrac{6}{5} \xrightarrow{\times 12} \dfrac{72}{60}$ (×12 top and bottom)

11. $\dfrac{3}{15} \xrightarrow{\div 3} \dfrac{1}{5}$ (÷3 top and bottom)

17. $\dfrac{80}{120} \xrightarrow{\div 10} \dfrac{8}{12} \xrightarrow{\div 4} \dfrac{2}{3}$

23. $\dfrac{144}{108} \xrightarrow{\div 2} \dfrac{72}{54} \xrightarrow{\div 6} \dfrac{12}{9} \xrightarrow{\div 3} \dfrac{4}{3}$

25. $\dfrac{34}{85} \xrightarrow{\div 17} \dfrac{2}{5}$

27. $\dfrac{52}{91} \xrightarrow{\div 13} \dfrac{4}{7}$

Exercise 3-3

Page 115 SET A	Set 116 SET B
1. 8	1. 6
2. 7	2. 9
* 3. 5	3. 6
4. 4	4. 6
* 5. 5 2/3	5. 6 1/3
6. 4 1/6	6. 6 1/4
7. 1 3/7	7. 1 5/6
8. 1 3/8	8. 2 2/5
* 9. 7 2/5	9. 5 2/7
10. 9 2/5	10. 8 2/7
*11. 7 13/16	11. 7 7/19
12. 8 4/17	12. 11 7/13
13. 65 2/7	13. 78 1/7
14. 56 4/7	14. 84 5/7
15. 9/1	15. 7/1
16. 8/1	16. 6/1
*17. 3/2	17. 5/3
18. 8/5	18. 7/4
*19. 21/8	19. 17/6
20. 20/7	20. 23/8
21. 37/4	21. 33/4
22. 25/3	22. 28/3
*23. 75/2	23. 49/2
24. 313/6	24. 231/5
*25. 65/6	25. 165/8
26. 188/9	26. 97/9
*27. 871/7	27. 996/7
28. 629/3	28. 859/8
*29. 780/17	29. 758/19
30. 627/32	30. 561/32

*Selected Solutions from Set A

3. $3\overline{)15} = 5$

5. $3\overline{)17} = 5\ 2/3$, remainder 2

9. $5\overline{)37} = 7\ 2/5$, 35, remainder 2

11. $16\overline{)125} = 7\ 13/16$, 112, remainder 13

17. $1\dfrac{1}{2} = \dfrac{3}{2}$

19. $2\dfrac{5}{8} = \dfrac{21}{8}$

23. $37\dfrac{1}{2} = \dfrac{75}{2}$

25. $10\dfrac{5}{6} = \dfrac{65}{6}$

27. $124\dfrac{3}{7} = \dfrac{871}{7}$

29. $45\dfrac{15}{17} = \dfrac{780}{17}$

Exercise 3-4

Page 119   SET A   Page 120   SET B
* 1. 3/8   1. 3/16
  2. 2/15   2. 1/6
  3. 35/72   3. 30/77
  4. 63/80   4. 21/40
* 5. 3/5   5. 4/5
  6. 2/3   6. 3/4
  7. 3 1/3   7. 3 3/4
  8. 2 4/7   8. 4 2/3
* 9. 5/21   9. 7/15
 10. 5/21   10. 5/14
 11. 1 13/42   11. 2 8/9
 12. 1 7/15   12. 1 5/21
*13. 3/28   13. 4/15
 14. 1/6   14. 7/15
*15. 1/5   15. 1/5
 16. 1/7   16. 1/7
*17. 1 5/21   17. 22/27
 18. 1 5/28   18. 14/15
 19. 32/105   19. 1
 20. 21/40   20. 4/75
*21. 2/5   21. 14/135
 22. 1/9   22. 15/56
*23. 55/126   23. 7/72
 24. 11/12   24. 5/84
 25. 35/576   25. 392/405
 26. 147/2560   26. 28/405

*Selected Solutions from Set A

1. $\dfrac{1}{2} \times \dfrac{3}{4} = \dfrac{3}{8}$ 5. $3 \times \dfrac{1}{5} = \dfrac{3}{5}$

9. $\dfrac{\cancel{3}^1}{7} \times \dfrac{5}{\cancel{9}_3} = \dfrac{5}{21}$ 13. $\dfrac{\cancel{12}^{\,3}}{\cancel{35}_{\,7}} \times \dfrac{\cancel{5}^{\,1}}{\cancel{16}_{\,4}} = \dfrac{3}{28}$

15. $3 \times \dfrac{\cancel{4}^{\,1}}{\cancel{25}_{\,5}} \times \dfrac{\cancel{5}^{\,1}}{\cancel{12}_{\,3}} = \dfrac{3}{15} = \dfrac{1}{5}$

17. $\dfrac{13}{\cancel{35}_{\,7}} \times \dfrac{\cancel{14}^{\,2}}{\cancel{21}_{\,3}} \times \cancel{8}^{\,1} = \dfrac{26}{21} = 1\dfrac{5}{21}$

21. $\dfrac{1}{\cancel{3}} \times \dfrac{\cancel{6}^{\,2}}{\cancel{13}} \times \dfrac{\cancel{39}^{\,3}}{15} = \dfrac{6}{15} = \dfrac{2}{5}$

23. $\dfrac{\cancel{20}^{\,5}}{\cancel{21}_{\,7}} \times \dfrac{\cancel{19}^{\,1}}{\cancel{26}_{\,9}} \times \dfrac{\cancel{33}^{\,11}}{\cancel{38}_{\,2}} = \dfrac{55}{126}$

Exercise 3-5

Page 123   SET A   Page 124   SET B
* 1. 2/3   1. 5/6
  2. 3/4   2. 3/5
* 3. 1 1/9   3. 1 3/25
  4. 35/36   4. 15/16
* 5. 10/21   5. 15/28
  6. 28/45   6. 21/44
  7. 5/7   7. 1 1/4
  8. 1 1/3   8. 3/5
* 9. 1/6   9. 1/12
 10. 1/10   10. 1/16
 11. 3   11. 5
 12. 1/4   12. 3
*13. 20 1/4   13. 13 1/3
 14. 11 1/5   14. 12 3/5
*15. 11/42   15. 5/24
 16. 5/18   16. 7/30
*17. 8/9   17. 2 4/9
 18. 9/20   18. 5/12
 19. 3 1/9   19. 1 1/7
 20. 1 1/2   20. 1/2

*Selected Solutions from Set A

1. $\dfrac{1}{3} \div \dfrac{1}{2} = \dfrac{1}{3} \times \dfrac{2}{1} = \dfrac{2}{3}$

3. $\dfrac{2}{3} \div \dfrac{3}{5} = \dfrac{2}{3} \times \dfrac{5}{3} = \dfrac{10}{9} = 1\dfrac{1}{9}$

5. $\dfrac{6}{9} \div \dfrac{7}{5} = \dfrac{\cancel{6}^{\,2}}{\cancel{9}_{\,3}} \times \dfrac{5}{7} = \dfrac{10}{21}$

9. $\dfrac{2}{3} \div 4 = \dfrac{2}{3} \div \dfrac{4}{1} = \dfrac{\cancel{2}^{\,1}}{3} \times \dfrac{1}{\cancel{4}_{\,2}} = \dfrac{1}{6}$

13. $18 \div \dfrac{8}{9} = \dfrac{\cancel{18}^{\,9}}{1} \times \dfrac{9}{\cancel{8}_{\,4}} = \dfrac{81}{4} = 20\dfrac{1}{4}$

15. $\dfrac{5}{18} \div \dfrac{35}{33} = \dfrac{\cancel{5}^{\,1}}{\cancel{18}_{\,6}} \times \dfrac{\cancel{33}^{\,11}}{\cancel{35}_{\,7}} = \dfrac{11}{42}$

17. $\dfrac{20}{42} \div \dfrac{15}{28} = \dfrac{\cancel{20}^{\,4}}{\cancel{42}_{\,3}} \times \dfrac{\cancel{28}^{\,2}}{\cancel{15}_{\,3}} = \dfrac{8}{9}$

Exercise 3-6

Page 127 SET A   Page 128 SET B

* 1. 4 7/12   1. 6 5/12
  2. 6 3/4   2. 12 3/4
* 3. 86   3. 105
  4. 91 1/5   4. 66 1/4
* 5. 109 5/7   5. 54 6/11
  6. 496   6. 100
* 7. 1 13/63   7. 1 21/37
  8. 65/76   8. 25/36
* 9. 6 5/32   9. 8 27/32
  10. 2 5/32   10. 8 1/8
*11. 2/3   11. 6/7
  12. 6   12. 12

*Selected Solutions from Set A

1. $2\frac{3}{4} \times 1\frac{2}{3} = \frac{11}{4} \times \frac{5}{3} = \frac{55}{12} = 4\frac{7}{12}$

3. $5\frac{3}{8} \times 16 = \frac{43}{\cancel{8}_1} \times \frac{\cancel{16}^2}{1} = \frac{86}{1} = 86$

5. $26\frac{2}{3} \times 4\frac{4}{35} = \frac{\cancel{80}^{16}}{\cancel{3}_1} \times \frac{\cancel{144}^{48}}{\cancel{35}_7} = \frac{768}{7} = 109\frac{5}{7}$

7. $7\frac{3}{5} \div 6\frac{3}{10} = \frac{38}{5} \div \frac{63}{10} = \frac{38}{\cancel{5}_1} \times \frac{\cancel{10}^2}{63}$

   $= \frac{76}{63} = 1\frac{13}{63}$

9. $24\frac{5}{8} \div 4 = \frac{197}{8} \div \frac{4}{1} = \frac{197}{8} \times \frac{1}{4}$

   $= \frac{197}{32} = 6\frac{5}{32}$

11. $5 \div 7\frac{1}{2} = \frac{5}{1} \div \frac{15}{2} = \frac{\cancel{5}^1}{1} \times \frac{2}{\cancel{15}_3} = \frac{2}{3}$

Exercise 3-7

Page 131 SET A   Page 132 SET B

* 1. 268 1/2   1. 13
  2. 147   2. 28 1/2
* 3. 1708   3. 380 2/3
  4. 1578 3/4   4. 488
* 5. 2 13/16   5. 18 27/32
  6. 3 1/3   6. 10 5/16
* 7. 9 5/32   7. 53 21/32
  8. 16 7/10   8. 81 9/16
* 9. 1839 voters   9. 930 students
  10. 6250 bulbs   10. 42 2/3 oz
*11. $39   11. 3562 students
  12. 628 students   12. $117
*13. 219 pennies   13. 308 people
  14. 16 players   14. 171 workers

*Selected Solutions from Set A

1. $\frac{1}{2}$ of $537 = \frac{1}{2} \times \frac{537}{1} = 268\frac{1}{2}$

3. $\frac{2}{5}$ of $4270 = \frac{2}{\cancel{5}_1} \times \frac{\cancel{4270}^{854}}{1} = 1708$

5. $\frac{3}{8}$ of $7\frac{1}{2} = \frac{3}{8} \times \frac{15}{2} = \frac{45}{16} = 2\frac{13}{16}$

7. $\frac{1}{4}$ of $36\frac{5}{8} = \frac{1}{4} \times \frac{293}{8} = \frac{293}{32} = 9\frac{5}{32}$

9. Find 3/4 of the total voters:

   $\frac{3}{4}$ of $2452 = \frac{3}{\cancel{4}_1} \times \frac{\cancel{2452}^{613}}{1} = 1839$

11. Discount:

    1/4 of $52 = 1/4 × $52 = $13

    Sale price: $52 = $13 = $39

13. Find the number of pennies:

    3/4 of 876 = 3/4 × 876 = 657

    2/3 are dated 1983 or later.

    2/3 of 657 = 2/3 × 657 = 438

    Number after 1983: 657 − 438 = 219

Exercise 3-8

Page 135 SET A   Page 136 SET B

* 1. 2/7   1. 2/9
  2. 2/3   2. 2/5
* 3. 1/9   3. 1/7
  4. 2/7   4. 2/9
* 5. 1/2   5. 1/2
  6. 1/2   6. 1/2
* 7. 1/9   7. 1/3
  8. 2/5   8. 2/9
* 9. 3   9. 5
  10. 6   10. 4
*11. 3   11. 2
  12. 3 2/3   12. 2
*13. 1   13. 1
  14. 1   14. 1
*15. 7/16   15. 9/16
  16. 9/16   16. 3/4
*17. 1 1/6   17. 31/32
  18. 15/16   18. 1 1/18

*Selected Solutions from Set A

1. $\frac{1}{7} + \frac{1}{7} = \frac{1+1}{7} = \frac{2}{7}$

3. $\frac{3}{9} - \frac{2}{9} = \frac{3-2}{9} = \frac{1}{9}$

5. $\frac{3}{8} + \frac{1}{8} = \frac{3+1}{8} = \frac{4}{8} = \frac{1}{2}$

7. $\frac{7}{18} - \frac{5}{18} = \frac{7-5}{18} = \frac{2}{18} = \frac{1}{9}$

9. $\frac{4}{3} + \frac{5}{3} = \frac{4+5}{3} = \frac{9}{3} = 3$

11. $\frac{19}{5} - \frac{4}{5} = \frac{19-4}{5} = \frac{15}{5} = 3$

13. $\frac{3}{8} + \frac{2}{8} + \frac{3}{8} = \frac{3+2+3}{8} = \frac{8}{8} = 1$

15. $\frac{11}{32} + \frac{5}{32} - \frac{2}{32} = \frac{11+5-2}{32} = \frac{14}{32} = \frac{7}{16}$

17. $\frac{25}{54} + \frac{21}{54} + \frac{17}{54} = \frac{25+21+17}{54}$
$= \frac{63}{54} = \frac{7}{6} = 1\frac{1}{6}$

Exercise 3-9

Page 139 SET A   Page 140 SET B

* 1. 3/4   1. 1/2
  2. 5/6   2. 3/4
* 3. 1/8   3. 3/8
  4. 3/8   4. 5/8
* 5. 5/12   5. 1/12
  6. 5/12   6. 1/12
* 7. 9/10   7. 3/4
  8. 9/10   8. 3/4
* 9. 1/30   9. 1/20
  10. 1/40   10. 7/30
*11. 4   11. 3
  12. 5   12. 4
*13. 1 9/32   13. 29/32
  14. 27/32   14. 31/32
*15. 2 11/12   15. 1 1/4
  16. 2 5/12   16. 2 3/4

*Selected Solutions from Set A

1. $\frac{1}{2} = \frac{2}{4}$ 3. $\frac{3}{4} = \frac{6}{8}$
   $+\frac{1}{4} = \frac{1}{4}$ $-\frac{5}{8} = \frac{5}{8}$
   $\frac{3}{4}$ $\frac{1}{8}$

5. $\frac{3}{4} = \frac{9}{12}$ 7. $\frac{3}{20} = \frac{3}{20}$
   $-\frac{1}{3} = \frac{4}{12}$ $+\frac{3}{4} = \frac{15}{20}$
   $\frac{5}{12}$ $\frac{18}{20} = \frac{9}{10}$

9. $\frac{5}{6} = \frac{25}{30}$ 11. $\frac{7}{10} = \frac{7}{10}$
   $-\frac{4}{5} = \frac{24}{30}$ $+\frac{5}{2} = \frac{25}{10}$
   $\frac{1}{30}$ $+\frac{4}{5} = \frac{8}{10}$
                                            $\frac{40}{10} = 4$

15. $\frac{1}{4} = \frac{3}{12}$ 13. $\frac{3}{8} = \frac{12}{32}$
    $+\frac{3}{6} = \frac{6}{12}$ $+\frac{15}{32} = \frac{15}{32}$
    $+\frac{9}{2} = \frac{54}{12}$ $+\frac{7}{16} = \frac{14}{32}$
    $-\frac{7}{3} = \frac{28}{12}$
    $\frac{35}{12} = 2\frac{11}{12}$ $\frac{41}{32} = 1\frac{9}{32}$

Exercise 3-10

Page 143 SET A   Page 144 SET B

* 1. 7/18   1. 19/24
  2. 13/24   2. 11/18
* 3. 1/12   3. 13/30
  4. 11/30   4. 1/12
* 5. 13/36   5. 23/48
  6. 25/48   6. 23/36
* 7. 19/72   7. 83/240
  8. 101/240   8. 23/72
* 9. 13/36   9. 7/12
  10. 55/84   10. 127/252
*11. 41/192   11. 127/960
  12. 79/252   12. 17/84

*Selected Solutions from Set A

1. LCD = 18

$$\frac{1}{6} = \frac{3}{18}$$
$$+\frac{2}{9} = \frac{4}{18}$$
$$\frac{7}{18}$$

3. LCD = 60

$$\frac{11}{20} = \frac{33}{60}$$
$$-\frac{7}{15} = \frac{28}{60}$$
$$\frac{5}{60} = \frac{1}{12}$$

5. LCD = 36

$$\frac{5}{18} = \frac{10}{36}$$
$$+\frac{1}{12} = \frac{3}{36}$$
$$\frac{13}{36}$$

7. LCD = 72

$$\frac{13}{18} = \frac{52}{72}$$
$$-\frac{11}{24} = \frac{33}{72}$$
$$\frac{19}{72}$$

9. LCD = 252

$$\frac{5}{42} = \frac{30}{252}$$
$$+\frac{7}{36} = \frac{49}{252}$$
$$+\frac{1}{21} = \frac{12}{252}$$
$$\frac{91}{252} \overset{\div 7}{=} \frac{13}{36}$$

11. LCD = 960

$$\frac{7}{64} = \frac{105}{960}$$
$$+\frac{11}{48} = \frac{220}{960}$$
$$-\frac{5}{40} = \frac{120}{960}$$
$$\frac{205}{960} = \frac{41}{192}$$

LCD for Problem 9:

$42 = 2 \times 3 \times 7$
$36 = 2 \times 2 \times 3 \times 3$
$21 = 3 \times 7$

LCD $= 2 \times 2 \times 3 \times 3 \times 7 = 252$

Exercise 3-11

Page 147 SET A   Page 148 SET B

* 1. 7   1. 8
  2. 11 3/8   2. 14 1/2
* 3. 16 7/16   3. 16 7/16
  4. 23 1/24   4. 14 1/18
  5. 128 23/24   5. 106 1/24
  6. 141 67/90   6. 136 31/42
* 7. 2 3/4   7. 2 3/4
  8. 4 1/2   8. 2 4/7
* 9. 72 5/6   9. 62 5/6
  10. 54 13/30   10. 114 35/54
*11. 4 4/9   11. 2 2/3
  12. 5 3/7   12. 3 4/11

*Selected Solutions from Set A

1. $4\frac{5}{6} = \frac{29}{6}$

$+2\frac{1}{6} = \frac{13}{6}$

$\frac{42}{6} = 7$

3. $9\frac{13}{16} = 9\frac{13}{16}$

$+6\frac{5}{8} = 6\frac{10}{16}$

$15\frac{23}{16} = 15 + 1\frac{7}{16} = 16\frac{7}{16}$

7. $8\frac{5}{8} = \frac{69}{8}$

$-5\frac{7}{8} = \frac{47}{8}$

$\frac{22}{8} = \frac{11}{4} = 2\frac{3}{4}$

11. $9 = 8\frac{9}{9}$

$-4\frac{5}{9} = 4\frac{5}{9}$

$4\frac{4}{9}$

9. $94\frac{5}{14} = 94\frac{15}{42} = 93\frac{57}{42}$

$-21\frac{11}{21} = -21\frac{22}{42} = -21\frac{22}{42}$

$72\frac{35}{42} = 72\frac{5}{6}$

Exercise 3-12

Page 151 SET A   Page 152 SET B

* 1. 1 1/8   1. 9/10
  2. 20/21   2. 1 1/20
  3. 1 1/6   3. 2 1/2
  4. 1 1/4   4. 5/6
* 5. 111/122   5. 116/195
  6. 145/194   6. 201/308
  7. 1 4/7   7. 1 5/14
  8. 1 7/13   8. 4 3/5
* 9. 12/17   9. 12/17
  10. 15/22   10. 15/22
*11. 61/384   11. 19/90
  12. 41/288   12. 55/144
  13. 2 19/34   13. 12 2/7
  14. 2 2/9   14. 4 16/21

*Selected Solutions from Set A

1. $\dfrac{\tfrac{3}{4}}{\tfrac{2}{3}} = \dfrac{3}{4} \div \dfrac{2}{3} = \dfrac{3}{4} \times \dfrac{3}{2} = \dfrac{9}{8} = 1\dfrac{1}{8}$

5. $\dfrac{3\tfrac{7}{10}}{4\tfrac{1}{15}} = \dfrac{\tfrac{37}{10}}{\tfrac{61}{15}} = \dfrac{37}{10} \div \dfrac{61}{15}$

$= \dfrac{37}{\cancel{10}} \times \dfrac{\cancel{15}^{3}}{61} = \dfrac{111}{122}$

9. $\dfrac{\tfrac{3}{8} + \tfrac{5}{8}}{\tfrac{2}{3} + \tfrac{3}{4}} = \dfrac{\tfrac{8}{8}}{\tfrac{8}{12} + \tfrac{9}{12}} = \dfrac{1}{\tfrac{17}{12}}$

$= 1 \div \dfrac{17}{12} = 1 \times \dfrac{12}{17} = \dfrac{12}{17}$

11. $\dfrac{1\tfrac{5}{16} + 2\tfrac{1}{2}}{24} = \dfrac{\tfrac{21}{16} + \tfrac{5}{2}}{24}$

$= \dfrac{\tfrac{21}{16} + \tfrac{40}{16}}{24} = \dfrac{\tfrac{61}{16}}{\tfrac{24}{1}}$

$= \dfrac{61}{16} \div \dfrac{24}{1} = \dfrac{61}{16} \times \dfrac{1}{24} = \dfrac{61}{384}$

Exercise 3-13

Page 155 SET A   Page 156 SET B

* 1. =   1. =
  2. =   2. =
  3. >   3. >
  4. >   4. >
* 5. <   5. <
  6. <   6. <
* 7. <   7. >
  8. >   8. <
* 9. 9/16, 1/2, 3/8   9. 5/8, 1/2, 7/16
  10. 1 2/3, 1 3/5,   10. 1 3/4, 1 2/3,
      1 1/2   1 2/5
*11. 3/2, 5/4, 4/5,   11. 3/2, 4/3, 4/5,
      2/3   3/4
  12. 9/4, 15/7, 2,   12. 7/3, 15/7, 2,
      11/6   11/6
*13. 11/12, 13/15,   13. 7/15, 5/12,
      11/18   7/18

*Selected Solutions from Set A

1. LCD = 6; $\dfrac{1}{2} = \dfrac{3}{6}$

5. LCD = 42; $\dfrac{5}{6} = \dfrac{35}{42}$, $\dfrac{6}{7} = \dfrac{36}{42}$

So 5/6 < 6/7

7. LCD = 12; $\dfrac{5}{4} = \dfrac{15}{12}$, $\dfrac{4}{3} = \dfrac{16}{12}$

So 5/4 < 4/3

9. LCD = 16; $\dfrac{1}{2} = \dfrac{8}{16}$, $\dfrac{3}{8} = \dfrac{6}{16}$, $\dfrac{9}{16} = \dfrac{9}{16}$

But 9/16 > 8/16 > 6/16
So 9/16 > 1/2 > 3/8

11. LCD = 60

$\dfrac{5}{4} = \dfrac{75}{60}$ $\dfrac{3}{2} = \dfrac{90}{60}$

$\dfrac{4}{5} = \dfrac{48}{60}$ $\dfrac{2}{3} = \dfrac{40}{60}$

But 90/60 > 75/60 > 48/60 > 40/60
So   3/2 > 5/4 > 4/5 > 2/3

13. LCD = 180
$\dfrac{11}{12} = \dfrac{165}{180}$, $\dfrac{13}{15} = \dfrac{156}{180}$, $\dfrac{11}{18} = \dfrac{110}{180}$

But 165/180 > 156/180 > 110/180
So   11/12 > 13/15 > 11/18

Exercise 3-14

Page 159 SET A

* 1. 1/4" bit
  2. 1/2" staple
* 3. 1/5
  4. 13 1/2 lb
* 5. 14
  6. 20 ft
* 7. 101 3/4 lb
  8. 1 4/5 gal
  9. 77 5/8 in.
  10. 77¢
*11. 3 1/2 mi
  12. 1 1/3 oz
  13. 5 1/2 laps
  14. $7900
*15. 140 1/2 lb
  16. 2 5/8 lb

Page 160 SET B

1. 1/4
2. 4/5
3. 34 3/8 mi
4. 14 11/12
5. 154
6. 1/8
7. 4 1/8 lb
8. 3 1/4 lb
9. 69 1/8 in.
10. 391¢ = $3.91
11. 15 bottles
12. 7
13. $26,115
14. 32 1/2 hr
15. 410 5/8

*Selected Solutions from Set A

1. Change each to an LCD of 32 and compare: 5/16 = 10/32, 1/4 = 8/32, 9/32 = 9/32; so a 1/4" is smallest.

3. "No Shows": 26,000 − 20,800 = 5200

   Fraction: $\frac{5200}{26000} = \frac{52}{260} = \frac{1}{5}$

5. Divide the total length by 3 1/2.

   $49 \div 3\frac{1}{2} = 49 \div \frac{7}{2} = \cancel{49} \times \frac{2}{\cancel{7}} = 14$

7. Subtract the two weights.

   $236\frac{1}{2} - 134\frac{3}{4} = \frac{473}{2} - \frac{539}{4}$

   $= \frac{946}{4} - \frac{539}{4} = \frac{407}{4} = 101\frac{3}{4}$

11. Divide miles driven by total number of gallons used.

    $14\frac{1}{2} + 15\frac{1}{5} + 14\frac{1}{10} + 16\frac{1}{5} =$

    $14\frac{5}{10} + 15\frac{2}{10} + 14\frac{1}{10} + 16\frac{2}{10} = 60$ gal

    $1830 \div 60 = 30\frac{1}{2}$

15. Subtract the weight lost (6 × 3/4) from her original weight.

    $145 - \cancel{6}^3 \times \frac{3}{\cancel{4}_2} = 145 - \frac{9}{2} = 140\frac{1}{2}$

Chapter 3

Practice Test A
Page 163

1. 54
2. 107/6
3. 5/9
4. 1 1/4
5. 5 1/24
6. 3/4
7. 1/3
8. 1 17/45
9. 11/135
10. 5/6
11. 6 19/30
12. 1 1/3
13. 4/3, 7/6, 8/7
14. 21 games
15. 27¢

Practice Test B
Page 164

1. 41
2. 3 2/3
3. 15/56
4. 3/4
5. 9 43/54
6. 2/5
7. 1 1/5
8. 1 8/63
9. 13/30
10. 7/8
11. 7 17/21
12. 1 11/52
13. 9/11, 7/9, 2/3
14. 539 people
15. 12¢

CHAPTER 4

Page 170   Pretest

1. twenty-seven and sixteen thousandths
2. 3,907.620
3. 708.3235
4. 764.713
5. 6158.025
6. 6.5
7. 3.1
8. 26,370,000
9. 0.002182
10. 31/50
11. 3.0625
12. 0.5, 0.05, 0.049, 0.005
13. 16 8/45
14. $102.96
15. $249

Exercise 4-1

Page 173   SET A

1. five tenths
2. seventeen hundredths
3. thirty-nine thousandths
4. five and seven ten-thousandths
5. sixteen and thirty-five hundredths
6. four hundred twenty-six and nine tenths
7. six and one thousand two hundred thirty-six ten-thousandths
8. 0.7
9. 0.12
10. 9.003
11. 45.0006
12. 100.16
13. 356.207
14. 5023.3517
15. 2,080,000.001086
16. 2 tenths
17. 5 thousandths
18. 4 ten-thousandths
19. 6 tens

Page 174 SET B

1. seven tenths
2. twenty-nine hundredths
3. seventy-six thousandths
4. six and seven ten-thousandths
5. seventeen and eighty-two hundredths
6. four hundred twenty-nine and seven tenths
7. five and two thousand one hundred seventy-four ten-thousandths
8. 0.5
9. 0.16
10. 4.009
11. 61.0004
12. 1000.05
13. 999.609
14. 8011.1418
15. 3,004,000.000102
16. 4 hundredths
17. 6 thousandths
18. 1 hundred
19. 7 ten-thousandths

Exercise 4-2

Page 177 SET A   Page 178 SET B

*1. 36.7   1. 43.8
 2. 4.7   2. 116.8
 3. 125.5   3. 3.3
 4. 90.0   4. 10.0
*5. 9.0   5. 38.0
 6. 0.1   6. 0.3
 7. 18.72   7. 95.62
 8. 5.48   8. 0.37
*9. 792.04   9. 706.02
10. 0.90   10. 7.70
*11. 7.20   11. 24.30
12. 1.02   12. 100.03
*13. 562.018   13. 382.019
14. 562.0185   14. 382.0194
15. 560   15. 400
16. 600   16. 380
17. 46.96   17. 52.70
18. 46.964   18. 52.695
19. 46.9635   19. 52.6952

Selected Solutions from Set A

1.   36.72 ≗ 36.7
   tenths↑ ↳digit to the right
   place   is less than 5
Round off by discarding the digits
to the right of the tenths place.

5.   8.95 ≗ 9.0
   tenths↑ ↳digit to the right
   place   is 5
Round off by increasing the tenths
place by 1 and discarding the 5.

9.   792.038218 ≗ 792.04
   hundredths↑ ↳digit to the right
   place   is more than 5
Round off by increasing the hundredths
place by 1 and discarding the 8218.

11.   7.195 ≗ 7.20
   hundredths↑ ↳digit to the right
   place   is 5
Round off by increasing the hundredths
place by 1 and discarding the 5.

13.   562.01846 ≗ 562.018
   thousandths↑ ↳digit to the right
   place   is less than 5
Round off by discarding the
digits to the right of the 8.

Exercise 4-3

Page 181 SET A   Page 182 SET B

* 1. 320.54   1. 221.57
  2. 523.69   2. 193.63
* 3. 0.246   3. 0.0856
  4. 1.236   4. 0.0878
* 5. 145.88   5. 65.46
  6. 68.76   6. 89.46
* 7. 893.81   7. 600.15
  8. 1.166   8. 1.055
* 9. 153.023   9. 545.607
 10. 867.2907   10. 770.7505
*11. 908.1502   11. 148.8833
 12. 6104.2023   12. 2131.2073
*13. 1564.7699   13. 17330.121

*Selected Solutions from Set A

1.   12   3.   1
   287.04   0.05
     4.5   0.096
  + 29.   + 0.1
   320.54   0.246

5.   11 2   7.   11 1
    81.26   586.94
    43.99   + 306.87
    16.57   893.81
  + 4.06
   145.88

9.   11
    12.6
    14.
  +126.423
   153.023

11. 121 1
    809.
      3.65
     19.0702
  + 76.43
    908.1502

13.   22
     123.
      86.057
    1283.001
       4.7
  +   68.0119
    1564.7699

Exercise 4-4

Page 185 SET A   Page 186 SET B

* 1. 3.431   1. 3.433
  2. 9.24   2. 31.63
* 3. 21.45   3. 48.36
  4. 12.14   4. 38.23
* 5. 157.983   5. 31.872
  6. 278.965   6. 104.683
* 7. 33.693   7. 32.986
  8. 18.382   8. 30.867
* 9. 688.1852   9. 329.1972
 10. 448.1926   10. 181.0751
*11. 717.5   11. 823.2
 12. 405.3   12. 405.24
*13. 17.6876   13. 18.8765
 14. 414.0247   14. 688.4824
*15. 1504.4   15. 1712.85841
 16. 4088.2716   16. 1254.6

*Selected Solutions from Set A

                              6 10
1.   8.6 4 7   3.   4 7.7̸ 0̸
   - 5.2 1 6   - 2 6.2 5
     3.4 3 1   2 1.4 5

      10 15
5. 2 0̸ 5̸ 9 9 10
    3̸ 1̸ 6.0̸ 0̸ 0̸
  - 1 5 8.0 1 7
    1 5 7.9 8 3

                              17
7.   4 1110   9.   6 7̸ 15
     5̸ 2.0̸ 9 3   7̸ 8̸ 8̸.1 8 5 2
   - 1 8.4 0 0   -   9 7.0 0 0 0
     3 3.6 9 3   6 8 8.1 8 5 2

11.   4 9 10
      7 5̸ 0̸.0̸
    -   3 2.5
      7 1 7.5

13.   8 12
      8 9̸.2̸ 8 7 6
    - 7 1.6 0 0 0
      1 7.6 8 7 6

             11
15.   2 1̸ 10
      1 5 3̸ 2̸.0̸
    -     2 7.6
      1 5 0 4.4

Exercise 4-5

Page 189 SET A

* 1. 43.11
  2. 108.75
* 3. 57.368
  4. 30.906
* 5. 1805.44
  6. 2417.76
* 7. 0.0500350
  8. 0.140147
* 9. 65.36
  10. 0.0212
*11. 0.28076
  12. 265.02
*13. 299.20178
  14. 7.6155

Page 190 SET B

1. 65.44
2. 51.75
3. 57.267
4. 55.449
5. 686.72
6. 1443.05
7. 0.075042
8. 0.315072
9. 60.45
10. 0.0093
11. 0.63672
12. 309.452
13. 103.32816
14. 8.876525

*Selected Solutions from Set A

```
1.    1 4.3 7        3.      8.0 8
     x      3              x   7.1
      4 3.1 1                  8 0 8
                             5 6 5 6
                             5 7.3 6 8

5.      6 9 4.4
       x     2.6
        4 1 6 6 4
      1 3 8 8 8
      1 8 0 5.4 4

7.        2.0 0 1 4
         x    .0 2 5
          1 0 0 0 7 0
         4 0 0 2 8
         .0 5 0 0 3 5 0

9.    1 6.3 4
     x      4
      6 5.3 6

11.   7.0 1 9
     x    .0 4
      .2 8 0 7 6

13.       1 4 8.3 4
         x     2.0 1 7
          1 0 3 8 3 8
          1 4 8 3 4
        2 9 6 6 8 0
        2 9 9.2 0 1 7 8
```

Exercise 4-6

Page 193 SET A

* 1. 12.6
  2. 13.7
* 3. 8.14
  4. 9.31
* 5. 74
  6. 95
* 7. 0.0185
  8. 0.0175
* 9. 0.004
  10. 0.0035
*11. 6200
  12. 8400
*13. 39
  14. 58
*15. 0.34375
  16. 0.21875

Page 194 SET B

1. 16.9
2. 13.7
3. 7.23
4. 9.16
5. 25
6. 45
7. 0.0645
8. 0.0145
9. 0.0085
10. 0.0085
11. 4700
12. 9800
13. 63
14. 78
15. .40625
16. 0.4375

*Selected Solutions from Set A

```
1.       12.6           3.          8.14
      4⟌50.4            7.2⟌56.6̱08
        4                  57 6
        10                    1 0 0
         8                      7 2
         2 4                    2 88
         2 4                    2 88

5.           74.          7.      .0185
    0.65⟌48.10̱            84⟌1.5540
         45 5                  84
          2 60                 714
          2 60                 672
                               420
9.            .004             420
    4.09⟌.01̱636
          1 636        13.           39.
                          6.24⟌243.36̱
11.           6200.              187 2
    .0034⟌21.0800̱                 56 16
           20 4                   56 16
              68
              68       15.          .34375
               0          3.2⟌1.1̱00000
               0                 9 6
               0                 1 40
                                 1 28
                                   120
                                    96
                                   240
                                   224
                                    160
                                    160
```

Exercise 4-7

Page 197   SET A

* 1. 1.2
  2. 0.9
* 3. 8.4
  4. 6.5
* 5. 253.3
  6. 203.3
* 7. 1.41
  8. 1.45
* 9. 31.49
  10. 17.45
*11. 0.167
  12. 0.605
  13. 4.010
  14. 5.010

Page 198   SET B

  1. 0.8
  2. 1.9
  3. 4.7
  4. 5.5
  5. 443.3
  6. 346.7
  7. 0.37
  8. 0.75
  9. 30.77
  10. 71.07
  11. 0.583
  12. 0.467
  13. 3.010
  14. 6.010

*Selected Solutions from Set A

1. 1.21 ≟ 1.2
    7)8.50
      7
      ‾
      1 5
      1 4
      ‾
        10
         7
         ‾

3. 8.35 ≟ 8.4
    .36)3.00,60
         2 88
         ‾
           12 6
           10 8
           ‾
             1 80
             1 80
             ‾

5. 253.33 ≟ 253.3

           25 3.33
    2.4)608.0,00
        48
        ‾
        128
        120
        ‾
          8 0
          7 2
          ‾
            80
            72
            ‾
             80
             72
             ‾

7. 1.405 ≟ 1.41

        1.405
    6)8.430
      6
      ‾
      2 4
      2 4
      ‾
        30
        30
        ‾

9. 31.487 ≟ 31.49

              31.487
    .289)9.100,000
         8 67
         ‾
         430
         289
         ‾
         141 0
         115 6
         ‾
          25 40
          23 12
          ‾
           2 280
           2 023
           ‾

11. .1666 ≟ .167

             .1666
    66)11.0000
       6 6
       ‾
       4 40
       3 96
       ‾
         440
         396
         ‾
          440
          396
          ‾

Exercise 4-8

Page 201   SET A

* 1. 4360
  2. 52,300
* 3. 952,340
  4. 212,436
* 5. 3906
  6. 4536
* 7. 312,700
  8. 4,173,000
* 9. 0.543
  10. 0.756
*11. 0.35
  12. 0.4
*13. 0.00529
  14. 0.00746
*15. 0.003765
  16. 0.0004087

Page 202   SET B

  1. 817,000
  2. 6040
  3. 277,725
  4. 16,803,800
  5. 3888
  6. 3876
  7. 7,263,000
  8. 610,800
  9. 0.823
  10. 0.645
  11. 0.4
  12. 0.3
  13. 0.00472
  14. 0.00351
  15. 0.0005103
  16. 0.008274

*Selected Solutions from Set A

1.   4.3 6
    x 1 0 0 0
    ‾‾‾‾‾‾‾‾‾
    4 3 6 0.0 0

3.   4 7.6 1 7
    x   2 0 0 0 0
    ‾‾‾‾‾‾‾‾‾‾‾‾‾‾‾
    9 5 2 3 4 0.0 0 0

5.   0.0 9 3
    x   4 2 0 0 0
    ‾‾‾‾‾‾‾‾‾‾‾‾‾‾‾
        1 8 6 0 0 0
      3 7 2
    ‾‾‾‾‾‾‾‾‾‾‾‾‾‾‾
    3 9 0 6.0 0 0

7.   3.1 2 7
    x   1 0 0 0 0 0
    ‾‾‾‾‾‾‾‾‾‾‾‾‾‾‾‾‾‾‾‾‾
    3 1 2 7 0 0.0 0 0

9. 54.3 ÷ 100 = $\frac{54.3}{1.00}$ = .543

11. 700 ÷ 2000 = $\frac{700}{2,000}$ = $\frac{.7}{2}$ = .35

13. 1957.3 ÷ 370,000 = $\frac{1957.3}{37,0000}$ = $\frac{.19573}{37}$

        .00529
    37).19573
       185
       ‾
       107
        74
        ‾
        333
        333
        ‾

15. 10^5 = 10x10x10x10 = 100,000

    376.5 ÷ 10^5 = $\frac{00376.5}{1,00000}$ = .003765

Exercise 4-9

Page 205 SET A   Page 206 SET B

* 1. 0.6   1. 0.8
  2. 0.5   2. 0.75
  3. 0.625   3. 0.375
  4. 0.0625   4. 0.3125
* 5. 1.28125   5. 1.21875
  6. 1.15625   6. 1.09375
* 7. 0.545   7. 0.556
  8. 0.833   8. 0.571
* 9. 1.857   9. 2.231
 10. 2.444   10. 3.455
 11. 23.667   11. 44.167
 12. 37.778   12. 32.143
*13. 3/5   13. 1/5
 14. 2/5   14. 3/5
*15. 31/50   15. 43/50
 16. 37/50   16. 49/50
*17. 17/400   17. 3/80
 18. 19/400   18. 11/400
*19. 4 1/1000   19. 1 7/1000
 20. 3 9/1000   20. 2 3/1000

*Selected Solutions from Set A

1. $\frac{3}{5} = $ 5)3.0 .6
                    3 0

5. $\frac{41}{32} = $ 32)41.00000 1.28125
   32
    9 0
    6 4
    2 60
    2 56
       40
       32
       80
       64
      160
      160

7. .5454 ≐ .545
   $\frac{6}{11} = $ .5454 ≐ .54
          11)6.0000
             5 5
               50
               44
               60
               55
                50
                44

9. $1\frac{6}{7} = \frac{13}{7}$
   1.8571 ≐ 1.857

         1.8571
   7)13.0000
      7
      6 0
      5 6
        40
        35
        50
        49
        10
         7

13. $.6 = \frac{6}{10} = \frac{3}{5}$

15. $.62 = \frac{62}{100} = \frac{31}{50}$

17. $.0425 = \frac{425}{10000} = \frac{17}{400}$

19. $4.001 = 4\frac{1}{1000}$

Exercise 4-10

Page 209 SET A   Page 210 SET B

  1. 0.6   1. 0.9
  2. 0.75   2. 0.98
  3. 2.076   3. 3.1205
  4. 5.092   4. 12.706
* 5. 0.6   5. 0.7
  6. 0.162   6. 0.129
* 7. 3.402   7. 3.507
  8. 76.125   8. 84.306
* 9. 2/3   9. 0.4
 10. 3 1/4   10. 4 1/3
*11. 0.76>0.706>0.7   11. 0.37>0.307>
     >0.6>0.076   0.3>0.07>0.037
 12. 0.54>0.504>0.5   12. 0.901>0.9001>
     >0.4>0.054   0.9>0.19>0.091
 13. 3.12>3.102>2.3   13. 4.3>4.03>
     >2.13>2   3.4003>3.4>3.04
 14. 3/4>.7>.65   14. .82>4/5>
     >3/5>.076   .755>5/8>.5
*15. 1.8>1.625>   15. 2 7/8>2.78>
     1 1/2>1 1/3>   2.7>2 2/3>2.66
     1.258

*Selected Solutions from Set A

5.   .6 = .60   7. 3.4002 = 3.4002
     .06 = .06   3.402 = 3.4020
   but 60 > 06   but 3.4020 > 3.4002
   so .6 > .06   so 3.402 > 3.4002

9. $\frac{2}{3}$ = .666... so $\frac{2}{3}$ > .66

11. 0.76 = .760
    0.706 = .706
    0.7   = .700
    0.6   = .600
    0.076 = .076

    so .76 > .706 > .7 > .6 > .076

15. 1.8   = 1.800
    1.625 = 1.625

    $1\frac{1}{2}$ = 1.500

    $1\frac{1}{3}$ = 1.333

    1.258 = 1.258

    so 1.8 > 1.625 > $1\frac{1}{2}$ > $1\frac{1}{3}$ > 1.258

Exercise 4-11

Page 213 SET A   Page 214 SET B

* 1. 3.2   1. 6.4
  2. 12   2. 15
  3. 21   3. 29.9
  4. 11.7   4. 11.7
* 5. 5 9/20 ≐ 5.45   5. 7 1/25 ≐ 7.04
  6. 15 7/9 ≐ 15.78   6. 21 1/2 ≐ 21.5
  7. 7/12 ≐ 0.58   7. 1 2/3 or 1.67
  8. 1 13/20 or 1.65   8. 3 3/16 ≐ 3.19
* 9. 13.3357   9. 46.53
 10. 5.17   10. 20.59
 11. 12.25   11. 11.35
 12. 1608.25   12. 2006.35
*13. 5 1/3 ≐ 5.33   13. 11 1/3 ≐ 11.33
 14. 7 1/6 ≐ 7.17   14. 7 2/3 ≐ 7.67
 15. 4.8   15. 12.5
 16. 93.75   16. 0.625

*Selected Solutions from Set A

1. $14.4 \div 4\frac{1}{2} = $

$$4.5\overline{)14.4\,0} \quad \begin{array}{r} 3.2 \\ \underline{13\ 5} \\ 9\ 0 \\ \underline{9\ 0} \end{array}$$

5. $(1\frac{1}{2})^2 + 2\frac{2}{3} \times 1.2 =$

$(\frac{3}{2} \times \frac{3}{2}) + 2\frac{2}{3} \times 1\frac{1}{5} =$

$\frac{9}{4} + \frac{8}{3} \times \frac{\cancel{6}^2}{5} = \frac{9}{4} + \frac{16}{5} =$

$\frac{45}{20} + \frac{64}{20} = \frac{109}{20} = 5\frac{9}{20} = 5.45$

9. $\frac{1}{2} \times 3^3 - 5.3 \times .031 =$

$.5 \times 27 - 5.3 \times .031 =$

$13.5 - .1643 =$

13.3357

13. $\frac{1}{3} \times \sqrt{16} + 10 \div 2.5 =$

$\frac{1}{3} \times 4 + 10 \div 2\frac{1}{2}$

$\frac{4}{3} + 10 \div \frac{5}{2} = \frac{4}{3} + \cancel{10}^2 \times \frac{2}{\cancel{5}} =$

$\frac{4}{3} + \frac{4}{1} = \frac{4}{3} + \frac{12}{3} = \frac{16}{3} = 5\frac{1}{3} \doteq 5.33$

Exercise 4-12

Page 217 SET A   Page 218 SET B

* 1. $117.98   1. $141.06
  2. $143.23   2. $13.98
* 3. the $62.95   3. the 12.5 oz
     tire   can
  4. the 5.5 oz   4. $906.24
     tube
* 5. ≐ 28.16   5. ≐ 19.26
  6. ≐ 23.99   6. ≐ $.11
  7. $250.48   7. $303.97
  8. $136.50   8. 1.625 in
* 9. $5.44   9. $14.69
 10. ≐ $46.09   10. $711.75
*11. $693.50   11. $250.60
 12. $169.50   12. $5.93

*Selected Solutions from Set A

1. Subtract the total of the checks
   from balance plus deposits.

   checks: $ 47.35   bal: $ 22.56
             10.00   dep: + 415.25
            239.50   437.81
          + 22.98   checks: - 319.83
           $319.83 $117.98

3. Determine the cost per mile of
   each tire. The one that costs
   less per mile is a better buy.
   Remember: "per" means "divide."
   $62.95 ÷ 50,000 = .0012590
   $57.50 ÷ 40,000 = .0014375

5. Divide number of miles driven
   (difference in odometer readings)
   by the number of gallons used.
   39762.4
   -39247.1
   ───────
     515.3   515.3 ÷ 18.3 = 28.16

9. Add the cost for the first 3 min.
   ($.85) and the cost for the next
   27 min. (27 x $.17).
   27 x $.17 = $4.59
              + .85
              ─────
              $5.44

11. Multiply the cost for each day
    (2 x $.95 = $1.90) by 365.
    365 x $1.90 = $693.50

Exercise 4-13

Page 221 SET A   Page 222 SET B

* 1. p=26, a=36   1. p=38, a=84
  2. p=60, a=225   2. p=72, a=324
* 3. $283.80   3. $179.85
  4. $399.80   4. $711.36
* 5. $909.00   5. $1410.16
  6. $67.49   6. $50.25
* 7. $65.66   7. $91.28
  8. $27.72   8. $206.40
* 9. $9588   9. $39,381.12
  10. $513.77   10. $73.41
*11. $.13   11. 4.8 acres;
                           $6710.40

*Selected Solutions from Set A

1. perimeter = 9+4+9+4 = 26 ft
   area = 9x4 = 36 sq ft

3. Convert feet to yards.
   Multiply area (sq yd) by $23.65.
   area = 3 x 4 = 12 sq yd
   cost = 12 x $23.65 = $283.80

5. Add the cost of the fence and sod.
   fence: perimeter times $4.15
   sod: area times $.39
   fence = (30+33+30+33) x $4.15
         = 126 x $4.15 = $522.90
   sod   = (30 x 33) x $.39 = $386.10
   total = $522.90 + $386.10 =$909.00

7. Multiply the total area of both
   rooms by $.49.
   areas: 9 x 11 = 99
           5 x 7 = 35
           total = 134 sq ft
   cost: 134 x $.49 = $65.66

9. Multiply the area by $39.95.
   area: 12 x 20 = 240 sq ft
   cost: 240 x $39.95 = $9588.00

11. Divide the cost by the number of
    square feet in 5.5 acres.
    sq ft: 5.5 x 43,560 = 239,580
    cost per sq ft:
      $32,000 ÷ 239,580 $\doteq$ $.13

Chapter 4
Practice Test A
Page 225

1. 407.007
2. 746.33
3. 321.4217
4. 38.004
5. 809.4572
6. 70.5
7. 11.15
8. 2,431,000
9. 0.0001911
10. 13/500
11. 2.3125
12. 1.27, 1.2, 1.1027, 1.0127
13. 37 17/18 $\doteq$ 37.94
14. $61.54
15. $359.20

Chapter 4
Practice Test B
Page 226

1. eighty-four and fifty-six
   hundredths
2. 56.4
3. 944.5118
4. 38.042
5. 1270.2
6. 2.65
7. 7.22
8. 36,866,000
9. 0.000953
10. 19/50
11. 5.375
12. 0.306, 0.3, 0.036, 0.0306
13. 52 19/27 $\doteq$ 52.7
14. $6.73
15. $1059.40

Unit II Exam
Page 230

1. 4 2/3
2. one hundred fifty-two and
   thirty-seven thousandths
3. 0.58   10. 260,400
4. 7/36   11. 88.76
5. 1 1/2   12. 14 students
6. 1 13/120   13. 3 5/6 ft
7. 7 13/15   14. $4237.44
8. 9/10   15. 4675 sq in.
9. 196.376

UNIT III

Page 232 **BRAIN BUSTER**

height $\stackrel{\bullet}{=}$ 5.3 miles

CHAPTER 5

Page 234 Pretest

1. 3/5
2. 24
3. 195
4. $\stackrel{\bullet}{=}$ 3.94
5. $\stackrel{\bullet}{=}$ 8.76
6. .045; 9/200
7. 57/500; 11.4%
8. 44.7
9. $\stackrel{\bullet}{=}$ $5.20
10. 16%
11. 75
12. $\stackrel{\bullet}{=}$ $196.17
13. $1.82
14. $\stackrel{\bullet}{=}$ 44.9%
15. $80.54

Note: Unless otherwise indicated in this unit, when the answers have more than three decimal digits, they will be rounded off to the nearest hundredth.

Exercise 5-1

Page 237 SET A   Page 238 SET B

* 1. 2/3	1. 9/20
2. 2/1	2. 9/11
* 3. 1/3	3. 11/20
4. 2/3	4. 1/2
* 5. 1/4	5. 8/15
6. 9/11	6. 1/4
* 7. 10/7	7. 3/5
8. 7/10	8. 5/3
* 9. 14/45	9. 1/3
10. 45/79	10. 15/23
*11. 2/5	11. 2/5
12. 1/12	12. 1/12
*13. 4/15	13. 1/4
14. 91/68	14. 4/3
*15. 5/14	15. 42/79
16. 32/1	16. 64/7
*17. 3/16	17. 1/8

*Selected Solutions from Set A

1. 8 out of 12 = 8/12 = 2/3

3. 4 out of 12 = 4/12 = 1/3

5. The total number of balls is 12 + 10 + 18 = 40. Red balls to total = 10/40 = 1/4.

7. 20 nickels = 100¢; 7 dimes = 70¢; 100/70 = 10/7

9. 7 dimes = 70¢; 9 quarters = 225¢; 70/225 = 14/45

11. 18:45 = 18/45 = 2/5

13. 10 yd = 30 ft, so 8 ft to 10 yd = 8:30 = 8/30 = 4/15

15. $\dfrac{3\frac{3}{4}}{10\frac{1}{2}} = \dfrac{\frac{15}{4}}{\frac{21}{2}} = \dfrac{15}{4} \div \dfrac{21}{2} = \dfrac{\cancel{15}^{5}}{\cancel{4}_{2}} \times \dfrac{\cancel{2}^{1}}{\cancel{21}_{7}} = \dfrac{5}{14}$

17. 3 yd = 108 in

$\dfrac{20\frac{1}{4}}{108} = \dfrac{\frac{81}{4}}{\frac{108}{1}} = \dfrac{81}{4} \div \dfrac{108}{1} = \dfrac{\cancel{81}^{3}}{4} \times \dfrac{1}{\cancel{108}_{4}} = \dfrac{3}{16}$

Exercise 5-2

Page 241 SET A   Page 242 SET B

* 1. yes   1. no
  2. yes   2. yes
* 3. yes   3. yes
  4. yes   4. yes
* 5. no   5. no
  6. no   6. no
* 7. 3   7. 2
  8. 4   8. 3
* 9. 35   9. 48
  10. 56   10. 63
*11. 120   11. 90
  12. 450   12. 60
*13. 4   13. 95
  14. 13   14. 91
*15. 42   15. 56
  16. 28   16. 20
*17. 171   17. 153
  18. 40   18. 39

*Selected Solutions from Set A

1. cross products $12 \times 14 = 168$
   are equal: $21 \times 8 = 168$

3. cross products $57 \times 76 = 4332$
   are equal: $38 \times 114 = 4332$

5. cross products $73 \times 33 = 2409$
   are not equal: $53 \times 53 = 2809$

7. $\frac{N}{4} = \frac{6}{8}$

   $8 \times N = 4 \times 6$
   $\frac{8 \times N}{8} = \frac{24}{8}$
   $N = 3$

9. $\frac{20}{N} = \frac{4}{7}$

   $4 \times N = 20 \times 7$
   $\frac{4 \times N}{4} = \frac{140}{4}$
   $N = 35$

11. $\frac{18}{15} = \frac{N}{100}$

    $15 \times N = 18 \times 100$
    $\frac{15 \times N}{15} = \frac{1800}{15}$
    $N = 120$

13. $\frac{65}{52} = \frac{5}{N}$

    $65 \times N = 52 \times 5$
    $\frac{65 \times N}{65} = \frac{260}{65}$
    $N = 4$

15. $\frac{N}{18} = \frac{63}{27}$

    $27 \times N = 18 \times 63$
    $\frac{27 \times N}{27} = \frac{1134}{27}$
    $N = 42$

17. $\frac{9}{14} = \frac{N}{266}$

    $14 \times N = 9 \times 266$
    $\frac{14 \times N}{14} = \frac{2394}{14}$
    $N = 171$

Exercise 5-3

Page 245 SET A   Page 246 SET B

* 1. yes   1. yes
  2. yes   2. yes
* 3. yes   3. yes
  4. yes   4. yes
* 5. 1.12   5. .54
  6. 1.35   6. 2.31
* 7. 11.2   7. $\doteq$ 6.86
  8. 15.75   8. $\doteq$ 5.83
* 9. $\doteq$ 2.47   9. $\doteq$ 8.28
  10. $\doteq$ .30 10. $\doteq$ 14.44
*11. $\doteq$ 58.89 11. $\doteq$ 38.18
  12. $\doteq$ 65.71 12. $\doteq$ 40.77
*13. 76   13. 28
  14. 18   14. 144
*15. 680   15. $\doteq$ 228.07
  16. = 717.95   16. $\doteq$ 220.93
  17. 2 1/2 or 2.5 17. 1/5 or .2
  18. 3 1/2 or 3.5 18. 5/18 $\doteq$ .28

*Selected Solutions from Set A

1. cross products $2\frac{2}{3} \times \frac{3}{4} = 2$
   are equal:

   $1\frac{1}{5} \times 1\frac{2}{3} = 2$

3. cross products $1.4 \times 1.6 = 2.24$
   are equal: $5.6 \times .4 = 2.24$

5. $\frac{A}{16} = \frac{7}{100}$

   $100 \times A = 16 \times 7$
   $\frac{100 \times A}{100} = \frac{112}{100}$
   $A = 1.12$

7. $\frac{5}{8} = \frac{7}{N}$

   $5 \times N = 8 \times 7$
   $\frac{5 \times N}{5} = \frac{56}{5}$
   $N = 11.2$

9. $\frac{Y}{.35} = \frac{1.2}{.17}$

   $.17 \times Y = 1.2 \times .35$
   $\frac{.17 \times Y}{.17} = \frac{.42}{.17}$
   $Y \doteq 2.47$

11. $\frac{5.3}{B} = \frac{9}{100}$

    $9 \times B = 5.3 \times 100$
    $\frac{9 \times B}{9} = \frac{530}{9}$
    $B \doteq 58.89$

13. $\frac{F}{\frac{1}{2}} = \frac{38}{\frac{1}{4}}$

    $\frac{1}{4} \times F = \frac{1}{2} \times 38$
    $\frac{\frac{1}{4} \times F}{\frac{1}{4}} = \frac{19}{\frac{1}{4}}$
    $F = 76$

15. $\frac{17}{2.5} = \frac{F}{100}$

    $2.5 \times P = 17 \times 100$
    $\frac{2.5 \times P}{2.5} = \frac{1700}{2.5}$
    $P = 680$

Exercise 5-4

Page 249 SET A   Page 250 SET B
* 1. 44.4 ft   1. 123 ft
  2. ≐ 37.08 ft   2. ≐ 44.29 ft
* 3. ≐ 12.54 mi   3. 6 mi
  4. ≐ 9.33 mi   4. ≐ 21.40 in
* 5. ≐ 8.57 lb   5. ≐ 4.67 oz
  6. 17 gal   6. 474
* 7. ≐ $48,546.51   7. ≐ $20,824.07
  8. ≐ 335.42 mi   8. 44 qt jars
* 9. ≐ 48.39 min   9. ≐ 37.13 gal
 10. $7.00   10. ≐ $8.59
*11. ≐ $3.96   11. ≐ $1.24
 12. ≐ $3.71   12. 2,310,000 ft
*13. ≐ 36.57 oz   13. 1587 acres;
 14. 32 games   3703 acres
*15. 1400 acres;   14. 25 people
     4900 acres   15. 77 game

*Selected Solutions from Set A

1. $\dfrac{\text{height}}{\text{shadow}} = \dfrac{\text{height}}{\text{shadow}}$ 3. $\dfrac{\text{miles}}{\text{inches}} = \dfrac{\text{miles}}{\text{inches}}$

   $\dfrac{6}{5} = \dfrac{H}{37}$ $\dfrac{27}{7} = \dfrac{M}{3\,1/4}$

   $\dfrac{5 \times H}{5} = \dfrac{222}{5}$ $\dfrac{2 \times M}{7} = \dfrac{87.75}{7}$

   $H = 44.4$ $M \doteq 12.54$

5. $\dfrac{\text{lb}}{\text{sq ft}} = \dfrac{\text{lb}}{\text{sq ft}}$ 7. $\dfrac{\text{tax}}{\text{price}} = \dfrac{\text{tax}}{\text{price}}$

   $\dfrac{6}{1400} = \dfrac{L}{2000}$ $\dfrac{688}{40000} = \dfrac{835}{P}$

   $\dfrac{1400 \times L}{1400} = \dfrac{12000}{1400}$ $\dfrac{688 \times P}{688} = \dfrac{33400000}{688}$

   $L \doteq 8.57$ $P \doteq 48,546.51$

9. $\dfrac{\text{miles}}{\text{time}} = \dfrac{\text{miles}}{\text{time}}$ 11. $\dfrac{\text{oz}}{\text{cost}} = \dfrac{\text{oz}}{\text{cost}}$

   $\dfrac{6.2}{40} = \dfrac{7.5}{T}$ $\dfrac{12\,1/2}{2.95} = \dfrac{16.76}{C}$

   $\dfrac{6.2 \times T}{6.2} = \dfrac{300}{6.2}$ $\dfrac{12.5 \times C}{12.5} = \dfrac{49.442}{12.5}$

   $T \doteq 48.39$ $C \doteq 3.96$

13. $\dfrac{\text{mix}}{\text{total}} = \dfrac{\text{mix}}{\text{total}}$ 15. $\dfrac{\text{part}}{\text{total}} = \dfrac{\text{part}}{\text{total}}$

    $\dfrac{2}{7} = \dfrac{M}{128}$ $\dfrac{2}{9} = \dfrac{P}{6300}$

    $\dfrac{7 \times M}{7} = \dfrac{256}{7}$ $\dfrac{9 \times P}{9} = \dfrac{12600}{9}$

    $M \doteq 36.57$ $P = 1400$
                                other $= 4900$

Exercise 5-5

Page 253 SET A   Page 254 SET B
* 1. 17/100   1. 19/100
  2. 23/100   2. 31/100
* 3. 1/20   3. 2/25
  4. 3/50   4. 1/50
* 5. 21/250   5. 19/250
  6. 31/500   6. 6/125
* 7. 29/300   7. 19/300
  8. 47/600   8. 11/120
* 9. 1 9/25   9. 2 3/25
 10. 2 3/20   10. 1 13/25
*11. 7/16   11. 229/400
 12. 109/400   12. 351/400
*13. .45   13. .36
 14. .58   14. .45
*15. .05   15. .04
 16. .08   16. .03
*17. 2.36   17. 1.57
 18. 1.89   18. 1.39
*19. .265   19. .158
 20. .204   20. .163
*21. .0825   21. .0475
 22. .0975   22. .0625

*Selected Solutions from Set A

1. $17\% = \dfrac{17}{100}$ 3. $5\% = \dfrac{5}{100} = \dfrac{1}{20}$

5. $8.4\% = .084 = \dfrac{84}{1000} = \dfrac{21}{250}$

7. $9\tfrac{2}{3}\% = \dfrac{9\tfrac{2}{3}}{100} = \dfrac{\tfrac{29}{3}}{100} = \dfrac{29}{3} \times \dfrac{1}{100} = \dfrac{29}{300}$

9. $136\% = \dfrac{136}{100} = 1\dfrac{36}{100} = 1\dfrac{9}{25}$

11. $43\tfrac{3}{4}\% = \dfrac{43\tfrac{3}{4}}{100} = \dfrac{\tfrac{175}{4}}{100} = \dfrac{\cancel{175}}{4} \times \dfrac{1}{\cancel{100}} = \dfrac{7}{16}$

13. ⤺ 45% = .45   15. ⤺ 5% = .05

17. 2⤺36% = 2.36   19. ⤺ 26.5% = .265

21. $8\tfrac{1}{4}\% = $ ⤺ 8.25% = .0825

Exercise 5-6

Page 257   SET A   Page 258   SET B

* 1. 0.74; 74%   1. .26; 26%
  2. 0.12; 12%   2. .08; 8%
* 3. 1/50; 2%   3. 1/25; 4%
  4. 2/25; 8%   4. 3/50; 6%
* 5. 57/1000; .057   5. 61/1000; .061
  6. 19/1000; .019   6. 43/1000; .043
* 7. 0.2; 20%   7. 0.6; 60%
  8. 0.7; 70%   8. 0.4; 40%
* 9. 187/400; 46.75%   9. 19/80; 23.75%
  10. 177/400; 44.25%   10. 97/400; 24.25%
*11. $\doteq$ 2.17; $\doteq$ 217%   11. $\doteq$ 1.67; $\doteq$ 167%
  12. $\doteq$ 1.83; $\doteq$ 183% 12. $\doteq$ 2.33; $\doteq$ 233%
*13. 2 3/8; 237.5%   13. 1 1/8; 112.5%
  14. 1 1/40; 102.5%   14. 2 3/40; 207.5%
  15. 53/800; .06625   15. 43/800; .05375

*Selected Solutions from Set A

1. $\frac{37}{50}$ = $50 \overline{)37.00}$ .74 = .74⌒ = 74%
   $\quad\quad\quad\quad\underline{35\ 0}$
   $\quad\quad\quad\quad\ \ 2\ 00$
   $\quad\quad\quad\quad\ \ \underline{2\ 00}$

3. .02⌒ = 2% = $\frac{2}{100}$ = $\frac{1}{50}$

5. ⌒5.7% = .057 = $\frac{57}{1000}$

7. $\frac{1}{5}$ = $5\overline{)1.0}$ .2 = .20⌒ = 20%
   $\quad\quad\ \underline{1\ 0}$

9. .4675 = $\frac{4675}{10000}$ = $\frac{187}{400}$
   .46⌒75 = 46.75%

11. $2\frac{1}{6}$ = $\frac{13}{6}$ = $6\overline{)13.000}$ 2.166 = 2.17⌒ = 217%
    $\quad\quad\quad\quad\quad\underline{12}$
    $\quad\quad\quad\quad\quad\ 1\ 0$
    $\quad\quad\quad\quad\quad\ \ \ \underline{6}$
    $\quad\quad\quad\quad\quad\ \ \ 40$
    $\quad\quad\quad\quad\quad\ \ \ \underline{36}$
    $\quad\quad\quad\quad\quad\ \ \ 40$

13. 2.375 = $2\frac{375}{1000}$ = $2\frac{3}{8}$
    2.37⌒5 = 237.5%

Exercise 5-7

Page 261   SET A   Page 262 SET B
* 1. 15   1. 11
  2. 39   2. 26
* 3. 0.935   3. 1.104
  4. 1.062   4. 0.552
* 5. $\doteq$ $11.81   5. $\doteq$ $59,520.89
  6. $\doteq$ $17,100.78 6. $\doteq$ $25.97
* 7. 98.4   7. 73
  8. 135.1   8. 116.9
* 9. $\doteq$ $60.30   9. $\doteq$ $528.37
  10. $\doteq$ $670.97   10. $\doteq$ $61.38
*11. 4.715   11. 1.885
  12. $\doteq$ 3.87   12. 5.67
*13. $\doteq$ $14.25   13. $\doteq$ $15.55
  14. $\doteq$ $52.80   14. $35.90
*15. $80.85   15. $\doteq$ $95.63
  16. $4062.50   16. $\doteq$ $14,642.78
*17. $1043.55   17. $1447.25
  18. $90,141.60   18. $30.59 a share

*Selected Solutions from Set A

1. 30% of 50 = .30 x 50 = 15

3. 5% of 18.7 = .05 x 18.7 = 0.935

5. 0.8% of $1476.59 = .008 x $1476.59
   $\doteq$ $11.81

7. 16.4% of 600 = .164 x 600 = 98.4

9. 78.32% of $76.99 = .7832 x $76.99
   $\doteq$ $60.30

11. 5 3/4% of 82 = .0575 x 82 = 4.715

13. 7.5% of $189.95 = .075 x $189.95
    $\doteq$ $14.25

15. 8 1/4% of $980 = .0825 x $980
    = $80.85

17. 15% of $6957 = .15 x $6957
    = $1043.55

Exercise 5-8

Page 265 SET A

* 1. 60%
  2. 80%
* 3. ≐ 33.33
  4. 45
* 5. 3
  6. 3.5
  7. 26%
  8. 21%
* 9. 225
  10. 187.5
  11. 47.36
  12. 125.13
*13. 360%
  14. 600%
*15. ≐ 423.53
  16. ≐ 385.71
*17. ≐ 8.03
  18. ≐ 3.17
  19. 120%

Page 266 SET B

  1. 30%
  2. 40%
  3. 80
  4. 45
  5. 3.2
  6. 4.2
  7. 22%
  8. 24%
  9. ≐ 566.67
  10. 600
  11. 144.32
  12. 130.91
  13. 250%
  14. 640%
  15. ≐ 547.83
  16. ≐ 415.38
  17. ≐ 3.23
  18. ≐ 5.54
  19. 50%

*Selected Solutions from Set A

1. P is unknown
   B=15, A=9
   $$\frac{9}{15} = \frac{P}{100}$$
   $$\frac{15 \times P}{15} = \frac{900}{15}$$
   $$P = 60$$

3. B is unknown
   P=60, A=20
   $$\frac{20}{B} = \frac{60}{100}$$
   $$\frac{60 \times B}{60} = \frac{2000}{60}$$
   $$B ≐ 33.33$$

5. A is unknown
   P=6, B=50
   $$\frac{A}{50} = \frac{6}{100}$$
   $$\frac{100 \times A}{100} = \frac{300}{100}$$
   $$A = 3$$

9. B is unknown
   P=8, A=18
   $$\frac{18}{B} = \frac{8}{100}$$
   $$\frac{8 \times B}{8} = \frac{1800}{8}$$
   $$B = 225$$

13. P is unknown
    B=17.5, A=63
    $$\frac{63}{17.5} = \frac{P}{100}$$
    $$\frac{17.5 \times P}{17.5} = \frac{6300}{17.5}$$
    $$P = 360$$

15. B is unknown
    P=5 2/3, A=24
    $$\frac{24}{B} = \frac{5\ 2/3}{100}$$
    $$\frac{5\ 2/3 \times B}{5\ 2/3} = \frac{2400}{5\ 2/3}$$
    $$B = \frac{2400}{17/3} ≐ 423.53$$

17. A is unknown
    P=17.6, B=45.6
    $$\frac{A}{45.6} = \frac{17.6}{100}$$
    $$\frac{100 \times A}{100} = \frac{802.56}{100}$$
    $$A ≐ 8.03$$

Exercise 5-9

Page 269 SET A

* 1. $723.98
  2. $107.99
* 3. $671.84
  4. $1494.79
  5. $8491.80
  6. $2587.50
* 7. 14,896 voters
  8. $26,160.82
* 9. $733.50
  10. 85%
  11. $562.59
  12. 70%
*13. 30%
  14. 15%
*15. 625 people
  16. 2630 people
  17. 6.25%

Page 270 SET B

  1. $1042.51
  2. $475.29
  3. $67.31
  4. $2545.16
  5. $8194.26
  6. $8032.50
  7. 7755 voters
  8. $19,522.80
  9. $14,542.23
  10. 40%
  11. $498.12
  12. 67.5%
  13. 18%
  14. 9.6%
  15. 45 students
  16. 1200 bulbs
  17. 12%

*Selected Solutions from Set A

1. Add sales tax to original price.
   tax: .065 x 679.79 ≐ $44.19
   total: $679.79 + 44.19 = $723.98

3. Subtract discount (.12 x $763.45)
   from the cost.
   $763.45 - $91.61 = $671.84

7. Find 32% of 46,550.
   .32 x 46,550 = 14,896

9. Add raise (12.5% of 652) to
   original salary.
   12.5% of 652 = .125 x 652 = $81.50
   new salary = 652 + 81.50 = $733.50

13. Amount of decrease (200-140) = 60.
    Base price is 200, percent unknown.
    $$\frac{60}{200} = \frac{P}{100}$$
    $$200 \times P = 6000$$
    $$P = 30$$

15. Restate the problem:
    325 is 52% of what number?
    $$\frac{325}{B} = \frac{52}{100}$$
    $$52 \times B = 32{,}500$$
    $$B = 625$$

Exercise 5-10

Page 273 SET A

* 1. $150.00
  2. $336.00
* 3. $630.00
  4. $1000.00
* 5. $3828.00
  6. $7080.00
* 7. $2253.25
  8. $2135.36
* 9. $9720.00
  10. $10,972.80
*11. $24.79
  12. $273.68
*13. $43.12
  14. $343.39
  15. $228.31

Page 274 SET B

1. $256.00
2. $156.00
3. $1980.00
4. $108.00
5. $3553.68
6. $6106.70
7. $998.50
8. $4865.78
9. $8346.00
10. $11,692.80
11. $35.88
12. $264.03
13. $62.82
14. $249.34
15. $227.30

*Selected Solutions from Set A

1. $I = P \times R \times T$
   $= 500 \times .15 \times 2 = \$150.00$

3. $I = P \times R \times T$
   $= 2000 \times .09 \times 3.5 = \$630.00$

5. $I = P \times R \times T$
   $= 7500 \times .1276 \times 4 = \$3828.00$

7. T = 42 months = 3.5 years
   $I = 4655 \times .1383 \times 3.5 \doteq \$2253.25$

9. T = 5 years = 60 months
   $I = 10800 \times .015 \times 60 = \$9720.00$

11. Calculate interest:
    $I = 700 \times .175 \times 4 = \$490$
    Add interest to amount borrowed:
    $700 + 490 = \$1190.00$
    Divide total by 48 months:
    $\$1190 \div 48 \doteq \$24.79$

13. Calculate amount after discount:
    $\$987.48 - (.25 \times 987.48) = \$740.61$
    Calculate amount after sales tax:
    $\$740.61 + (.065 \times 740.61) \doteq \$788.75$
    Calculate interest:
    $I = 788.75 \times .156 \times 2 \doteq \$246.09$
    Add interest to amount borrowed:
    $\$788.75 + 246.09 \doteq \$1034.84$
    Divide total by 24 months:
    $\$1034.84 \div 24 \doteq \$43.12$

Chapter 5 Practice Test A

Page 277

1. 4/15
2. 56
3. 700
4. $\doteq$ 10.71
5. 62.5
6. .064; 8/125
7. 33/250; 13.2%
8. 70.56
9. 60%
10. $\doteq$ 271.43
11. $\doteq$ 71.28
12. $68.11
13. 26 2/3 ft
14. $3077
15. $76.99

Chapter 5 Practice Test B

Page 278

1. 10/21
2. 35
3. 550
4. $\doteq$ 10.52
5. 43.75
6. .082; 41/500
7. 73/500; 14.6%
8. 20.76
9. 36%
10. $\doteq$ 188.89
11. $\doteq$ $49.86
12. $90.05
13. 38.7 oz
14. $564.90
15. $139.84

Chapter 6 Pretest

Page 284

1. 4.075
2. 14,256 mi
3. 12.5 c
4. 56,000 oz
5. 1000 liters
6. 100 cg
7. 900 cm
8. 1.159 m
9. 57.6 mg
10. 127 cm
11. 67 kg
12. 8.48 gal
13. 120 yd
14. 70° C
15. $637.50

Exercise 6-1

Page 287 SET A   Page 288 SET B

* 1. 1.1 cm   1. 3/16 in.
  2. 2.8 cm   2. 1 7/16 in.
* 3. 4.5 cm   3. 2 3/8 in.
  4. 5.6 cm   4. 3 5/8 in.
* 5. 7.2 cm   5. 4 1/4 in.
  6. 9 cm   6. 5 3/4 in.
* 7. 30 mph   7. 102 lb
  8. 37.5 mph   8. 116 lb
* 9. 52.5 mph   9. 134 lb
  10. 60 mph   10. 146 lb
*11. 480 ml   11. 28°
  12. 360 ml   12. 20°
*13. 240 ml   13. 12°
  14. 120 ml   14. 3°

*Selected Solutions from Set A

1. basic interval: 1 ÷ 10 = 0.1 cm

   A is 1 basic interval above 1 cm.
   A measures 1 + 0.1 = 1.1 cm

3. C is 5 basic intervals above 4 cm
   C measures 4 + 5 x 0.1 = 4.5 cm

5. E is 2 basic intervals above 7 cm
   E measures 7 + 2 x 0.1 = 7.2 cm

7. basic interval: 10 ÷ 4 = 2.5 mph

   G is 2 basic intervals above 25 mph.
   G measures 25 + 2 x 2.5 = 30 mph

9. I is 3 basic intervals above 45 mph.
   I measures 45 + 3 x 2.5 = 52.5 mph

11. basic interval: 100 ÷ 5 = 20 ml

    K is 4 basic intervals above 400 ml.
    K measures 400 + 4 x 20 = 480 ml

13. M is 2 basic intervals above 200 ml.
    M measures 200 + 2 x 20 = 240 ml

Exercise 6-2

Page 291 SET A   Page 292 SET B

* 1. 108 in.   1. 40 fl oz
  2. 240 oz   2. 14 pt
* 3. 24 qt   3. 6 3/4 ft
  4. 9 c   4. 6 lb
* 5. 37.8 ft   5. 9500 lb
  6. 19.2 qt   6. 5385.6 in.
* 7. 4.76 T   7. 16.9 yd
  8. 11.6 ft   8. 9 3/8 gal
* 9. 22,000 ft   9. 294 oz
  10. 30 2/3 c   10. 332 in
*11. 1 1/12 lb   11. 3 1/4 yd
  12. 1 17/32 ft   12. 3 3/8 c
  13. 300 c   13. 2.3 T
  14. 1.6 T   14. 92 c
*15. 112,000 oz   15. 976 oz
  16. 21.15 qt   16. 1.025 gal
*17. 5456 yd   17. 9 1/8 qt
  18. 2.41 mi   18. 86,400 oz

*Selected Solutions from Set A

1. $9 \text{ ft} \times \dfrac{12 \text{ in.}}{1 \text{ ft}} = \dfrac{9 \times 12 \text{ in.}}{1} = 108 \text{ in.}$

3. $48 \text{ pt} \times \dfrac{1 \text{ qt}}{2 \text{ pt}} = \dfrac{48 \times 1 \text{ qt}}{2} = 24 \text{ qt}$

5. $12.6 \text{ yd} \times \dfrac{3 \text{ ft}}{1 \text{ yd}} = \dfrac{12.6 \times 3 \text{ ft}}{1} = 37.8 \text{ ft}$

7. $9516.8 \text{ lb} \times \dfrac{1 \text{ T}}{2000 \text{ lb}} = \dfrac{9516.8 \times 1 \text{T}}{2000}$

   $= 4.76 \text{ T}$

9. $4\tfrac{1}{6} \text{ mi} \times \dfrac{5280 \text{ ft}}{1 \text{ mi}} = \dfrac{4\tfrac{1}{6} \times 5280 \text{ ft}}{1}$

   $= 22,000 \text{ ft}$

11. $17\tfrac{1}{3} \text{ oz} \times \dfrac{1 \text{ lb}}{16 \text{ oz}} = \dfrac{17\tfrac{1}{3} \times 1 \text{ lb}}{16} = 1\tfrac{1}{12} \text{ lb}$

15. $3\tfrac{1}{2} \text{ T} \times \dfrac{2000 \text{ lb}}{1 \text{ T}} = 7000 \text{ lb}$

    $7000 \text{ lb} \times \dfrac{16 \text{ oz}}{1 \text{ lb}} = 112,000 \text{ oz}$

17. $3.1 \text{ mi} \times \dfrac{5280 \text{ ft}}{1 \text{ mi}} = 16,368 \text{ ft}$

    $16,368 \text{ ft} \times \dfrac{1 \text{ yd}}{3 \text{ ft}} = 5456 \text{ yd}$

Exercise 6-3

Page 295 SET A   Page 296 SET B

* 1. kilometer   1. kilogram
  2. kiloliter   2. kilometer
* 3. centigram   3. centiliter
  4. centimeter   4. centigram
* 5. milliliter   5. millimeter
  6. milligram   6. milliliter
* 7. 1000 m   7. 1000 liters
  8. 1000 g   8. 1000 m
* 9. .01 liter   9. .01 g
  10. .01 m   10. .01 liter
*11. .001 g   11. .001 m
  12. .001 liter   12. .001 g
*13. 100 cm   13. 100 cl
  14. 100 cg   14. 100 cm
*15. 1000 ml   15. 1000 mg
  16. 1000 mm   16. 1000 ml
*17. kg, g, cg   17. g, cg, mg
  18. m, cm, mm   18. km, m, cm
  19. km, cm, mm   19. kl, cl, ml
  20. kl, cl, ml   20. kg, cg, mg

*Selected Solutions from Set A

1. k means kilo; m means meter;
   so km means kilometer.

3. c means centi; g means gram;
   so cg means centigram.

5. m means milli; l means liter;
   so ml means milliliter.

7. k means kilo (1000);
   so 1 km = 1000 m.

9. c means centi (.01);
   so 1 cl = .01 liter.

11. m means milli (.001);
    so 1 mg = .001 g.

13. cm = 1/100 m;
    so 1 m = 100 cm.

15. ml = 1/1000 liter;
    so 1 liter = 1000 ml.

17. kg = 1000 g; cg = .01 g; so the
    order is kg, g, cg.

Exercise 6-4

Page 301 SET A   Page 302 SET B

* 1. 3000 g   1. 4000 liters
  2. 8000 liters   2. 9000 g
  3. 52 liters   3. 53 m
  4. 71 m   4. 41 liters
* 5. 500 cm   5. 900 cg
  6. 700 cg   6. 200 cm
* 7. 5 liters   7. 3 g
  8. 7 g   8. 2 liters
  9. 4600 ml   9. 5900 mm
  10. 7300 mm   10. 2800 ml
  11. 1.697 m   11. 3.127 g
  12. 1.555 g   12. 2.095 m
*13. 8900 mg   13. 4300 ml
  14. 7600 ml   14. 5900 mg
  15. .57 kl   15. .61 km
  16. .86 km   16. .23 kl
*17. 400 mm   17. 700 ml
  18. 500 ml   18. 600 mm
*19. .76 cl   19. .29 cg
  20. .83 cg   20. .32 cl
*21. 420,000 cg   21. 410,000 cm
  22. 370,000 cm   22. 640,000 cg
*23. .0165 liter   23. .28 g
  24. 1.8 m   24. .14 m

*Selected Solutions from Set A

1. $3 \text{ kg} \times \dfrac{1000 \text{ g}}{1 \text{ kg}} = 3000$ g

5. $5 \text{ m} \times \dfrac{1 \text{ cm}}{.01 \text{ m}} = 500$ cm

7. 5,000 ml = 5 liters (ml to liter
   is a 3 place movement to the left).

13. 8.9 g = 8900 mg (g to mg is a
    3 place movement to the right).

17. 40 cm = 400 mm (cm to mm is a
    1 place movement to the right).

19. .7.6 ml = .76 cl (ml to cl is a
    1 place movement to the left).

21. 4.2 kg = 420,000 cg (kg to cg
    is a 5 place movement to the right).

23. $16.5 \text{ ml} \times \dfrac{.001 \text{ liter}}{1 \text{ ml}} = .0165$ liter

Exercise 6-5

Page 307 SET A   Page 308 SET B

* 1. 22.86 cm   1. 141.75 g
  2. 8.48 qt   2. 19.8 lb
* 3. 15 kg   3. 15 m
  4. 5 m   4. 30 kg
* 5. 157.6 in.   5. 49.68 mi
  6. 3178 g   6. 5.3 qt
* 7. 6 oz   7. 3.65 lb
  8. 30 km   8. 3.45 liters
  9. 975.15 in.   9. ≗ 135.255 cm
  10. 129.69 lb   10. 168.19 lb
*11. 67 kg   11. ≗ 2.15 m
  12. ≗ 24.70 in.   12. ≗ 11.27 oz
*13. 1.76 oz   13. 54.82 m
  14. ≗ 3.05 m   14. ≗ 3.66 liters
*15. ≗ 16,510 ml   15. 8.48 gal
  16. 2.835 cg   16. ≗ 377.36 ml
  17. 200 liter drum   17. 5 ft woman
  18. 64 kg man   18. 7/16 in. bit

*Selected Solutions from Set A

1. $9 \text{ in.} \times \dfrac{2.54 \text{ cm}}{1 \text{ in.}} = 22.86$ cm

3. $33 \text{ lb} \times \dfrac{1 \text{ kg}}{2.2 \text{ lb}} = 15$ kg

5. $4 \text{ m} \times \dfrac{39.4 \text{ in.}}{1 \text{ m}} = 157.6$ in.

7. $170.1 \text{ g} \times \dfrac{1 \text{ oz}}{28.35 \text{ g}} = 6$ oz

11. $147.4 \text{ lb} \times \dfrac{1 \text{ kg}}{2.2 \text{ lb}} = 67$ kg

13. Change kg to lb:
$.05 \text{ kg} \times \dfrac{2.2 \text{ lb}}{1 \text{ kg}} = 0.11$ lb

Change lb to oz:
$0.11 \text{ lb} \times \dfrac{16 \text{ oz}}{1 \text{ lb}} = 1.76$ oz

15. Change fl oz to qt:
$560 \text{ fl oz} \times \dfrac{1 \text{ qt}}{32 \text{ fl oz}} = 17.5$ qt

Change qt to liters:
$17.5 \text{ qt} \times \dfrac{1 \text{ liter}}{1.06 \text{ qt}} \doteq 16.51$ liters

Change liters to ml:
16.51 liters ≗ 16,510 ml

Exercise 6-6

Page 311 SET A   Page 312 SET B

* 1. 158°F   1. 140°F
  2. 113°F   2. 185°F
* 3. 40°C   3. 20°C
  4. 10°C   4. 35°C
* 5. 125.6°F   5. 109.4°F
  6. 179.6°F   6. 163.4°F
* 7. ≗ 176.67 C   7. ≗ 204.44°C
  8. ≗ 37.78°C   8. 135°C
* 9. 162.68°F   9. 200.12°F
  10. 180.86°F   10. 167.54°F
*11. 200°C   11. 300°C
  12. They are =.   12. They are =.
*13. 4°C   13. 36°F

*Selected Solutions from Set A

1. $F = \dfrac{9}{5} \times 70 + 32$
   $= 126 + 32 = 158°$

3. $C = \dfrac{5}{9} \times (104 - 32)$
   $= \dfrac{5}{9} \times 72 = 40°$

5. $F = \dfrac{9}{5} \times 52 + 32$
   $= 93.6 + 32 = 125.6°$

7. $C = \dfrac{5}{9} \times (350 - 32)$
   $= \dfrac{5}{9} \times 318 \doteq 176.67°$

9. $F = \dfrac{9}{5} \times 72.6 + 32$
   $= 130.68 + 32 = 162.68°$

11. Change 200°C to °F:
$F = \dfrac{9}{5} \times 200 + 32 = 392°$

So 200°C is hotter than 390°F.

13. Change 39°F to °C:
$C = \dfrac{5}{9} \times (39 - 32) \doteq 3.89°$

So 4°C is hotter than 39°F.

Exercise 6-7

Page 317 SET A

1. 81°
2. 45°
* 3. 62.25°
4. ≟ 48.89%
* 5. ≟ 31.34%
6. 21-44
7. over 64
8. 145,280,000
* 9. 156,630,000
*10. 24,970,000

Page 318 SET B

1. 25'0"
2. 20'6"
3. 22'
4. 12%
5. ≟ 19.05%
6. Asia; 17,077,500 sq mi
7. Australia; 2,990,000 sq mi
8. 7,705,000 sq mi
9. 2,472,500 sq mi
10. 16,272,500 sq mi

*Selected Solutions from Set A

3. To find the average, add up the temperatures and divide by 12.
   45+50+54+60+67+74+81+78+74+64+54+46 = 747, and 747 ÷ 12 = 62.25

5. Amount of decrease (67-46) = 21; base temp. = 67°; percent unknown.

   $\frac{A}{B} = \frac{P}{100}$ $\frac{21}{67} = \frac{P}{100}$

   $67 \times P = 2100$

   $P \doteq 31.34$

9. The percent under 45 is 69%:

   21-44:   33%
   14-20:   14%
   under 14: 22%
   total: 69%

   69% of 227,000,000
   = .69 × 227,000,000
   =   156,630,000

10. under 14: 22% of 227,000,000
              = .22 × 227,000,000
              =   49,940,000

    over 64: 11% of 227,000,000
              = .11 × 227,000,000
              =   24,970,000

    difference: 49,940,000
                -24,970,000
                 24,970,000

Chapter 6 Practice Test A

Page 321

1. 2.05
2. 17,400 lb
3. 13.5 ft
4. 3 gal
5. 1000 g
6. 100 cm
7. 8000 mg
8. 1.697 liters
9. 82.4 mm
10. 3.105 mi
11. 40 in.
12. ≟ 1.41 oz
13. 440 yd
14. 40°C
15. ≟ 216.67%

Chapter 6 Practice Test B

Page 322

1. 3.25
2. 150 in.
3. 576 oz
4. 140 fl oz
5. 1000 m
6. 1000 ml
7. 500 cm
8. 1.555 g
9. 0.623 cm
10. 1970 in.
11. 8 km
12. ≟ 9.55 liters
13. 1000 kg
14. 113°F
15. 80%

Unit III Exam

Page 326

1. 49
2. ≟ 1.27
3. 0.45; 9/20
4. 75%
5. 86.5
6. 158,400 in.
7. 5.6 g
8. ≟ 1415.09 ml
9. 140 F
10. 30 oz
11. 4 31/32 in.
12. $30,720
13. 420 mi
14. 31.25%
15. $192.72

UNIT IV

Page 328 BRAIN BUSTER

27 triangles

CHAPTER 7

Page 330 Pretest

1. 8
2. >
3. <
4. −7
5. −31
6. −12
7. −2
8. −42
9. 45
10. 11
11. 9
12. −22
13. 38
14. −189
15. −3

Exercise 7-1

Page 333 SET A   Page 334 SET B

* 1. 8   1. 4
  2. 6   2. 8
* 3. 14   3. 4
  4. 2   4. 4
* 5. 0   5. 0
  6. 0   6. 0
* 7. 62   7. 60
  8. 46   8. 60
* 9. 23.5   9. 44.25
 10. 38.6   10. 13
*11. 11 1/4   11. 1/2
 12. 9   12. 14
*13. 522   13. 75

*Selected Solutions from Set A

1. A + B − C =   3. $\frac{4\ B}{A}$ =
   2 + 7 − 1 =
         8                $\frac{4 \cdot 7}{2}$ = 14

5. 2 B C − 7 A =   7. A(3 B + 5 A) =
   2·7·1 − 7·2 =   2(3·7 + 5·2) =
   14 − 14 =   2(21 + 10) =
         0   2 · 31 = 62

9. x y + $\frac{z}{x}$ − $\frac{y}{2}$ =

   (1.2)(5) + $\frac{24}{1.2}$ − $\frac{5}{2}$ =

   6.0 + 20 − 2.5 =
   26 − 2.5 =
       23.5

11. M(3 N − 2 M) =

    $\frac{3}{4}(3 \cdot 5\frac{1}{2} − 2 \cdot \frac{3}{4})$ =

    $\frac{3}{4}(3 \cdot \frac{11}{2} − \frac{6}{4})$ =

    $\frac{3}{4}(\frac{33}{2} − \frac{3}{2}) = \frac{3}{4}(\frac{30}{2})$ =

    $\frac{3}{4}(15) = \frac{45}{4} = 11\frac{1}{4}$

13. 3 h f + h ($\frac{h}{f}$ − 6 g) =

    $3 \cdot 15 \cdot \frac{2}{5} + 15(\frac{15}{2/5} − 6(.65))$ =

    18 + 15($\frac{75}{2}$ − 3.9) =

    18 + 15(33.6) = 18 + 504 = 522

Exercise 7-2

Page 337 SET A   Page 338 SET B

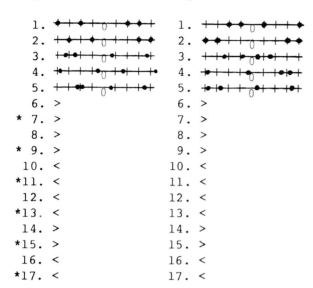

6. >   6. >
*7. >   7. >
8. >   8. >
*9. >   9. >
10. <   10. <
*11. <   11. <
12. <   12. <
*13. <   13. <
14. >   14. >
*15. >   15. >
16. <   16. <
*17. <   17. <

*Selected Solutions from Set A

7. 4 > 2, since 4 is farther to the right on the number line.

9. 0 > -2, since 0 is farther to the right on the number line.

11. -7 < -2, since -7 is farther to the left on the number line.

13. -6 < -2, since -6 is farther to the left on the number line.

15. $-7\frac{1}{2} > -7\frac{3}{4}$, since $-7\frac{1}{2}$ is farther to the right on the number line.

17. -2.7 < -2.6, since -2.7 is farther to the left on the number line.

Exercise 7-3

Page 341 SET A   Page 342 SET B

1. 12   1. 12
2. 14   2. 14
*3. 3   3. 3
4. 2   4. 1
5. -4   5. -6
6. -6   6. -8
*7. -10   7. -15
8. -12   8. -14
*9. 2   9. 7
10. 3   10. 11
*11. -18   11. -16
12. -18   12. -17
13. -9   13. -11
14. 4   14. -10
15. 0   15. 0
16. 0   16. 0
*17. 38   17. 38
18. 35   18. 54
*19. -48   19. -73
20. -50   20. -38
21. -239   21. -211
22. -281   22. -314
23. -132   23. -88
24. -262   24. -133

*Selected Solutions from Set A

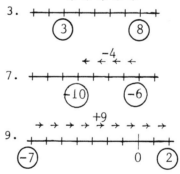

11. -3 + (-15) = -18; both signs are the same, so add the number parts and use the "-" sign.

17. 75 + (-37) = 38; signs are different, so subtract the number parts and use the sign of the larger number part--the "+" of 75.

19. -124 + 76 = -48; signs are different, so subtract the number parts and use the sign of the larger number part--the "-" of -124.

Exercise 7-4

Page 345 SET A   Page 346 SET B

1.	3	1.	3
2.	5	2.	5
*3.	-4	3.	-6
4.	-5	4.	-6
*5.	-10	5.	-13
6.	-12	6.	-14
*7.	8	7.	9
8.	10	8.	11
*9.	-2	9.	-2
10.	-5	10.	-2
*11.	0	11.	0
12.	0	12.	0
*13.	27	13.	16
14.	14	14.	14
*15.	-8	15.	-4
16.	-2	16.	-8
*17.	-90	17.	-92
18.	-93	18.	-64
*19.	166	19.	145
20.	160	20.	131
*21.	-13	21.	-4
22.	69	22.	68
*23.	-53	23.	-93
24.	-85	24.	-146

*Selected Solutions from Set A

3. $6 - 10 = 6 + (-10) = -4$

5. $-7 - 3 = -7 + (-3) = -10$

7. $5 - (-3) = 5 + 3 = 8$

9. $-7 - (-5) = -7 + 5 = -2$

11. $-7 - (-7) = -7 + 7 = 0$

13. $52 - 6 - 19 = 52 + (-6) + (-19)$
    $= 46 + (-19) = 27$

15. $20 - 11 - 17 = 20 + (-11) + (-17)$
    $= 9 + (-17) = -8$

17. $53 - 143 = 53 + (-143) = -90$

19. $72 - (-94) = 72 + 94 = 166$

21. $-51 - (-38) = -51 + 38 = -13$

23. $-125 - (-72) = -125 + 72 = -53$

Exercise 7-5

Page 349 SET A   Page 350 SET B

1.	15	1.	21
2.	24	2.	24
*3.	-21	3.	-40
4.	-18	4.	-27
*5.	-32	5.	-42
6.	-30	6.	-63
*7.	48	7.	56
8.	35	8.	36
*9.	72	9.	45
10.	63	10.	32
11.	4	11.	5
12.	9	12.	4
*13.	-8	13.	-4
14.	-6	14.	-5
*15.	-7	15.	-7
16.	-7	16.	-6
*17.	8	17.	9
18.	4	18.	6
19.	5	19.	9
20.	6	20.	7
21.	-13	21.	-12
22.	-12	22.	-15

*Selected Solutions from Set A

3. $3(-7) = -21$, since $(+) \cdot (-) = (-)$ and $3 \cdot 7 = 21$.

5. $-4 \cdot 8 = -32$, since $(-) \cdot (+) = (-)$ and $4 \cdot 8 = 32$.

7. $(-8)(-6) = +48$, since $(-) \cdot (-) = (+)$ and $8 \cdot 6 = 48$.

9. $-9 \cdot (-8) = +72$, since $(-) \cdot (-) = (+)$ and $9 \cdot 8 = 72$.

13. $\frac{-40}{5} = -8$, since the numbers have different signs and $\frac{40}{5} = 8$.

15. $\frac{56}{-8} = -7$, since the numbers have different signs and $\frac{56}{8} = 7$.

17. $\frac{-72}{-9} = +8$, since the numbers have the same signs and $\frac{72}{9} = 8$.

Exercise 7-6

Page 353 SET A	Page 354 SET B
* 1. 10	1. -6
2. 26	2. 1
* 3. -27	3. -28
4. -26	4. -18
* 5. -20	5. 14
6. 18	6. -24
* 7. -11	7. -10
8. -13	8. -8
* 9. -1	9. 8
10. 6	10. 1
*11. 2	11. 0
12. 9	12. -20
*13. 19	13. 17
14. 2	14. 37
*15. -2	15. -3
16. 4	16. 4
*17. 1/4	17. -2/5
18. -1/2	18. -2/3
19. 5	19. 3
20. 2	20. -14

Selected Solutions from Set A

1. $4^2 + (2)(-3) = 16 + (2)(-3)$
   $= 16 + (-6) = 10$

3. $(-2)(6) - 3\cdot 5 = -12 - 15$
   $= -12 + (-15) = -27$

5. $-5(-3 + 7) = -5(4) = -20$

7. $8 - \frac{8}{2} - 5\cdot 3 = 8 - 4 - 15$
   $= 8+(-4)+(-15) = -11$

9. $-3 - (4 - 6) = -3 - (-2)$
   $= 3 + 2 = -1$

11. $(-5 + 2\cdot 2)(7 - 9) = (-5 + 4)(7 - 9)$
    $= (-1)(-2) = 2$

13. $5 + 2[3-(-4)] = 5 + 2[3+4]$
    $= 5 + 2\cdot 7$
    $= 5 + 14 = 19$

15. $\frac{-4 - 2(8-5)}{5} = \frac{-4 - 2(3)}{5}$

    $= \frac{-4-6}{5} = \frac{-10}{5} = -2$

17. $\frac{8 + (-6)}{5 - (-3)} = \frac{2}{5+3} = \frac{2}{8} = \frac{1}{4}$

Exercise 7-7

Page 357 SET A	Page 358 SET B
* 1. A = 2	1. X = 4
2. B = 4	2. Y = 5
* 3. X = 11	3. C = 7
4. Y = 11	4. D = 10
* 5. P = 4	5. r = 6
6. z = 3	6. s = 6
* 7. r = 20	7. t = 15
8. s = 18	8. V = 8
* 9. t = -9	9. A = -4
10. v = -8	10. B = -6
*11. T = 2	11. R = 4
12. N = 5	12. T = 3
*13. y = 2	13. n = 2
14. k = 3	14. m = 3
*15. m = 0	15. Z = 0
16. n = 0	16. r = 0
*17. no	17. no
18. no	18. no
*19. yes	19. yes
20. yes	20. no

Selected Solutions from Set A

1. $A + 3 = 5$ $A = 2$
   since $2 + 3 = 5$.

3. $X - 4 = 7$ $X = 11$
   since $11 - 4 = 7$.

5. $3P = 12$ $P = 4$
   since $3\cdot 4 = 12$.

7. $\frac{r}{5} = 4$ $r = 20$

   since $\frac{20}{5} = 4$.

9. $-5 = t + 4$ $t = -9$
   since $-5 = -9 + 4$.

11. $-16 = -8T$ $T = 2$
    since $-16 = -8\cdot 2$.

13. $3y + 2 = 8$ $y = 2$
    since $3\cdot 2 + 2 = 8$.

15. $5 = 5 - 2m$ $m = 0$
    since $5 = 5 - 2\cdot 0$.

17. No; $\frac{-3(-2)}{2} - 6 = 3 - 6 = -3$.

19. Yes; $5(-2) + (-7) = -10+(-7)=-17$.

Exercise 7-8

Page 361   SET A   Page 362   SET B

* 1. S = 2   1. Z = 9
  2. S = 6   2. X = 8
* 3. t = 12   3. K = 12
  4. t = 9   4. K = 10
* 5. y = −4   5. X = −4
  6. y = −5   6. Z = −4
* 7. X = −2   7. R = −3
  8. X = −1   8. R = −6
* 9. X = 13   9. S = 19
  10. K = 29   10. S = 26
*11. K = 32   11. t = 31
  12. K = 42   12. t = 41
*13. Z = −3   13. y = −3
  14. Z = −4   14. y = −9
*15. p = 0.8   15. N = 1.9
  16. p = 2.1   16. N = 2.1
*17. R = 12   17. M = 8
  18. R = 21   18. M = 15
*19. A = 2/15   19. B = −1/12

*Selected Solutions from Set A

1. S + 7 = 9   3. t − 7 = 5
       −7 −7   +7 +7
    S   = 2   t   = 12

5. 1 = y + 5   7. −6 = X − 4
    −5   −5   +4   +4
    −4 = y   −2 = X

9. X + 9 = 22   11. K − 15 = 17
       −9   −9   +15 +15
    X   = 13   K   = 32

13. 12 = Z + 15   15. −2.5 + p = 1.7
    −15   −15   +2.5   +2.5
    −3 = Z   p = 0.8

17. R − (−8) = 20
    R + 8 = 20
       −8   −8
    R   = 12

19. A + 2/3 = 4/5
       − 2/3 − 2/3
    A = 4/5 − 2/3 = 12/15 − 10/15

    A = 2/15

Exercise 7-9

Page 365   SET A   Page 366   SET B

* 1. m = 9   1. m = 8
  2. m = 7   2. m = 7
* 3. Z = −8   3. X = −8
  4. X = −12   4. X = −7
* 5. N = 35   5. N = 21
  6. N = 24   6. N = 36
* 7. R = 648   7. R = 63
  8. R = 208   8. R = 48
* 9. K = −8   9. K = −6
  10. K = −8   10. K = −9
*11. t = −32   11. t = −27
  12. t = −36   12. t = −55
*13. y = 15   13. y = 13
  14. y = 14   14. y = 14
*15. v = 7   15. r = 9
  16. v = 6   16. r = 8
*17. m = −6.5   17. m = −2.5
  18. m = −3.5   18. m = −4.5
*19. T = −936   19. T = −897

*Selected Solutions from Set A

1. 5m/5 = 45/5   3. −8Z/−8 = 64/−8
       m = 9   Z = −8

5. 5·N/5 = 7·5   7. −9·R/−9 = −72(−9)
    N = 35   R = 648

9. −48/6 = 6K/6   11. −8 = (1/4)t
    −8 = K   4(−8) = 4·(1/4)t
                               −32 = t

13. −90/−6 = −6y/−6   15. 85v/85 = 595/85
    15 = y   v = 7

17. −3.2m/−3.2 = 20.8/−3.2
    m = 6.5

19.   78 = T/−12

    (−12)·78 = (T/−12)·(−12)
    −936 = T

Exercise 7-10

Page 369 SET A   Page 370 SET B

* 1. X = 2   1. X = 6
  2. X = 4   2. X = 5
* 3. y = −4   3. y = −6
  4. y = −6   4. y = −8
* 5. z = 3   5. z = 2
  6. z = 5   6. z = 3
* 7. K = 4   7. k = 2
  8. K = 3   8. k = 1
* 9. N = 12   9. N = 8
  10. N = 16   10. N = 25
*11. m = 42   11. m = 24
  12. m = 28   12. m = 24
*13. R = −20   13. R = −18
  14. R = −54   14. R = −35
*15. t = 84   15. t = 80
  16. t = 36   16. t = 45
*17. X = 36   17. X = 48
  18. X = 27   18. X = 20
  19. N = 7.7   19. N = 6.5

*Selected Solutions from Set A

1. $2X + 5 = 9$
   $-5 -5$
   $\frac{2X}{2} = \frac{4}{2}$
   $X = 2$

3. $40 = -8y + 8$
   $-8 -8$
   $\frac{32}{-8} = \frac{-8y}{-8}$
   $-4 = y$

5. $-3 + 5z = 12$
   $+3 +3$
   $\frac{5z}{5} = \frac{15}{5}$
   $z = 3$

7. $-18 = -3K - 6$
   $+6 +6$
   $\frac{-12}{-3} = \frac{-3K}{-3}$
   $4 = K$

9. $\frac{N}{3} + 6 = 10$
   $-6 -6$
   $3 \cdot \frac{N}{3} = 4 \cdot 3$
   $N = 12$

11. $\frac{1}{6}m - 4 = 3$
    $+4 +4$
    $6 \cdot \frac{1}{6}m = 7 \cdot 6$
    $m = 42$

13. $-12 = \frac{R}{5} - 8$
    $+8 +8$
    $5 \cdot (-4) = \frac{R}{5} \cdot 5$
    $-20 = R$

15. $4 + \frac{t}{-7} = -8$
    $-4 -4$
    $-7 \cdot \frac{t}{-7} = -12(-7)$
    $t = 84$

17. $4 \cdot \frac{3X}{4} = 27 \cdot 4$
    $\frac{3X}{3} = \frac{108}{3} \rightarrow X = 36$

Exercise 7-11

Page 375 SET A   Page 376 SET B

* 1. X + 6   1. X + 12
  2. X + 3   2. X + (−9)
* 3. 2X − 5   3. 3X − 3
  4. 2X − 6   4. 2X − 7
* 5. 6X   5. −7X
  6. 9X   6. 39X
* 7. 3X + 4   7. 2X + 10
  8. 5X + 2   8. 5X + 9
* 9. X/7   9. X/68
  10. X/4   10. X/−4
*11. 37   11. 39
  12. 22   12. 18
*13. 7   13. 4
  14. 14   14. 44
*15. −16   15. 36
  16. −72   16. −34
*17. 15¢   17. 28¢
  18. 17 years   18. 18 years
*19. 160 pounds   19. 60 inches
  20. 20   20. −45

*Selected Solutions from Set A

1. *Sum* means add, X + 6

3. *Twice* means 2 times; *fewer* means subtract.   2X − 5

5. *Product* means multiply.   6X

7. *Four more than* means plus 4.   3X+4

9. *Quotient* means divide.   X/7

11. N = the number
    $N + 8 = 45$
    $-8 -8$
    $N = 37$

13. N = the number
    $5N - 6 = 29$
    $+6 +6$
    $\frac{5N}{5} = \frac{35}{5}$
    $N = 7$

15. N = the number
    $\frac{-7N}{-7} = \frac{112}{-7}$
    $N = -16$

17. P = pencil's cost
    $50 + 3P = 95$
    $-50 -50$
    $\frac{3P}{3} = \frac{45}{3}$
    $P = 15$

19. K = Kevin's wt.
    $3K - 2(100) = 280$
    $3K - 200 = 280$
    $+200 +200$
    $\frac{3K}{3} = \frac{480}{3}$
    $K = 160$

Chapter 7

Practice Test A

Page 379

1. 2
2. <
3. >
4. -4
5. -42
6. -11
7. 6
8. -40
9. -7
10. -6
11. X = 24
12. P = 36
13. N = 38
14. r = 57
15. -5

Practice Test B

Page 380

1. 19
2. >
3. >
4. 3
5. -53
6. -15
7. 14
8. -54
9. 32
10. -32
11. X = 26
12. P = 36
13. n = 39
14. n = -8
15. 48

CHAPTER 8

Page 386   **Pretest**

1. $\overline{AE}$, $\overline{BD}$, $\overline{AB}$, $\overline{AC}$, $\overline{BC}$, $\overline{ED}$, $\overline{DC}$, $\overline{EC}$
2. AB = 12.5
3. acute angle
4. supplement = 134°, complement = 44°

5.
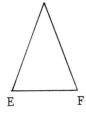

6.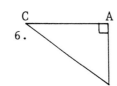

7.
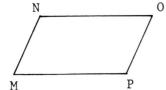

8. pentagon
9. C = 78.5 cm, A = 490.625 sq cm
10. A = 28, P = 24

Exercise 8-1

Page 389   SET A

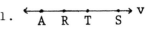

1. Y P F w (line)
2. J C X K (line)

3. J, V, G (in parallelogram)

4. d, W (in parallelogram with arrow)

5. g, h (two parallel lines)

6.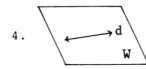

7.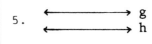

8.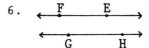

9. $\overline{MA}$, $\overline{MT}$, $\overline{MH}$,
   $\overline{AT}$, $\overline{AH}$, $\overline{TH}$

10. $\overline{SR}$, $\overline{SE}$, $\overline{RE}$,
    $\overline{SU}$, $\overline{SP}$, $\overline{UP}$,
    $\overline{RU}$, $\overline{EP}$

11. EA=5, AS=5,
    AY=10

12. NI=24, IC=12,
    CE=12

Page 390   SET B

1. A R T S v (line)
2. M B E N (line)

3. p, q, R (in parallelogram)

4. A, L, T, Q (in parallelogram)

5. a, b, c (three parallel rays)

6.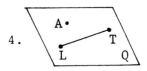

7.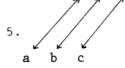

8.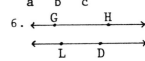

9. $\overline{GR}$, $\overline{RE}$, $\overline{AE}$, $\overline{GA}$,
   $\overline{GT}$, $\overline{GE}$, $\overline{TE}$, $\overline{RT}$,
   $\overline{RA}$, $\overline{TA}$

10. $\overline{LE}$, $\overline{LK}$, $\overline{EK}$,
    $\overline{LI}$, $\overline{IE}$, $\overline{IK}$

11. HO=3, OP=3,
    PE=6

12. LU=7, UC=7,
    CK=7

Exercise 8-2

Page 393 SET A

1.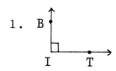
2.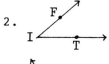
3.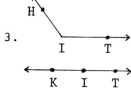
4.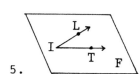
5.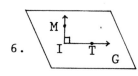
6.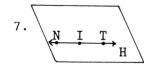
7.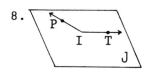
8.

9. acute
10. obtuse
11. straight
12. right
13. comp.: 45°
    supp.: 135°
14. comp.: 30°
    supp.: 120°
15. comp.: 67½°
    supp.: 157½°
16. comp.: 74.5°
    supp.: 164.5°

Page 394 SET B

1.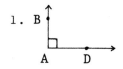
2.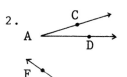
3.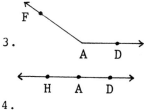
4.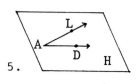
5.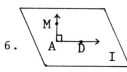
6.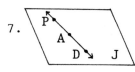
7.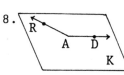
8.

9. acute
10. obtuse
11. right
12. straight
13. comp.: 75°
    supp.: 165°
14. comp.: 60°
    supp.: 150°
15. comp.: 72½°
    supp.: 162½°
16. comp.: 59.5°
    supp.: 149.5°

Exercise 8-3

Page 397 SET A

1.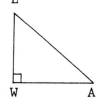
2.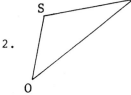
3.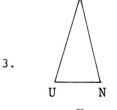
4.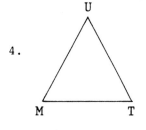

5. △ FRA, △ FAC, △ ENC
6. △ PAS, △ PAN, △ PNS, △ PAI, △ ANI
7. right
8. acute
9. obtuse
10. acute
11. isosceles
12. scalene
13. equilateral
14. isosceles

Page 398 SET B

1.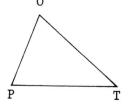
2.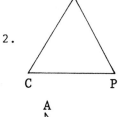
3.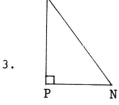
4.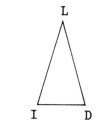

5. △ JSE, △ JSN, △ JSO, △ JEN, △ JEO, △ JNO,
6. △ TSW, △ TSO, △ TWO, △ TAZ, △ TAK, △ TZK
7. obtuse
8. right
9. right
10. acute
11. equilateral
12. isosceles
13. isosceles
14. scalene

Exercise 8-4

Page 401 SET A   Page 402 SET B

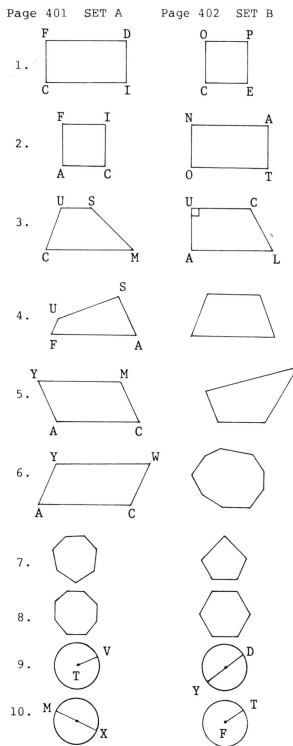

11. $\overline{ED}, \overline{EI}, \overline{EA}, \overline{EN}$ 11. $\overline{AD}, \overline{AE}, \overline{AF}$
12. $\overline{DN}$ 12. $\overline{FE}$
13. AE = 4   13. AD = 5
14. DN = 10   14. FE = 8
15. EI = $6\frac{1}{2}$ 15. AD = $12\frac{1}{2}$

Exercise 8-5

Page 405 SET A   Page 406 SET B

1. P = 28 ft   1. P = 26 cm
   A = 45 sq ft   A = 40 sq cm
2. P = 22 ft   2. P = 30 cm
   A = 28 sq ft   A = 54 sq cm
3. P = 34 in   3. P = 25 in
   A = 72.25 sq in.   A $\doteq$ 39.06 sq in.
4. P = 46 in   4. P = 45 in
   A = 132.25 sq in.   A $\doteq$ 126.56 sq in.
5. C = 42.076 cm   5. C = 79.756 m
   A $\doteq$ 140.95 sq cm A $\doteq$ 506.45 sq m
6. C = 52.124 cm   6. C = 97.968 m
   A $\doteq$ 216.31 sq cm A $\doteq$ 764.15 sq m
*7. P = 54   7. P = 24
   A = 126   A = 24
8. P = 60   8. P = 54
   A = 120   A = 126
*9. P = 25   9. P = 9
   A = 31.25   A = 3.6
10. P = 20   10. P = 9.2
    A = 19.5   A = 4.59
*11. P = 17.99   11. P = 32.13
    A $\doteq$ 19.23   A = 63.585
12. P = 12.85   12. P = 24.99
    A $\doteq$ 9.81   A = 38.465
13. P = 18   13. P = 44
    A = 15   A = 112
14. P = 64   14. P = 34
    A = 222   A = 66

*Selected Solutions from Set A

7. perimeter = sum of the sides
   = 13 + 20 + 21 = 54

   area = $\frac{1}{2}$bh = $\frac{1}{2}$(21)(12) = 126

9. perimeter = sum of the sides
   = $6\frac{1}{4}$ + $6\frac{1}{4}$ + $6\frac{1}{4}$ + $6\frac{1}{4}$ = 25

   area = bh = ($6\frac{1}{4}$)(5) = 6.25×5 = 31.25

11. circumference of the half-circle =
    $\frac{1}{2}$(2πr) = $\frac{1}{2}$(2)(3.14)(3.5) = 10.99

    perimeter = circumference of the
    half-circle + length of diameter
    = 10.99 + 7 = 17.99

    area of the half-circle =
    $\frac{1}{2}$(πr^2) = $\frac{1}{2}$(3.14)(3.5)^2 = $\frac{1}{2}$(38.465)
    = 19.2325 $\doteq$ 19.23

Chapter 8
Page 409 — Practice Test A

1. $\overline{AG}$, $\overline{GI}$, $\overline{IN}$, $\overline{NY}$, $\overline{YI}$, $\overline{IA}$, $\overline{GY}$, $\overline{AN}$
2. GI = $8\frac{1}{2}$
3. obtuse angle
4. supplement = 143°, complement = 53°

5.   6.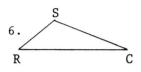

7.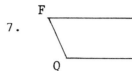

8. octagon
9. C = 116.18 cm, A = 1074.665 sq cm
10. A = 18, P = 20

Page 410 — Chapter 8 Practice Test B

1. $\overline{SX}$, $\overline{ST}$, $\overline{XT}$, $\overline{SF}$, $\overline{FX}$, $\overline{FT}$
2. ST = 38
3. right angle
4. supplement = 151°, complement = 61°

5. 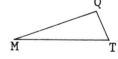   6.

7.

8. hexagon
9. C = 135.02 cm, A = 1451.465 sq cm
10. A = 336, P = 84

Unit IV Exam
Page 413

1. 20
2. -9
3. -8
4. 4
5. 45
6. -5
7. -49
8. -10
9. 7
10. -12
11. acute
12. equilateral
13. comp: 48°
    supp: 138°
14. A = 64 sq in.
    P = 32 in.
15. A = 28.26 sq cm
    C = 18.84 cm

Final Exam
Pages 414-415

1. 20,380,000
2. 27,491
3. 567
4. 92
5. 2 41/56
6. 1 5/84
7. 53 5/6
8. 266.58
9. 98.045
10. 6.05
11. 441.9
12. -29
13. 0.486
14. 3.5%
15. 5.8 kg
16. 571 in. (rounded to the nearest inch)
17. 87.5
18. 12
19. 89
20. $94.48
21. $170.06
22. $40.04
23. $699.72
24. $393.86
25. A = 324 sq cm
    C = 62.8 cm

Index

Acute angle, 392, 407
Acute triangle, 396, 407
Addend, 13, 41
Addition
   carrying in, 15-16
   checking, 15
   of decimals, 179-180
   of fractions, 133-142
   of mixed numbers, 145-146
   of signed numbers, 339-340
   of whole numbers, 13-16
Addition Property of Equality, 359
Algebraic expressions, 332, 377
   evaluating, 332
Angle, 391, 407
   acute, 392, 407
   complementary, 392, 407
   obtuse, 392, 407
   right, 392, 408
   straight, 392, 408
   supplementary, 392, 408
Area, 219-220, 224, 403-404
   of plane figures, 403-404
   of rectangles, 219-220
Average, 87-88, 91

Bar graph, 313, 319
Base
   of exponents, 33, 41
   of percent proportions, 263-268
   of plane figures, 403-404
Borrowing, 53

Canceling, 109, 117, 162
Carrying
   in addition, 15-16
   in multiplication, 21-22
Celsius scale, 309-310
Centigrade scale, 309-310
Circle, 400, 407
   area of, 404
   circumference of, 404
   diameter of, 400
   radius of, 400
Circle graph, 315, 319
Circumference, 404, 407
Complementary angles, 392, 407
Complex fraction, 149-150, 162
Composite number, 83, 91
Conversion chart
   metric, 303, 320
   U.S., 289, 320

Conversions
   between the U.S. and metric
      systems, 303-306, 320
   in the metric system, 297-300, 320
   in the U.S. system, 289-290, 320
Cross multiply, 239-240, 275
Cross products, 239-240, 275

Decimal digit, 187, 224
Decimal numbers, 171-172, 224
   addition of, 179-180
   comparing, 207-208
   converting to fractions, 204
   converting to percents, 255-256
   division of, 191-196
   multiplication of, 187-188
   reading and writing, 217-218
   repeating, 195
   rounding off, 175-176
   subtraction of, 183-184
   word problems involving, 215-216
Decimal place, 175, 224
Decimal point, 171, 224
Denominator, 103, 162
Diameter, 400, 407
Difference, 51, 91
Digits, 5, 41
Dividend, 57, 91
Divisibility, tests for, 83
Division
   by zero, 62
   checking, 71
   of decimals, 191-195
   of fractions, 121-122
   of mixed numbers, 126
   of signed numbers, 348
   of whole numbers, 57-76
   remainders in, 61, 91
   zeros in the divisor, 199-200
   zeros in the quotient, 75-76
Division Property of Equality, 363
Divisor, 57, 91

Equals (=), 34
Equation, 355, 377
   left side, 355
   right side, 355
   solution of, 355-368
Equilateral triangle, 395, 407
Even number, 109
Exponent, 33, 41

Factor, 19, 41

Factoring into primes, 84, 91
Fahrenheit scale, 309-310
Fraction(s), 103, 162
    addition of, 133-142
    canceling, 117-118, 162
    comparing, 153-154
    complex, 149-150, 162
    converting to decimals, 203
    converting to percents, 255-256
    division of, 121-122
    equivalent, 107
    improper, 104, 162
    least common denominator, 137-142, 162
    like, 133, 162
    line, 103, 162
    multiplication of, 117-118
    of numbers, 129-130
    proper, 104, 162
    raising to higher terms, 107-108, 162
    reducing, 107-110, 162
    subtraction of, 133-142
    unit, 289, 320
    word problems involving, 157-158

Geometry, 387
Gram, 293, 319
Graph(s)
    bar, 313, 319
    circle, 315, 319
    line, 314, 319
    statistical, 313-316, 319
Greater than (>), 34

Hexagon, 399, 407

Improper fraction, 104, 162
Interest, 271-276
Intersecting lines, 388, 407
Isosceles triangle, 395, 407

Least common denominator, 137-142, 162
Length
    metric units of, 293-294, 320
    U.S. units of, 289-290, 320
Less than (<), 34
Like fractions, 133, 162
Line, 387
    intersecting, 388, 407
    parallel, 388, 407

Line graph, 314, 319
Line segment, 388, 407
Liter, 293, 320
Lowest terms, 109

Measuring devices, 285-286
Meter, 293, 320
Metric system, 293-294
    basic units of, 293-294
    conversion chart, 303, 320
    conversions to U.S. units, 303-306
    conversions within the, 297-300
    prefixes in the, 293, 319-320
Midpoint, 388, 407
Minuend, 51, 91
Mixed numbers, 113-114, 162
    addition of, 145-146
    division of, 126
    multiplication of, 125
    subtraction of, 145-146
Multiplication
    checking, 26
    of decimals, 187-188
    of fractions, 117-118
    of mixed numbers, 125
    of signed numbers, 347-348
    of whole numbers, 19-22, 25-26
    zeros in, 29-30, 199-200
Multiplication Property of Equality, 363-364

Number(s)
    decimal, 171-172, 224
    even, 83
    fractional, 103-104
    line, 335-336, 377
    mixed, 113-114, 162
    negative, 335
    positive, 335
    prime, 83, 92
    signed, 335, 377
    whole, 1, 5-6
Numerator, 103, 162

Obtuse angle, 392, 407
Obtuse triangle, 396, 408
Octagon, 399, 408
Order of operations, 79-81, 91, 351-352

Parallel lines, 388, 408
Parallelogram, 399, 408
Pentagon, 399, 408
Percent, 251, 276
    converting to a fraction, 251-252
    of a number, 259-260
    proportion, 263-264, 276
    word problems involving, 267-268
Percentage problems, 263-264
Perimeter
    of plane figures, 403-404
    of rectangles, 219-220, 224
Pi, 404, 408
Place-value
    of decimal numbers, 171-172
    of whole numbers, 5-6, 41
Plane, 387
Plane figure, 399, 408
Point, 387
Powers, 33, 41
Prime numbers 83, 92
Principle, 271-272, 276
Product, 19, 41
Proper fraction, 104, 162
Proportion, 239-240, 275
    solution of, 239-244
    word problems involving, 247-248

Quadrilateral, 399, 408
    parallelogram, 399, 408
    rectangle, 399, 408
    square, 399, 408
    trapezoid, 399, 408
Quotient, 57, 92

Radius, 400, 408
Raising a fraction to higher terms,
    107-108, 162
Rate of interest, 271-272
Ratio, 235-236, 276
Ray, 391, 408
Reading measuring devices, 285-286
Reciprocal, 121, 162
Rectangle, 219, 399, 408
    area of, 219-220, 403
    perimeter of, 219-220, 403
Reducing a fraction, 107-108, 162
Remainder, 61, 92
Right angle, 392, 408
Right triangle, 396, 408
Rounding off
    decimals, 175-176
    whole numbers, 9-10

Scalene triangle, 395, 408
Signed numbers, 335, 377
    addition of, 339-340
    comparing, 335-336
    division of, 347-348
    multiplication of, 347-348
    order of operations, 351-352
    subtraction of, 343-344
Signs of comparison (>, <, =), 34
Simple interest, 271-272, 276
Square
    of a number, 33, 41
    plane figure, 399, 408
Square root, 33, 41
Squaring, 33, 41
Statistical graphs, 313-316
Straight angle, 392, 408
Subtraction
    borrowing in, 53
    checking, 54
    of decimals, 183-184
    of fractions, 133-142
    of mixed numbers, 145-146
    of signed numbers, 343-344
    of whole numbers, 51-54
Subtrahend, 51, 92
Sum, 13, 41
Supplementary angles, 392, 408

Temperature
    Celsius or Centigrade, 309-310
    Fahrenheit, 309-310
Trapezoid, 399, 408
Triangle, 395, 408
    acute, 396, 407
    equilateral, 395, 407
    isosceles, 395, 407
    obtuse, 396, 408
    right, 396, 408
    scalene, 395, 408

Unit-fraction, 289-290, 297, 320
U.S. system of measurement, 289-290
    conversion chart, 289, 320
    conversion to metrics, 303-306
    conversions within the, 289-290

Variable, 331, 377
Volume
    metric units of, 293-294, 320
    U.S. units of, 289-290, 320

Weight
  metric units of, 293-294, 320
  U.S. units of, 289-290, 320
Whole numbers, 1
  addition of, 13-16
  comparing, 34
  division of, 57-76
  multiplication of, 19-26
  reading and writing, 5-6
  rounding off, 9-10
  square root of, 34, 41
  squaring, 33, 41
  subtraction of, 51-54
  word problems involving, 37-38, 87-88
Word problems
  involving algebra, 371-374
  involving decimals, 215-216
  involving fractions, 129-130
  involving graphs, 313-316
  involving percents, 267-268
  involving perimeters and areas, 219-220
  involving proportions, 247-248
  involving whole numbers, 37-38, 87-88

Zero(s)
  dividing by, 62
  in multiplication, 29-30, 199-200
  in the divisor, 199-200
  in the quotient, 75-76